THE ART AND CRAFT
OF PROBLEM SOLVING

THE ART AND CRAFT OF PROBLEM SOLVING

Paul Zeitz

University of San Francisco

JOHN WILEY & SONS, INC.

New York   Chichester   Weinheim   Brisbane   Singapore   Toronto

ACQUISITIONS EDITOR	Ruth Baruth
MARKETING MANAGER	Audra Silverie
PRODUCTION EDITOR	Ken Santor
COVER DESIGNER	Madelyn Lesure
COVER PHOTO	Telegraph Colour Library/FPG International

This book was set in 10/12 Times Roman by the author and printed and bound by Quebecor-Fairfield, Inc. The cover was printed by Phoenix Color Corporation.

This book is printed on acid-free paper.

The paper in this book was manufactured by a mill whose forest management programs include sustained yield harvesting of its timberlands. Sustained yield harvesting principles ensure that the numbers of trees cut each year does not exceed the amount of new growth.

ISBN: 0-471-13571-2

Printed in the United States of America

10 9 8 7 6 5 4 3 2 1

To Eric

The explorer is the person who is lost.

—Tim Cahill, *Jaguars Ripped My Flesh*

Preface

Why This Book?

This is a book about mathematical problem solving, for college-level novices. By this I mean bright people who know some mathematics (ideally, at least some calculus), who enjoy mathematics, who have at least a vague notion of proof, but who have spent most of their time doing exercises rather than problems.

An *exercise* is a question that tests the student's mastery of a narrowly focused technique, usually one that was recently "covered." Exercises may be hard or easy, but they are never puzzling, for it is always immediately clear how to proceed. Getting the solution may involve hairy technical work, but the path towards solution is always apparent. In contrast, a *problem* is a question that cannot be answered immediately. Problems are often open-ended, paradoxical, and sometimes unsolvable, and require investigation before one can come close to a solution. Problems and problem solving are at the heart of mathematics. Research mathematicians do nothing but open-ended problem solving. In industry, being able to solve a poorly-defined problem is much more important to an employer than being able to, say, invert a matrix. A computer can do the latter, but not the former.

A good problem solver is not just more employable. Someone who learns how to solve mathematical problems enters the mainstream culture of mathematics; he or she develops great confidence and can inspire others. Best of all, problem solvers have fun; the adept problem solver knows how to play with mathematics, and understands and appreciates beautiful mathematics.

An analogy: The average (non-problem solver) math student is like someone who goes to a gym three times a week to do lots of repetitions with low weights on various exercise machines. In contrast, the problem solver goes on a long, hard backpacking trip. Both people get stronger. The problem solver gets hot, cold, wet, tired, and hungry. The problem solver gets lost, and has to find his or her way. The problem solver gets blisters. The problem solver climbs to the top of mountains, sees hitherto undreamed of vistas. The problem solver arrives at places of amazing beauty, and experiences ecstasy which is amplified by the effort expended to get there. When the problem solver returns home, he or she is energized by the adventure, and cannot stop telling others about their wonderful experiences. Meanwhile, the gym rat has gotten steadily stronger, but has not had much fun, and has little to share with others.

While the majority of American math students are not problem solvers, there does exist an elite problem solving culture. Its members were raised with math clubs, and often participated in math contests, and learned the important "folklore" problems and ideas that most mathematicians take for granted. This culture is prevalent in parts of Eastern Europe and exists in small pockets in the United States. I grew up in New York City and attended Stuyvesant High School, where I was captain of the math team, and consequently had a problem solver's education. I was and am deeply involved with problem solving contests. In high school, I was a member of the first USA team to participate in the International Mathematical Olympiad (IMO) and twenty years later, as a college professor, have coached several of the most recent IMO teams, including one which in 1994 achieved the only perfect performance in the history of the IMO.

But most people don't grow up in this problem solving culture. My experiences as a high school and college teacher, mostly with students who did not grow up as problem solvers, have convinced me that problem solving is something that is easy for any bright math student to learn. As a missionary for the problem solving culture, *The Art and Craft of Problem Solving* is a first approximation of my attempt to spread the gospel. I decided to write this book because I could not find any suitable text that worked for my students at the University of San Francisco. There are many nice books with lots of good mathematics out there, but I have found that mathematics itself is not enough. *The Art and Craft of Problem Solving* is guided by several principles:

- Problem solving can be taught, and can be learned.
- Success at solving problems is crucially dependent on psychological factors. Attributes like confidence, concentration, and courage are vitally important.
- No holds-barred investigation is at least as important as rigorous argument.
- The non-psychological aspects of problem solving are a mix of strategic principles, more focused tactical approaches, and narrowly defined technical tools.
- Knowledge of folklore (for example, the pigeonhole principle, or Conway's Checker problem) is as important as mastery of technical tools.

Reading This Book

Consequently, although this book is organized like a standard math textbook, its tone is much less formal: it tries to play the role of a friendly coach, teaching not just by exposition, but by exhortation, example, and challenge. There are few prerequisites—only a smattering of calculus is assumed—and while my target audience is college math majors, the book is certainly accessible to advanced high school students and to people reading on their own, especially teachers (at any level).

The book is divided into two parts. Part I is an overview of problem-solving methodology, and is the core of the book. Part II contains four chapters, that can be read independently of one another, which outline algebra, combinatorics, number theory, and calculus from the problem solver's point of view. In order to keep the book's length manageable, there is no geometry chapter. Geometric ideas are diffused throughout the book, and concentrated in a few places (for example, Section 4.2). Nevertheless, the book is a bit light on geometry. Luckily, a number of great geometry books have already been written. At the elementary level, *Geometry Revisited* [3] and *Geometry and the Imagination* [13] have no equals.

The structure of each section within each chapter is simple: exposition, examples, and problems—lots and lots—some easy, some hard, some very hard. The purpose of the book is to teach problem solving, and this can only be accomplished by grappling with many problems, solving some and learning from others that not every problem is meant to be solved, and that any time spent thinking honestly about a problem is time well spent.[1]

My goal is that reading this book and working on some of its 660 problems should be like the backpacking trip described above. The reader will definitely get lost for some of the time, and will get very, very sore. But at the conclusion of the trip, the reader will be toughened and happy and ready for more adventures.

And he or she will have learned a lot about mathematics—not a specific branch of mathematics, but mathematics, pure and simple. Indeed, a recurring theme throughout the book is the unity of mathematics. Many of the specific problem solving methods involve the idea of recasting from one branch of math to another; for example, a geometric interpretation of an algebraic inequality.

Teaching With This Book

In a one-semester course, virtually all of Part I should be studied, although not all of it will be mastered. In addition, the instructor can choose selected sections from Part II. For example, a course at the freshman or sophomore level might concentrate on chapters 1-6, while more advanced classes would omit much of chapter 5 (except the last section) and chapter 6, concentrating instead on chapters 7 and 8.

This book is aimed at beginning students, and I don't assume that the instructor is expert, either. The *Instructor's Resource Manual* contains solution sketches to most of the problems as well as some ideas about how to teach a problem solving course. For more information, please visit the Wiley web site at www.wiley.com.

Acknowledgements

Deborah Hughes Hallet has been the guardian angel of my career for nearly twenty years. Without her kindness and encouragement, this book would not exist, nor would I be a teacher of mathematics. I owe it to you, Deb. Thanks!

I have had the good fortune to work at the University of San Francisco, where I am surrounded by friendly and supportive colleagues and staff members, students who love learning, and administrators who strive to help the faculty. In particular, I'd like to single out a few people for heartfelt thanks:

- My dean, Stanley Nel, has helped me generously in concrete ways, with computer upgrades and travel funding. But more importantly, he has taken an active interest in my work from the very beginning. His enthusiasm and the knowledge that he supports my efforts has helped keep me going for the past four years.

- Tristan Needham has been my mentor, colleague, and friend since I came to USF in 1992. I could never have finished this book without his advice and hard

---

[1]An appendix at the end of the book contains hints to some of the problems and more help is available on the internet. Visit www.wiley.com.

labor on my behalf. Tristan's wisdom spans the spectrum from the tiniest LaTeX details to deep insights about the history and foundations of mathematics. In many ways that I am still just beginning to understand, Tristan has taught me what it means to really understand a mathematical truth.

- Nancy Campagna, Marvella Luey, Tonya Miller, and Laleh Shahideh have generously and creatively helped me with administrative problems so many times and in so many ways that I don't know where to begin. Suffice to say that without their help and friendship, my life at USF would often have become grim and chaotic.

- Not a day goes by without Wing Ng, our multitalented department secretary, helping me to solve problems involving things such as copier misfeeds to software installation to page layout. Her ingenuity and altruism have immensely enhanced my productivity.

Many of the ideas for this book come from my experiences teaching students in two vastly different arenas: a problem-solving seminar at USF and the training program for the USA team for the IMO. I thank all of my students for giving me the opportunity to share mathematics.

My colleagues in the math competitions world have taught me alot about problem solving. In particular, I'd like to thank Titu Andreescu, Jeremy Bem, Doug Jungreis, Kiran Kedlaya, Jim Propp, and Alexander Soifer for many helpful conversations.

Bob Bekes, John Chuchel, Dennis DeTurk, Tim Sipka, Robert Stolarsky, Agnes Tuska, and Graeme West reviewed earlier versions of this book. They made many useful comments and found many errors. The book is much improved because of their careful reading. Whatever errors remain, I of course assume all responsibility.

This book was written on a Macintosh computer, using LaTeX running on the wonderful Textures® 2.0 program, which is miles ahead of any other TeX system. I urge anyone contemplating writing a book using TeX or LaTeX to consider this program (www.bluesky.com). Another piece of software which helped me immensely was Eric Scheide's indexer program, which automates much of the LaTeX indexing process. His program easily saved me a week's tedium. Contact scheide@usfca.edu for more information.

Ruth Baruth, my editor at Wiley, has helped me transform a vague idea into a book in a surprisingly short time, by expertly mixing generous encouragement, creative suggestions, and gentle prodding. I sincerely thank her for her help, and look forward to more books in the future.

My wife and son have endured a lot during the writing of this book. This is not the place to thank them for their patience, but for me to apologize for my neglect. It is certainly true that I could have gotten a lot more work done, and done the work that I did do with less guilt, if I didn't have a family making demands on my time. But without my family, nothing—not even the beauty of mathematics—would have any meaning at all.

Paul Zeitz

San Francisco
November, 1998

Contents

Part I

In General

Chapter 1

What this Book Is About and How to Read It

1.1 "Exercises" vs. "Problems"

This is a book about mathematical problem solving. We make three assumptions about you, our reader:

- You enjoy math.
- You know high-school math pretty well, and have at least begun the study of "higher mathematics" such as calculus and linear algebra.
- You want to become better at solving math problems.

First, what is a **problem**? We distinguish between **problems** and **exercises**. An exercise is a question that you know how to resolve immediately. Whether you get it right or not depends on how expertly you apply specific techniques, but you don't need to puzzle out what techniques to use. In contrast, a problem demands much thought and resourcefulness before the right approach is found. For example, here is an exercise.

Example 1.1.1 Compute 5436^3 without a calculator.

You have no doubt about how to proceed—just multiply, carefully. The next question is more subtle.

Example 1.1.2 Write

$$\frac{1}{1 \cdot 2} + \frac{1}{2 \cdot 3} + \frac{1}{3 \cdot 4} + \cdots + \frac{1}{99 \cdot 100}$$

as a fraction in lowest terms.

At first glance, it is another tedious exercise, for you can just carefully add up all 99 terms, and hope that you get the right answer. But a little investigation yields something intriguing. Adding the first few terms, and simplifying, we discover that

$$\frac{1}{1 \cdot 2} + \frac{1}{2 \cdot 3} = \frac{2}{3},$$

3

$$\frac{1}{1\cdot 2}+\frac{1}{2\cdot 3}+\frac{1}{3\cdot 4}=\frac{3}{4},$$

$$\frac{1}{1\cdot 2}+\frac{1}{2\cdot 3}+\frac{1}{3\cdot 4}+\frac{1}{4\cdot 5}=\frac{4}{5},$$

which leads to the **conjecture** that for all positive integers n,

$$\frac{1}{1\cdot 2}+\frac{1}{2\cdot 3}+\frac{1}{3\cdot 4}+\cdots+\frac{1}{n(n+1)}=\frac{n}{n+1}.$$

So now we are confronted with a **problem**: is this conjecture true, and if so, how do we *prove* that it is true? If we are experienced in such matters, this is still a mere exercise, in the technique of mathematical induction (see page 50). But if we are not experienced, it is a problem, not an exercise. To solve it, we need to spend some time, trying out different approaches. The harder the problem, the more time we need. Often the first approach fails. Sometimes the first dozen approaches fail!

Here is another question, the famous "Census-Taker Problem." A few people might think of this as an exercise, but for most, it is a problem.

Example 1.1.3 A census-taker knocks on a door, and asks the woman inside how many children she has and how old they are.

"I have three daughters, their ages are whole numbers, and the product of the ages is 36," says the mother.

"That's not enough information," responds the census-taker.

"I'd tell you the sum of their ages, but you'd still be stumped."

"I wish you'd tell me something more."

"Okay, my oldest daughter Annie likes dogs."

What are the ages of the three daughters?

After the first reading, it seems impossible — there isn't enough information to determine the ages. That's why it is a problem, and a fun one, at that. (The answer is at the end of this chapter, on page 14, if you get stumped.)

If the Census-Taker Problem is too easy, try this next one (see page 84 for solution):

Example 1.1.4 I invite 10 couples to a party at my house. I ask everyone present, including my wife, how many people they shook hands with. It turns out that everyone shook hands with a different number of people. If we assume that no one shook hands with his or her partner, how many people did my wife shake hands with? (I did not ask myself any questions.)

A good problem is mysterious and interesting. It is mysterious, because at first you don't know how to solve it. If it is not interesting, you won't think about it much. If it is interesting, though, you will want to put a lot of time and effort into understanding it.

This book will help you to investigate and solve problems. If you are an inexperienced problem solver, you may often give up quickly. This happens for several reasons.

- You may just not know how to begin.

- You may make some initial progress, but then cannot proceed further.
- You try a few things, nothing works, so you give up.

An experienced problem solver, in contrast, is rarely at a loss for how to begin investigating a problem. He or she[1] confidently tries a number of approaches to get started. This may not solve the problem, but some progress is made. Then more specific techniques come into play. Eventually, at least some of the time, the problem is resolved. The experienced problem solver operates on three different levels:

Strategy: Mathematical and psychological ideas for starting and pursuing problems.

Tactics: Diverse mathematical methods that work in many different settings.

Tools: Narrowly focused techniques and "tricks" for specific situations.

1.2   The Three Levels of Problem Solving

Some branches of mathematics have very long histories, with many standard symbols and words. Problem solving is not one of them.[2] We use the terms **strategy**, **tactics** and **tools** to denote three different levels of problem solving. Since these are not standard definitions, it is important that we understand exactly what they mean.

A Mountaineering Analogy

You are standing at the base of a mountain, hoping to climb to the summit. Your first *strategy* may be to take several small trips to various easier peaks nearby, so as to observe the target mountain from different angles. After this, you may consider a somewhat more focused strategy, perhaps to try climbing the mountain via a particular ridge. Now the *tactical* considerations begin: how to actually achieve the chosen strategy. For example, suppose that strategy suggests climbing the south ridge of the peak, but there are snowfields and rivers in our path. Different tactics are needed to negotiate each of these obstacles. For the snowfield, our tactic may be to travel early in the morning, while the snow is hard. For the river, our tactic may be scouting the banks for the safest crossing. Finally, we move onto the most tightly focused level, that of *tools*: specific techniques to accomplish specialized tasks. For example, to cross the snowfield we may set up a particular system of ropes for safety and walk with ice axes. The river crossing may require the party to strip from the waist down and hold hands for balance. These are all tools. They are very specific. You would never summarize, "To climb the mountain we had to take our pants off and hold hands," because it was a minor — though essential — component of the entire climb. On the other hand, strategic and sometimes tactical ideas are often described in your summary: "We decided to reach the summit via the south ridge and had to cross a difficult snowfield and a dangerous river to get to the ridge."

---

[1]We will henceforth avoid the awkward "he or she" construction by alternating genders in subsequent chapters.

[2]In fact, there does not even exist a standard name for the theory of problem solving, although George Pólya and others have tried to popularize the term **heuristics** (see, for example, [20]).

As we climb a mountain, we may encounter obstacles. Some of these obstacles are easy to negotiate, for they are mere exercises (of course this depends on the climber's ability and experience). But one obstacle may present a difficult miniature problem, whose solution clears the way for the entire climb. For example, the path to the summit may be easy walking, except for one 10-foot section of steep ice. Climbers call negotiating the key obstacle the **crux move**. We shall use this term for mathematical problems as well. A crux move may take place at the strategic, tactical or tool level; some problems have several crux moves; many have none.

From Mountaineering to Mathematics

Let's approach mathematical problems with these mountaineering ideas. When confronted with a problem, you cannot immediately solve it, for otherwise, it is not a problem but a mere exercise. You must begin a process of **investigation**. This investigation can take many forms. One method, by no means a terrible one, is to just randomly try whatever comes into your head. If you have a fertile imagination, and a good store of methods, and a lot of time to spare, you may eventually solve the problem. However, if you are a beginner, it is best to cultivate a more organized approach. First, think strategically. Don't try immediately to solve the problem, but instead think about it on a less focused level. The goal of strategic thinking is to come up with a plan that may only barely have mathematical content, but which leads to an "improved" situation, not unlike the mountaineer's strategy, "If we get to the south ridge, it looks like we will be able to get to the summit."

Strategies help us get started, and help us continue. But they are just vague outlines of the actual work that needs to be done. The concrete tasks to accomplish our strategic plans are done at the lower levels of tactic and tool.

Here is an example which shows the three levels in action, from a 1926 Hungarian contest.

Example 1.2.1 Prove that the product of four consecutive natural numbers cannot be the square of an integer.

Solution: Our initial strategy is to familiarize ourselves with the statement of the problem, i.e., to **get oriented**. We first note that the question asks us to prove something. Problems are usually of two types—those that ask you to prove something and those that ask you to find something. The Census-Taker Problem (1.1.3) is an example of the latter type.

Next, observe that the problem is asking us to prove that something *cannot* happen. We divide the problem into **hypothesis** (also called "the given") and **conclusion** (whatever the problem is asking you to find or prove). The hypothesis is:

Let n be a natural number.

The conclusion is:

$n(n+1)(n+2)(n+3)$ *cannot be the square of an integer.*

Formulating the hypothesis and conclusion isn't a triviality, since many problems don't state them precisely. In this case, we had to introduce some notation. Sometimes our choice of notation can be critical.

Perhaps we should focus on the conclusion: how do you go about showing that something cannot be a square? This strategy, trying to think about what would immediately lead to the conclusion of our problem, is called looking at the **penultimate step**.[3] Unfortunately, our imagination fails us — we cannot think of any easy criteria for determining when a number cannot be a square. So we try another strategy, one of the best for beginning just about any problem: **get your hands dirty**. We try plugging in some numbers to experiment. If we are lucky, we may see a pattern. Let's try a few different values for n. Here's a table. We use the abbreviation $f(n) = n(n+1)(n+2)(n+3)$.

n	1	2	3	4	5	10
$f(n)$	24	120	360	840	1680	17160

Notice anything? The problem involves squares, so we are sensitized to look for squares. Just about everyone notices that the first two values of $f(n)$ are one less than a perfect square. A quick check verifies that additionally,

$$f(3) = 19^2 - 1, \quad f(4) = 29^2 - 1, \quad f(5) = 41^2 - 1, \quad f(10) = 131^2 - 1.$$

We confidently conjecture that $f(n)$ is one less than a perfect square for every n. Proving this conjecture is the penultimate step that we were looking for, *because a positive integer which is one less than a perfect square cannot be a perfect square* since the sequence $1, 4, 9, 16, \ldots$ of perfect squares contains no consecutive integers (the gaps between successive squares get bigger and bigger). Our new strategy is to prove the conjecture.

To do so, we need help at the tactical/tool level. We wish to prove that for each n, the product $n(n+1)(n+2)(n+3)$ is one less than a perfect square. In other words, $n(n+1)(n+2)(n+3) + 1$ must be a perfect square. How to show that an algebraic expression is always equal to a perfect square? One tactic: **factor** the expression! We need to manipulate the expression, always keeping in mind our goal of getting a square. So we focus on putting parts together that are almost the same. Notice that the product of n and $n+3$ is "almost" the same as the product of $n+1$ and $n+2$, in that their first two terms terms are both $n^2 + 3n$. After regrouping, we have

$$[n(n+3)][(n+1)(n+2)] + 1 = (n^2 + 3n)(n^2 + 3n + 2) + 1. \tag{1}$$

Rather than multiply out the two almost-identical terms, we introduce a little **symmetry** to bring squares into focus:

$$(n^2 + 3n)(n^2 + 3n + 2) + 1 = \left((n^2 + 3n + 1) - 1\right)\left((n^2 + 3n + 1) + 1\right) + 1.$$

Now we use the "difference of two squares" factorization (a tool!) and we have

$$\left((n^2 + 3n + 1) - 1\right)\left((n^2 + 3n + 1) + 1\right) + 1 = (n^2 + 3n + 1)^2 - 1 + 1$$
$$= (n^2 + 3n + 1)^2.$$

---

[3]The word "penultimate" means "next to last."

We have shown that $f(n)$ is one less than a perfect square for all integers n, namely

$$f(n) = (n^2 + 3n + 1)^2 - 1,$$

and we are done.   ∎

Let us look back and analyze this problem in terms of the three levels. Our first strategy was orientation, reading the problem carefully and classifying it in a preliminary way. Then we decided on a strategy to look at the penultimate step, which did not work at first, but the strategy of numerical experimentation lead to a conjecture. Successfully proving this involved the tactic of factoring, coupled with a use of symmetry and the tool of recognizing a common factorization.

The most important level was strategic. Getting to the conjecture was the crux move. At this point the problem metamorphosed into an exercise! For even if you did not have a good tactical grasp, you could have muddled through. One fine method is **substitution**: Let $u = n^2 + 3n$ in equation (1). Then the right-hand side becomes $u(u+2) + 1 = u^2 + 2u + 1 = (u+1)^2$. Another method is to multiply out (ugh!). We have

$$n(n+1)(n+2)(n+3) + 1 = n^4 + 6n^3 + 11n^2 + 6n + 1.$$

If this is going to be the square of something, it will be the square of the quadratic polynomial $n^2 + an + 1$ or $n^2 + an - 1$. Trying the first case, we equate

$$n^4 + 6n^3 + 11n^2 + 6n + 1 = (n^2 + an + 1)^2 = n^4 + 2an^3 + (a^2 + 2)n^2 + 2an + 1$$

and we see that $a = 3$ works; i.e., $n(n+1)(n+2)(n+3) + 1 = (n^2 + 3n + 1)^2$. This was a bit less elegant than the first way we solved the problem, but it is a fine method. Indeed, it teaches us a useful *tool*: the method of **undetermined coefficients**.

1.3   A Problem Sampler

The problems in this book are classified into three large families: **recreational**, **contest** and **open-ended**. Within each family, problems split into two basic kinds: problems **"to find"** and problems **"to prove."**[4] Problems "to find" ask for a specific piece of information, while problems "to prove" require a more general argument. Sometimes the distinction is blurry. For example, Example 1.1.4 above is a problem "to find," but its solution may involve a very general argument.

What follows is a descriptive sampler of each family.

Recreational Problems

Also known as "brain teasers," these problems usually involve little formal mathematics, but instead rely on creative use of basic strategic principles. They are excellent to work on, because no special knowledge is needed, and any time spent thinking about a

---

[4]These two terms are due to George Pólya [20].

recreational problem will help you later with more mathematically sophisticated problems. The Census-Taker problem (Example 1.1.3) is a good example of a recreational problem. A gold mine of excellent recreational problems is the work of Martin Gardner, who edited the "Mathematical Games" department for *Scientific American* for many years. Many of his articles have been collected into books. Two of the nicest are perhaps [7] and [6].

1.3.1 A monk climbs a mountain. He starts at 8am and reaches the summit at noon. He spends the night on the summit. The next morning, he leaves the summit at 8am and descends by the same route that he used the day before, reaching the bottom at noon. Prove that there is a time between 8am and noon at which the monk was at exactly the same spot on the mountain on both days. (Notice that we do not specify anything about the speed that the monk travels. For example, he could race at 1000 miles per hour for the first few minutes, then sit still for hours, then travel backward, etc. Nor does the monk have to travel at the same speeds going up as going down.)

1.3.2 You are in the downstairs lobby of a house. There are 3 switches, all in the "off" position. Upstairs, there is a room with a lightbulb that is turned off. One and only one of the three switches controls the bulb. You want to discover which switch controls the bulb, but you are only allowed to go upstairs once. How do you do it? (No fancy strings, telescopes, etc. allowed. You cannot see the upstairs room from downstairs. The lightbulb is a standard 100-watt bulb.)

1.3.3 You leave your house, travel one mile due south, then one mile due east, then one mile due north. You are now back at your house! Where do you live? There is more than one solution; find as many as possible.

Contest Problems

These problems are written for formal exams with time limits, often requiring specialized tools and/or ingenuity to solve. Several exams at the high school and undergraduate level involve sophisticated and interesting mathematics.

American High School Math Exam (AHSME)   A multiple-choice test, taken by nearly 500,000 self-selected high school students each year, similar to the hardest and most interesting problems on the SAT.

American Invitational Math Exam (AIME)   The top 2000 or so scorers on the AHSME qualify for this 3-hour, 15-question test. Both the AHSME and AIME feature problems "to find," since these tests are graded by machine.

USA Mathematical Olympiad (USAMO)   The top 150 AIME participants participate in this elite 3-and-a-half-hour, 5-question essay exam, featuring mostly challenging problems "to prove."

American Regions Mathematics League (ARML)   Every year, ARML sponsors a national contest between regional teams of high-school students. Some of the problems are quite challenging and interesting, roughly comparable to the harder questions on the AHSME and AIME and the easier USAMO problems.

Other national and regional olympiads   Many other nations conduct difficult problem solving contests. Eastern Europe in particular has a very rich contest tradition, including very interesting municipal contests, such as the Leningrad Mathematical Olympiad.[5] Recently China and Vietnam have developed very innovative and challenging examinations.

International Mathematical Olympiad (IMO)   The top 6 USAMO scorers are invited to form the USA team which competes in this international contest. It is a 9-hour, 6-question essay exam, spread over two days.[6] The IMO began in 1959, and takes place in a different country each year. At first it was a small event restricted to Iron Curtain countries, but recently the event has become quite inclusive, with 75 nations represented in 1996.

Putnam Exam   The most important problem solving contest for American undergraduates, a 12-question, 6-hour exam taken by several thousand students each December. The median score is often zero.

Problems in magazines   A number of mathematical journals have problem departments, in which readers are invited to propose problems and/or mail in solutions. The most interesting solutions are published, along with a list of those who solved the problem. Some of these problems can be extremely difficult, and many remain unsolved for years. Journals with good problem departments, in increasing order of difficulty level, are *Math Horizons, The College Mathematics Journal, Mathematics Magazine*, and *The American Mathematical Monthly*. All of these are published by the Mathematical Association of America. There is also a journal devoted entirely to interesting problems and problem solving, *Crux Mathematicorum*, published by the Canadian Mathematical Society.

Contest problems are very challenging. It is a significant accomplishment to solve a single such problem, even with no time limit. The samples below include problems of all difficulty levels.

1.3.4 (AHSME 1996) In the xy-plane, what is the length of the shortest path from $(0, 0)$ to $(12, 16)$ that does not go inside the circle $(x - 6)^2 + (y - 8)^2 = 25$?

1.3.5 (AHSME 1996) Given that $x^2 + y^2 = 14x + 6y + 6$, what is the largest possible value that $3x + 4y$ can have?

1.3.6 (AHSME 1994) When n standard 6-sided dice are rolled, the probability of obtaining a sum of 1994 is greater than zero and is the same as the probability of obtaining a sum of S. What is the smallest possible value of S?

1.3.7 (AIME 1994) Find the positive integer n for which

$$\lfloor \log_2 1 \rfloor + \lfloor \log_2 2 \rfloor + \lfloor \log_2 3 \rfloor + \cdots + \lfloor \log_2 n \rfloor = 1994,$$

---

[5]The Leningrad Mathematical Olympiad was renamed the St. Petersberg City Olympiad in the mid-1990's.

[6]Starting in 1996, the USAMO adopted a similar format: 6 questions, taken during two 3-hour-long morning and afternoon sessions.

where $\lfloor x \rfloor$ denotes the greatest integer less than or equal to x. (For example, $\lfloor \pi \rfloor = 3$.)

1.3.8 (AIME 1994) For any sequence of real numbers $A = (a_1, a_2, a_3, \ldots)$, define ΔA to be the sequence $(a_2 - a_1, a_3 - a_2, a_4 - a_3, \ldots)$, whose nth term is $a_{n+1} - a_n$. Suppose that all of the terms of the sequence $\Delta(\Delta A)$ are 1, and that $a_{19} = a_{94} = 0$. Find a_1.

1.3.9 (USAMO 1989) The 20 members of a local tennis club have scheduled exactly 14 two-person games among themselves, with each member playing in at least one game. Prove that within this schedule there must be a set of 6 games with 12 distinct players.

1.3.10 (USAMO 1995) A calculator is broken so that the only keys that still work are the sin, cos, tan, $\sin^{-1}$, $\cos^{-1}$, and $\tan^{-1}$ buttons. The display initially shows 0. Given any positive rational number q, show that pressing some finite sequence of buttons will yield q. Assume that the calculator does real number calculations with infinite precision. All functions are in terms of radians.

1.3.11 (Russia, 1995) Solve the equation

$$\cos(\cos(\cos(\cos x))) = \sin(\sin(\sin(\sin x))).$$

1.3.12 (IMO 1976) Determine, with proof, the largest number which is the product of positive integers whose sum is 1976.

1.3.13 (Putnam 1978) Let A be any set of 20 distinct integers chosen from the arithmetic progression $1, 4, 7, \ldots, 100$. Prove that there must be two distinct integers in A whose sum is 104.

1.3.14 (Putnam 1994) Let (a_n) be a sequence of positive reals such that, for all n, $a_n \leq a_{2n} + a_{2n+1}$. Prove that $\sum_{n=1}^{\infty} a_n$ diverges.

1.3.15 (Putnam 1994) Find the positive value of m such that the area in the first quadrant enclosed by the ellipse $\frac{x^2}{9} + y^2 = 1$, the x-axis, and the line $y = \frac{2}{3}x$ is equal to the area in the first quadrant enclosed by the ellipse $\frac{x^2}{9} + y^2 = 1$, the y-axis, and the line $y = mx$.

1.3.16 (Putnam 1990) Consider a paper punch that can be centered at any point of the plane and that, when operated, removes from the plane precisely those points whose distance from the center is irrational. How many punches are needed to remove every point?

Open-Ended Problems

These are mathematical questions that are sometimes vaguely worded, and possibly have no actual solution (unlike the two types of problems described above). Open-ended problems can be very exciting to work on, because you don't know what the outcome will be. A good open-ended problem is like a hike (or expedition!) in an uncharted region. Often partial solutions are all that you can get. (Of course, partial solutions are always OK, even if you know that the problem you are working on is a formal contest problem that has a complete solution.)

1.3.17 Here are the first few rows of **Pascal's Triangle**.

$$
\begin{array}{ccccccccccc}
& & & & & 1 & & & & & \\
& & & & 1 & & 1 & & & & \\
& & & 1 & & 2 & & 1 & & & \\
& & 1 & & 3 & & 3 & & 1 & & \\
& 1 & & 4 & & 6 & & 4 & & 1 & \\
1 & & 5 & & 10 & & 10 & & 5 & & 1,
\end{array}
$$

where the elements of each row are the sums of pairs of adjacent elements of the prior row. For example, $10 = 4 + 6$. The next row in the triangle will be

$$1, 6, 15, 20, 15, 6, 1.$$

There are many interesting patterns in Pascal's Triangle. Discover as many patterns and relationships as you can, and prove as much as possible. In particular, can you somehow extract the Fibonacci numbers (see next problem) from Pascal's Triangle (or vice versa)? Another question: is there a pattern or rule for the **parity** (evenness or oddness) of the elements of Pascal's Triangle?

1.3.18 The **Fibonacci numbers** f_n are defined by $f_0 = 0$, $f_1 = 1$ and $f_n = f_{n-1} + f_{n-2}$ for $n > 1$. For example, $f_2 = 1$, $f_3 = 2$, $f_4 = 3$, $f_5 = 5$, $f_6 = 8$, $f_7 = 13$, $f_8 = 21$. Play around with this sequence; try to discover as many patterns as you can, and try to prove your conjectures as best as you can. In particular, look at this amazing fact: for $n \geq 0$,

$$f_n = \frac{1}{\sqrt{5}} \left\{ \left(\frac{1+\sqrt{5}}{2} \right)^n - \left(\frac{1-\sqrt{5}}{2} \right)^n \right\}.$$

You should be able to prove this with mathematical induction (see pages 50–55 and problems 1–7), but the more interesting question is, where did this formula come from? Think about this, and other things that come up when you study the Fibonacci sequence.

1.3.19 An "ell" is an L-shaped tile made from three 1×1 squares (see picture). For what positive integers a, b is it possible to completely tile an $a \times b$ rectangle only using ells? ("Tiling" means that we cover the rectangle exactly with ells, with no overlaps.) For example, it is clear that you can tile a 2×3 rectangle with ells, but (draw a picture) you cannot tile a 3×3 with ells. After you understand rectangles, generalize in two directions: tiling ells in more elaborate shapes, tiling shapes with things other than ells.

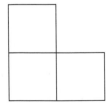

1.3.20 Imagine a long $1 \times L$ rectangle, where L is an integer. Clearly, one can pack this rectangle with L circles of diameter 1, and no more. (By "pack" we mean that touching is OK, but overlapping is not.) On the other hand, it is not immediately obvious that $2L$ circles is the maximum number possible for packing a $2 \times L$ rectangle. Investigate this, and generalize to $m \times L$ rectangles.

1.4 How to Read This Book

This book is not meant to be read from start to finish, but rather perused in a "non-linear" way. The book is designed to help you study two subjects: problem solving methodology *and* specific mathematical ideas. You will gradually learn more math and also become more adept at "problemsolvingology," and progress in one area will stimulate success in the other.

The book is divided into two parts. Part I, *In General*, gives an overview of strategies and tactics. Each strategy or tactic is discussed in a section which starts out with simple examples but ends with sophisticated problems. At some point, you may find that the text gets harder to understand, because it requires more mathematical experience. You should read the beginning of each section carefully, but then start skimming (or skipping) as it gets harder. You can (and should) reread later.

Part II, *Specifics*, is devoted to mathematical ideas at the tactical or tool level, organized by mathematical subject and developed specifically from the problem solver's point of view. Depending on your interests and background, you will read all or just some of part II.

As you increase your *mathematical* knowledge (from part II), you will want to keep returning to part I to reread sections that you may have skimmed earlier. As your knowledge increases, you will better understand the problem solving strategies and tactics in part I.

Throughout the book, new terms and specific strategy, tactic and tool names are in **boldface**. From time to time,

> *When an important point is made, it is indented and printed in italics,
> like this.*

That means, "pay attention!" To signify the successful completion of a solution, we use the "Halmos" symbol, a filled-in square.[7] We used a Halmos at the end of example 1.2.1 on page 8, and this line ends with one.   ∎

Please read with pencil and paper by your side, and/or write in the margins! Mathematics is meant to be studied actively. Also—this requires great restraint—try to solve each example as you read it, before reading the solution in the text. At the very least, take a few moments to ponder the problem. Don't be tempted into immediately looking at the solution. The more actively you approach the material in this book, the faster you will master it. And you'll have more fun.

---

[7]Named after Paul Halmos, a mathematician and writer who popularized its use.

Of course, some of the problems presented are harder than other. Toward the end of each section (or subsection) we may discuss a "classic" problem, one that is usually too hard for the beginning reader to solve alone in a reasonable amount of time. These classics are included for several reasons: they illustrate important ideas; they are part of what we consider the essential "repertoire" for every young mathematician; and, most important, they are beautiful works of art, to be pondered and savored. This book is called *The Art and Craft of Problem Solving,* and while we devote many more pages to the *craft* aspect of problem solving, we don't want you to forget that problem solving, at its best, is a passionate, aesthetic endeavor. If you will indulge us in another analogy, pretend that you are learning jazz piano improvisation. It's vital that you practice scales, and work on your own improvisations, but you also need the instruction and inspiration that comes from listening to some great recordings.

Solution to the Census-Taker Problem

The product of the ages is 36, so there are only a few possible triples of ages. Here is a table of all the possibilities, with the sums of the ages below each triple.

(1,1,36)	(1,2,18)	(1,3,12)	(1,4,9)	(1,6,6)	(2,2,9)	(2,3,6)	(3,3,4)
38	21	16	14	13	13	11	10

Aha! Now we see what is going on. The mother's second statement ("I'd tell you the sum of their ages, but you'd still be stumped") gives us valuable information. It tells us that the ages are either $(1, 6, 6)$ or $(2, 2, 9)$, for in all other cases, knowledge of the sum would tell us unambiguously what the ages are! The final clue now makes sense; it tells us that there *is* an oldest daughter, eliminating the triple $(1, 6, 6)$. The daughters are thus 2, 2 and 9 years old. ∎

Chapter 2

Strategies for Investigating Problems

As we've seen, solving a problem is not unlike climbing a mountain. And for inexperienced climbers, the task may seem daunting. The mountain is so steep! There is no trail! You can't even see the summit! If the mountain is worth climbing, it will take effort, skill, and, perhaps, luck. Several abortive attempts (euphemistically called "reconnaissance trips") may be needed before the summit is reached.

Likewise, a good math problem, one that is interesting and worth solving, will not solve itself. You must expend effort to discover the combination of the right mathematical tactics with the proper strategies. "Strategy" is often non-mathematical. Some problem solving strategies will work on many kinds of problems, not just mathematical ones.

For beginners especially, strategy is very important. When faced with a new and seemingly difficult problem, often you don't know where to begin. Psychological strategies can help you get in the right frame of mind. Other strategies help you start the process of investigation. Once you have begun work, you may need an overall strategic framework to continue and complete your solution.

We begin with psychological strategies that apply to almost all problems. These are simple commonsense ideas. That doesn't mean they are easy to master. But once you start thinking about them, you will notice a rapid improvement in your ability to work at mathematical problems. Note that we are not promising improvement in *solving* problems. That will come with time. But first you have to learn to *really* work.

After psychological strategies, we examine several strategies that help you begin investigations. These too are very simple ideas, easy and often fun to apply. They may not help you to *solve* many problems at first, but they will enable you to make encouraging *progress*.

The solution to every problem involves two parts: the **investigation**, during which you discover what is going on, and the **argument**, in which you convince others of your discoveries. We discuss the most popular of the many methods of formal argument in this chapter. We conclude with a study of miscellaneous strategies that can be used at different stages of a mathematical investigation.

2.1   Psychological Strategies

Effective problem solvers stand out from the crowd. Their brains seem to work differently. They are tougher, yet also more sensitive and flexible. Few people possess these laudable attributes, but it is easy to begin acquiring them.

Mental Toughness: Learn from Pólya's Mouse

We will summarize our ideas with a little story, "Mice and Men," told by George Pólya, the great mathematician and teacher of problem solving ([21], p. 75).

> The landlady hurried into the backyard, put the mousetrap on the ground (it was an oldfashioned trap, a cage with a trapdoor) and called to her daughter to fetch the cat. The mouse in the trap seemed to understand the gist of these proceedings; he raced frantically in his cage, threw himself violently against the bars, now on this side and then on the other, and in the last moment he succeeded in squeezing himself through and disappeared in the neighbour's field. There must have been on that side one slightly wider opening between the bars of the mousetrap ... I silently congratulated the mouse. He solved a great problem, and gave a great example.
>
> That is the way to solve problems. We must try and try again until eventually we recognize the slight difference between the various openings on which everything depends. We must vary our trials so that we may explore all sides of the problem. Indeed, we cannot know in advance on which side is the only practicable opening where we can squeeze through.
>
> The fundamental method of mice and men is the same; to try, try again, and to *vary the trials* so that we do not miss the few favorable possibilities. It is true that men are usually better in solving problems than mice. A man need not throw himself bodily against the obstacle, he can do so mentally; a man can vary his trials more and learn more from the failure of his trials than a mouse.

The moral of the story, of course, is that a good problem solver doesn't give up. However, she doesn't just stupidly keep banging her head against a wall (or cage!), but instead *varies each attempt*. But this is too simplistic. If people *never* gave up on problems, the world would be a very strange and unpleasant place. Sometimes you just cannot solve a problem. You will have to give up, at least temporarily. All good problem solvers occasionally admit defeat. An important part of the problem solver's art is knowing when to give up.

But most beginners give up too soon, because they lack the **mental toughness** attributes of **confidence** and **concentration**. It is hard to work on a problem if you don't believe that you can solve it, and it is impossible to keep working past your "frustration threshold." The novice must improve her mental toughness in tandem with her mathematical skills in order to make significant progress.

It isn't hard to acquire a modest amount of mental toughness. As a beginner, you most likely lack some confidence and powers of concentration, but you can increase both simultaneously. You may think that building up confidence is a difficult and subtle thing, but we are not talking here about self-esteem or sexuality or anything very deep in your psyche. Math problems are easier to deal with. You are *already* pretty confident about your math ability or you would not be reading this. You build upon your preexisting confidence by working at first on "easy" problems, where "easy" means that *you* can solve it after expending a modest effort. As long as you work on problems rather than exercises, your brain gets a workout, and your subconscious gets used to success. Your confidence automatically rises.

As your confidence grows, so too will your frustration threshold, if you gradually increase the intellectual "load." Start with easy problems, to warm up, but then work on harder and harder problems that continually *challenge and stretch you to the limit.* As long as the problems are interesting enough, you won't mind working for longer and longer stretches on them. At first, you may burn out after 15 minutes of hard thinking. Eventually, you will be able to work for hours single-mindedly on a problem, and keep other problems simmering on your mental backburner for days or weeks.

That's all there is to it. There is one catch: developing mental toughness takes time, and maintaining it is a lifetime task. But what could be more fun than thinking about challenging problems as often as possible?

Here is a simple and amusing problem, actually used in a software job interview, that illustrates the importance of confidence in approaching the unknown.[1]

Example 2.1.1 Consider the following diagram. Can you connect each small box on the top with its same-letter mate on the bottom with paths that do not cross one another, nor leave the boundaries of the large box?

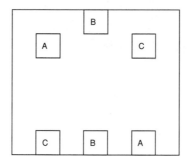

Solution: How to proceed? Either it is possible or it is not. The software company's personnel people were pretty crafty here; they wanted to see how quickly someone would give up. For certainly, it doesn't look possible. On the other hand, confidence dictates that

---

[1] We thank Denise Hunter for telling us about this problem.

Just because a problem seems impossible does not mean that it is impossible. Never admit defeat after a cursory glance. Begin optimistically; assume that the problem can be solved. Only after several failed attempts should you try to prove impossibility. If you cannot do so, then do not admit defeat. Go back to the problem later.

Now let us try to solve the problem. It is helpful to try to loosen up, and not worry about rules or constraints. **Wishful thinking** is always fun, and often useful. For example, in this problem, the main difficulty is that the top boxes labeled *A* and *C* are in the "wrong" places. So why not move them around to make the problem trivially easy? See the next diagram.

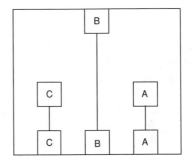

We have employed the all-important **make it easier** strategy:

If the given problem is too hard, solve an easier one.

Of course, we still haven't solved the original problem. Or have we? We can try to "push" the floating boxes back to their original positions, one at a time. First the *A* box:

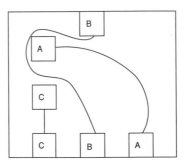

Now the *C* box,

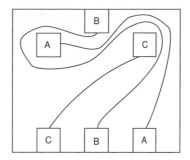

and suddenly the problem is solved!   ∎

There is a moral to the story, of course. Most people, when confronted with this problem, immediately declare that it is impossible. Good problem solvers do not, however. Remember, there is no time pressure. It might feel good to quickly "dispose" of a problem, by either solving it or declaring it to be unsolvable, but it is far better to take one's time to understand a problem. Avoid immediate declarations of impossibility; they are dishonest.

We solved this problem by using two strategic principles. First, we used the psychological strategy of cultivating an open, optimistic attitude. Second, we employed the enjoyable strategy of making the problem easier. We were lucky, for it turned out that the original problem was almost immediately equivalent to the modified easier version. That happened for a mathematical reason: the problem was a "topological" one. This "trick" of mutating a diagram into a "topologically equivalent" one is well worth remembering. It is not a strategy, but rather a *tool*, in our language.

Creativity

Most mathematicians are "Platonists," believing that the totality of their subject already "exists" and it is the job of human investigators to "discover" it, rather than create it. To the Platonist, problem solving is the art of seeing the solution that is already there. The good problem solver, then, is highly open and receptive to ideas that are floating around in plain view, yet invisible to most people.

This elusive receptiveness to new ideas is what we call **creativity**. Observing it in action is like watching a magic show, where wonderful things happen in surprising, hard-to-explain ways. Here is an example of a simple problem with a lovely, unexpected solution, one that appeared earlier as Problem 1.3.1 on page 9. Please think about the problem a bit before reading the solution!

Example 2.1.2 A monk climbs a mountain. He starts at 8am and reaches the summit at noon. He spends the night on the summit. The next morning, he leaves the summit at 8am and descends by the same route that he used the day before, reaching the bottom at noon. Prove that there is a time between 8am and noon at which the monk was at exactly the same spot on the mountain on both days. (Notice that we do not specify anything about the speed that the monk travels. For example, he could race at 1000 miles per hour for the first few minutes, then sit still for hours, then travel backward, etc. Nor does the monk have to travel at the same speeds going up as going down.)

Solution: Let the monk climb up the mountain in whatever way he does it. At the instant he begins his descent the next morning, have another monk start hiking up from the bottom, *traveling exactly as the first monk did the day before.* At some point, the two monks will meet on the trail. That is the time and place we want!   ∎

The extraordinary thing about this solution is the unexpected, clever insight of inventing a second monk. The idea seems to come from nowhere, yet it instantly resolves the problem, in a very pleasing way.[2]

That's creativity in action. The natural reaction to seeing such a brilliant, imaginative solution is to say, "Wow! How did she think of that? I could never have done it." Sometimes, in fact, seeing a creative solution can be inhibiting, for even though we admire it, we may not think that we could ever do it on our own. While it is true that some people do seem to be naturally more creative than others, we believe that almost everyone can learn to become more creative. Part of this process comes from cultivating a confident attitude, so that when you see a beautiful solution, you no longer think, "I could never have thought of that," but instead think, "Nice idea! It's similar to ones I've had. Let's put it to work!"

Learn to shamelessly appropriate new ideas and make them your own.

There's nothing wrong with that; the ideas are not patented. If they are beautiful ideas, you should excitedly master them and use them as often as you can, and try to stretch them to the limit by applying them in novel ways. Always be on the lookout for new ideas. Each new problem that you encounter should be analyzed for its "novel idea" content. The more you get used to appropriating and manipulating ideas, the more you will be able to come up with new ideas of your own.

One way to heighten your receptiveness to new ideas is to stay "loose," to cultivate a sort of mental **peripheral vision**. The receptor cells in the human retina are most densely packed near the center, but the most sensitive receptors are located on the periphery. This means that on a bright day, whatever you gaze at you can see very well. However, if it is dark, you will not be able to see things that you gaze at directly, but you will perceive, albeit fuzzily, objects on the periphery of your visual field (try exercise 2.1.10). Likewise, when you begin a problem solving investigation, you are "in the dark." Gazing directly at things won't help. You need to relax your vision and get ideas from the periphery. Like Pólya's mouse, constantly be on the lookout for twists and turns and tricks. Don't get locked into one method. Try to consciously break or bend the rules.

Here are a few simple examples, many of which are old classics. As always, don't jump immediately to the solution. Try to solve or at least think about each problem first!

Now is a good time to fold a sheet of paper in half or get a large index

---

[2]See page 59 for a more "conventional" solution to this problem.

card to hide solutions so that you don't succumb to temptation and read them before you have thought about the problems!

Example 2.1.3 Connect all nine points below with an unbroken path of four straight lines.

. . .

. . .

. . .

Solution: This problem is impossible unless you liberate yourself from the artificial boundary of the nine points. Once you decide to draw lines that extend past this boundary, it is pretty easy. Let the first line join three points, and make sure that each new line connects two more points.

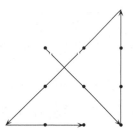

Example 2.1.4 Pat wants to take a 1.5-meter-long sword onto a train, but the conductor won't allow it as carry-on luggage. And the baggage person won't take any item whose greatest dimension exceeds 1 meter. What should Pat do?

Solution: This is unsolvable if we limit ourselves to 2-dimensional space. Once liberated from Flatland, we get a nice solution: The sword fits into a $1 \times 1 \times 1$- meter *box*, with a long diagonal of $\sqrt{1^2 + 1^2 + 1^2} = \sqrt{3} > 1.69$ meters. ∎

Example 2.1.5 What is the next letter in the sequence O, T, T, F, F, S, S, E ... ?

Solution: The sequence is a list of the first letters of the numerals one, two, three, four, ...; the answer is "N," for "nine." ∎

Example 2.1.6 Fill in the next column of the table.

1	3	9	3	11	18	13	19	27	55	
2	6	2	7	15	8	17	24	34	29	
3	1	5	12	5	13	21	21	23	30	

Solution: Trying to figure out this table one row at a time is pretty maddening. The values increase, decrease, repeat, etc., with no apparent pattern. But who said that the patterns had to be in rows? If you use peripheral vision to scan the table *as a whole* you will notice some familiar numbers. For example, there are lots of multiples of three. In fact, the first few multiples of three, in order, are hidden in the table.

1	3	**9**	3	11	**18**	13	19	**27**	55	
2	**6**	2	7	**15**	8	17	**24**	34	29	
3	1	5	**12**	5	13	**21**	21	23	**30**	

And once we see that the patterns are diagonal, it is easy to spot another sequence, the primes!

1	*3*	**9**	3	*11*	**18**	13	*19*	**27**	55	
2	**6**	2	7	**15**	8	*17*	**24**	34	*29*	
3	1	*5*	**12**	5	*13*	**21**	21	*23*	**30**	

The sequence that is left over is, of course, the Fibonacci numbers (1.3.18). So the next column of the table is $31, 33, 89$. ■

Example 2.1.7 Find the next member in this sequence.[3]

$$1, 11, 21, 1211, 111221, \ldots$$

Solution: If you interpret the elements of the sequence as numerical quantities, there seems to be no obvious pattern. But who said that they are numbers? If you look at the relationship between an element and its predecessor, and focus on "symbolic" content, we see a pattern. Each element "describes" the previous one. For example, the third element is 21, which can be described as "one 2 and one 1," i.e., 1211, which is the fourth element. This can be described as "one 1, one 2 and two 1s," i.e., 111221. So the next member is 312211 ("three 1s, two 2s and one 1"). ■

Example 2.1.8 Three women check into a motel room which advertises a rate of $27 per night. They each give $10 to the porter, and ask her to bring back 3 dollar bills. The porter returns to the desk, where she learns that the room is actually only $25 per night. She gives $25 to the motel desk clerk, returns to the room, and gives the guests back each one dollar, deciding not to tell them about the actual rate. Thus the porter has pocketed $2, while each guest has spent $10 - 1 = \$9$, a total of $2 + 3 \times 9 = \$29$. What happened to the other dollar?

---

[3] We thank Derek Vadala for bringing this problem to our attention. It appears in [29], p. 277.

Solution: This problem is deliberately trying to mislead the reader into thinking that the profit that the porter makes plus the amount that the guests spend *should* add up to $30. For example, try stretching things a bit: what if the actual room rate had been $0? Then the porter would pocket $27 and the guests would spend $27, which adds up to $54! The actual "invariant" here is not $30, but $27, the amount that the guests spend, and this will always equal the amount that the porter took ($2) plus the amount that went to the desk ($25). ■

Each example had a common theme: Don't let self-imposed, unnecessary restrictions limit your thinking. Whenever you encounter a problem, it is worth spending a minute (or more) asking the question, "Am I imposing rules that I don't need to? Can I change or **bend the rules** to my advantage?"

Nice guys may or may not finish last, but

> *Good, obedient boys and girls solve fewer problems than naughty and mischievous ones.*

Break or at least bend a few rules. It won't do anyone any harm, you'll have fun, and you'll start solving new problems.

We conclude this section with the lovely "Affirmative Action Problem," originally posed (in a different form) by Donald Newman. While mathematically more sophisticated than the monk problem, it too possesses a very brief and imaginative "one-liner" solution. The solution that we present is due to Jim Propp.

Example 2.1.9 Consider a network of finitely many balls, some of which are joined to one another by wires. We shall color the balls black and white, and call a network "integrated" if each white ball has at least as many black as white neighbors, and vice versa. The example below shows two different colorings of the same network. The one on the left is not integrated, because ball a has two white neighbors (c, d) and only one black neighbor (b). The network on the right is integrated.

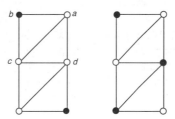

Given any network, is there a coloration that integrates it?

Solution: The answer is "yes." Let us call a wire "balanced" if it connects two differently colored balls. For example, the wire connecting a and b in the first network shown above is balanced, while the wire connecting a and c is not. Then our one-line solution is to

Maximize the balanced wires!

Now we need to explain our clever solution! Consider all the possible different colorings of a given network. There are finitely many colorings, so there must be one coloring (perhaps more than one) which produces the maximal number of balanced wires. We claim that *this* coloring is integrated. Assume, on the contrary, that it is not integrated. Then, there must be some ball, call it A, colored (without loss of generality) white, that has more white neighbors than black neighbors. Look at the wires emanating from A. The only balanced wires are the ones that connect A with black balls. More wires emanating from A are unbalanced than balanced. However, if we recolored A black, then more of the wires would be balanced rather than unbalanced. Since recoloring A only affects the wires that emanate from A, we have shown that recoloring A results in a coloration with more balanced wires than before. *That contradicts our assumption that our coloring already maximized the number of balanced wires!*

To recap, we showed that if a coloring is not integrated, then it cannot maximize balanced wires. Thus a coloring that maximizes balanced wires must be integrated! ∎

What are the novel ideas in this solution? That depends on how experienced you are, of course, but we can certainly isolate the stunning crux move: the idea of maximizing the number of balanced wires. The underlying idea, the **extreme principle**, is actually a popular "folklore" tactic used by experienced problem solvers (see Section 3.2 below). At first, seeing the extreme principle in action is like watching a karate expert break a board with seemingly effortless power. But once you master it for your own use, you will discover that breaking at least some boards isn't all that difficult. Another notable feature of this solution was the skillful use of argument by contradiction. Again, this is a fairly standard method of proof (see Section 2.3 below).

This doesn't mean that Jim Propp's solution wasn't clever. Indeed, it is one of the neatest one-liner arguments we've ever seen. But part of its charm is the simplicity of its ingredients, like origami, where a mere square of paper metamorphoses into surprising and beautiful shapes. Remember that the title of this book is *The Art and Craft of Problem Solving*. Craft goes a long way, and this is the route we emphasize, for without first developing craft, good art cannot happen. However, ultimately, the problem solving experience is an aesthetic one, as the Affirmative Action problem shows. The most interesting problems are often the most beautiful; their solutions are as pleasing as a good poem or painting.

OK, back to Earth! How do you become a board-breaking, paper-folding arts-and-crafts Master of Problem Solving? The answer is simple:

Toughen up, loosen up, and practice.

Toughen up by gradually increasing the amount and difficulty of your problem solving work. **Loosen up** by deliberately breaking rules and consciously opening yourself to new ideas (including shamelessly appropriating them!) Don't be afraid to play around, and try not to let failure inhibit you. Like Pólya's mouse, several failed attempts are perfectly fine, as long as you keep trying other approaches. And unlike Pólya's mouse, you won't die if you don't solve the problem. It's important to remember that. Problem solving isn't easy, but it should be fun, at least most of the time!

Finally, **practice** by working on lots and lots and lots of problems. Solving them is not as important. It is very healthy to have several unsolved problems banging around your conscious and unconscious mind. Here are a few to get you started.

Problems and Exercises

The first few (2.1.10–2.1.16) are mental training exercises. You needn't do them all, but please read each one, and work on a few (some of them require ongoing expenditures of time and energy, and you may consider keeping a journal to help you keep track). The remainder of the problems are mostly brain teasers, designed to loosen you up, mixed with a few open-ended questions to fire up your backburners.

2.1.10 Here are two fun experiments that you can do to see that your peripheral vision is both less acute yet more sensitive than your central vision.

1. On a clear night, gaze at the Pleiades constellation, which is also called the Seven Sisters, because it has seven prominent stars. Instead of looking directly at the constellation, try glancing at the Pleiades with your peripheral vision; i.e., try to "notice" it, while not quite looking at it. You should be able to see more stars!

2. Stare straight ahead at a wall while a friend slowly moves a card with a letter written on it into the periphery of your visual field. You will notice the movement of the card long before you can read the letter on it.

2.1.11 Many athletes benefit from "cross-training," the practice of working out regularly in another sport in order to enhance performance in the target sport. For example, bicycle racers may lift weights or jog. While we advocate devoting most of your energy to *math* problems, it may be helpful to diversify. Here a few suggestions.

(a) If English is your mother tongue, try working on word puzzles. Many daily newspapers carry the Jumble puzzle, in which you unscramble anagrams (permutations of the letters of a word). For example, **djauts** is **adjust**. Try to get to the point where the anagrams unscramble themselves unconsciously, almost instantaneously. This taps into your mind's amazing ability to make complicated associations. You may also find that it helps to read the original anagrams backward, upside down, or even arranged in a triangle, perhaps because this act of "restating" the problem loosens you up.

Another fun word puzzle is the cipher, in which you must decode a passage which has been encrypted with a single-letter substitution code (e.g., **A** goes to **L**, **B** goes to **G**, etc.) If you practice these until you can do the puzzle with little or no writing down, you will stimulate your association ability and enhance your deductive powers and concentration.

Standard crossword puzzles are OK, but not highly recommended, as they focus on fairly simple associations but with rather esoteric facts. Better than standard crossword puzzles, and orders of magnitude more challenging, are the "cryptic crosswords" found in the *Times* of London, the *Atlantic Monthly* and the *New Yorker*. These feature clues which are complicated puns and/or anagrams. For example, the July 21, 1997 *New Yorker* puzzle included the clue "Little devil wanders around, becomes a better person." The answer was "improves," because of the punnish spelling "imp" + "roves."

A good source of fun word problems is the Sunday morning puzzle on National Public Radio, hosted by Will Shortz. You can get these puzzles from the npr web site (http://www.npr.org/programs/wesun/puzzle.html), and/or listen on Sunday mornings, and/or look at collections such as [24].

(b) Learn to play a strategic game, such as Chess or Go. If you play cards, start concentrating on memorizing the hands as they are played.

(c) Take up a musical instrument, or if you used to play, start practicing again.

(d) Learn a "meditative" physical activity, such as yoga, tai chi, aikido, etc. Western sports like golf and bowling are OK, too.

(e) Read famous fictional and true accounts of problem solving and mental toughness. Some of our favorites are *The Gold Bug*, by Edgar Allan Poe (a tale of code-breaking); any Sherlock Holmes adventure, by Arthur Conan Doyle (masterful stories about deduction and concentration); *Zen in the Art of Archery*, by Eugen Herrigel (a Westerner goes to Japan to learn archery, and he *really* learns how to concentrate); *Endurance*, by Alfred Lansing (a true story of Antarctic shipwreck and the mental toughness needed to survive).

2.1.12 It doesn't matter when you work on problems, as long as you spend a lot of time on them, but do become aware of your routines. You may learn that, for example, you do your best thinking in the shower in the morning, or perhaps your best time is after midnight while listening to loud music, etc. Find a routine that works and then stick to it. (You may discover that walking or running is conducive to thought. Try this if you haven't before.)

2.1.13 Now that you have established one, occasionally shatter the routine. For example, if you tend to do your thinking in the morning in a quiet place, try to really concentrate on a problem at a concert at night, etc. This is a corollary of the "break rules" rule on page 23.

2.1.14 Here's a fun loosening-up exercise: pick a common object, for example, a brick, and list as quickly as possible as many uses for this object as you can. Try to be uninhibited and silly.

2.1.15 A nice source of amusing recreational problems are "lateral thinking" puzzlers. See, for example, [25]. These puzzles use mostly everyday ideas, but always require a good amount of peripheral vision to solve. Highly recommended, for warm-up work.

2.1.16 If you have trouble concentrating for long periods of time, try the following exercise: teach yourself some mental arithmetic. First, work out the squares from 1^2 to 32^2. Memorize this list. Then use the identity $x^2 - y^2 = (x - y)(x + y)$ to compute squares quickly in your head. For example, to compute 87^2, we reason as follows:

$$87^2 - 3^2 = (87 - 3)(87 + 3) = 84 \cdot 90 = 10(80 \cdot 9 + 4 \cdot 9) = 7560.$$

Hence $87^2 = 7560 + 3^2 = 7569$. Practice this method until you can reliably square any two-digit number quickly and accurately, in your head. Then try your hand at three-digit numbers. For example,

$$577^2 = 600 \cdot 554 + 23^2 = 332400 + 529 = 332929.$$

This should really impress your friends! This may seem like a silly exercise, but it will force you to focus, and the effort of relying on your mind's power of visualization or auditory memory may stimulate your receptiveness when you work on more serious problems.

2.1.17 In Example 2.1.9 on page 23, we proved that any coloring which maximizes the number of balanced wires will be integrated. Is this result "sharp;" i.e., must a coloring maximize balanced wires in order to be integrated?

2.1.18 The non-negative integers are divided into three groups as follows:

$$A = \{0, 3, 6, 8, 9, \ldots\}, \quad B = \{1, 4, 7, 11, 14, \ldots\}, \quad C = \{2, 5, 10, 13, \ldots\}.$$

Explain.

2.1.19 You are locked in a $50 \times 50 \times 50$-foot room which sits on 100- foot stilts. There is an open window at the corner of the room, near the floor, with a strong hook cemented into the floor by the window. So if you had a 100-foot rope, you could tie one end to the hook, and climb down the rope to freedom. (The stilts are not accessible from the window.) There are two 50-foot lengths of rope, each cemented into the ceiling, about 1 foot apart, near the center of the ceiling. You are a strong, agile rope climber, good at tying knots, and you have a sharp knife. You have no other tools (not even clothes). The rope is strong enough to hold your weight, but not if it is cut lengthwise. You can survive a fall of no more than 10 feet. How do you get out alive?

2.1.20 Compose your own "census-taker" problem. Invent a riddle which involves numerical information and clues which don't seem to be clues.

2.1.21 A group of jealous professors are locked up in a room. There is nothing else in the room but pencils and one tiny scrap of paper per person. The professors want to determine their average (mean, not median) salary so that each one can gloat or grieve over their personal situation compared to their peers. However, they are secretive people, and do not want to give away any personal salary information to anyone else. Can they determine the average salary in such a way that no professor can discover *any* fact about the salary of anyone but herself? For example, even facts such as "3 people earn more than \$40,000" or "no one earns more than \$90,000" are not allowed.

2.1.22 Bottle A contains a quart of milk and bottle B contains a quart of black coffee. Pour a small amount from B into A, mix well, and then pour back from A into B until both bottles again each contain a quart of liquid. What is the relationship between the fraction of coffee in A and the fraction of milk in B?

2.1.23 Indiana Jones needs to cross a flimsy rope bridge over a mile-long gorge. It is so dark that it is impossible to cross the bridge without a flashlight. Furthermore, the bridge is so weak that it can only support the weight of two people. The party has just one flashlight, which has a weak beam, so whenever two people cross, they are constrained to walk together, at the speed of the *slower* person. Indiana Jones can cross the bridge in 5 minutes. His girlfriend can cross in 10 minutes. His father needs

20 minutes, and his father's sidekick needs 25 minutes. They need to get everyone across safely in one hour to escape the bad guys. Can they do it?

2.1.24 You have already worked a little bit with Pascal's Triangle (Problem 1.3.17 on page 12). Find a way to get Fibonacci numbers (see Problem 1.3.18 on page 12) from Pascal's Triangle.

2.1.25 Example 1.1.2 involved the conjecture

$$\frac{1}{1\cdot 2} + \frac{1}{2\cdot 3} + \frac{1}{3\cdot 4} + \cdots + \frac{1}{n(n+1)} = \frac{n}{n+1}.$$

Experiment, and then conjecture more general formulas for sums where the denominators have products of 3 terms. Then generalize further.

2.1.26 It is possible to draw figure A below without lifting your pencil in such a way that you never draw the same line twice. However, no matter how hard you try, it seems impossible to draw figure B in this way. Can you find criteria which will allow you to quickly determine whether any given figure can or cannot be drawn in this way?

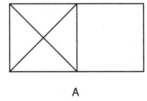

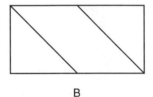

A   B

2.1.27 *Trick Questions.* All of the following problems seem to have obvious solutions, but the obvious solution is *not* correct. Ponder and solve!

(a) One day Martha said, "I have been alive during all or part of 5 decades." Rounded to the nearest year, what is the youngest she could have been?

(b) Of all the books at a certain library, if you select one at random, then there is a 90% chance that it has illustrations. Of all the illustrations in all the books, if you select one at random, then there is a 90% chance that it is in color. If the library has 10,000 books, then what is the minimum number of books that must contain colored illustrations?

(c) At least how many times must you flip a fair coin before there is at least a 50% probability that you will get at least 3 heads?

(d) Given two polyhedra (i.e., solids with polygonal faces), all of whose edges have length 1: a pyramid with a square base, and a tetrahedron (a tetrahedron is composed of four triangular faces). Suppose we glue the two polyhedra together along a triangular face (so that the attached faces exactly overlap). How many faces does the new solid have?

(e) The picture shows a square cut into two congruent polygons and another square cut into four congruent polygons.

For which positive integers n can a square be cut into n congruent polygons?

2.2   Strategies for Getting Started

The psychological ideas presented above may seem too vague. Perhaps you ask, "How can I learn to work hard on problems if I can't even get started?" You have already seen four very practical strategies that address this: the **penultimate step** and **get your hands dirty** strategies in Example 1.2.1, and the **wishful thinking** and **make it easier** strategies in Example 2.1.1. There is more to say about these and other ideas that help you begin a problem solving investigation.

As we said earlier, there are two parts to any successful solution: the investigation and the argument. Commonly, the investigation is obscured by the polished formal solution argument. But almost always, the investigation is the heart of the solution. Investigations are often tortuous, full of wrong turns and silly misconceptions. Once the problem is solved, it is easy to look over your prolonged investigation and wonder why it took you so long to see the light. But that is the nature of problem solving for almost everyone: you don't get rewarded with the flash of insight until you have paid your dues by prolonged, sometimes fruitless toil. Therefore,

Anything that stimulates investigation is good.

Here are some specific suggestions.

The First Step: Orientation

A few things need to be done at the beginning of every problem.

- Read the problem carefully. Pay attention to details such as positive vs. negative, finite vs. infinite, etc.
- Begin to classify: is it a "to find" or "to prove" problem? Is the problem similar to others you have seen?
- Carefully identify the hypothesis and the conclusion.
- Try some quick preliminary brainstorming:
  - Think about convenient notation.
  - Does a particular method of argument (see 2.3) seem plausible?
  - Can you guess a possible solution? Trust your intuition!
  - Are there key words or concepts that seem important? For example, might prime numbers or perfect squares or infinite sequences play an important role?

When you finish this (and don't rush!), go back and do it again. It pays to reread a problem several times. As you rethink classification, hypothesis and conclusion, ask yourself if you can **restate** what you have already formulated. For example, it may seem that the hypothesis is really trivial, and you just repeat it verbatim from the statement of the problem. But if you try to restate it, you may discover new information. Sometimes just reformulating hypothesis and conclusion with new notation helps (for example, Example 1.2.1 on page 6). Also, notice how restating helps one to solve the Census-Taker problem (Example 1.1.3 on page 4). More subtly, recall Example 2.1.7 on page 22, which involved the sequence 1, 11, 21, 1211, 111221, ... Normally, one reads a problem silently. But for many people, reciting the sequence *out loud* is just enough of a restatement to inspire the correct solution (as long as a number such as "1211" is read "one-two-one- one," not "one thousand, two hundred and eleven").

When looking at the conclusion of the problem, especially for a "to find" problem, sometimes it helps to "fantasize" an answer. Just make something up, and then reread the problem. Your fantasy answer is most likely false, and rereading the problem with this answer in mind may help you to see why the answer is wrong, which may point out some of the more important constraints of the problem.

Don't spend too much time on orientation. You are done once you have a clear idea of what the problem asks and what the given is. Promising guesses about answers or methodology are bonuses, and nothing you should expect. Usually this requires more intensive investigation.

I'm Oriented. Now What?

At this point, you understand what the problem is asking and you may have some ideas about what to do next. More often than not, this involves one or more of the four basic "startup" strategies that we have seen, **penultimate step**, **get your hands dirty**, **wishful thinking** and **make it easier**. Let's discuss these in more detail.

Get Your Hands Dirty: This is easy and fun to do. Stay loose and experiment. Plug in lots of numbers. Keep playing around until you see a pattern. Then play around some more, and try to figure out why the pattern you see is happening. It is a well-kept secret that much high-level mathematical research is the result of low-tech "plug and chug" methods. The great Carl Gauss, widely regarded as one of the greatest mathematicians in history (see page 75), was a big fan of this method. In one investigation, he painstakingly computed the number of integer solutions to $x^2 + y^2 \leq 90,000$.[4]

Penultimate Step: Once you know what the desired conclusion is, ask yourself, "What will yield the conclusion in a single step?" Sometimes a penultimate step is "obvious," once you start looking for one. And the more experienced you are, the more obvious the steps are. For example, suppose that A and B are weird, ugly expressions that seem to have no connection, yet you must show that $A = B$. One penultimate step would be to separately argue that $A \geq B$ AND $B \geq A$. Perhaps you want to show instead that $A \neq B$. A penultimate

---

[4]The answer is 282,697, in case you are interested. See [13], p. 33.

step would be to show that A is always even, while B is always odd. Always spend some time thinking very explicitly about possible penultimate steps. Of course, sometimes, the search for a penultimate step fails, and sometimes it helps one instead plan a proof strategy (see Section 2.3 below).

Wishful Thinking and **Make it Easier:**   These strategies combine psychology and mathematics to help break initial impasses in your work. Ask yourself, "What is it about the problem that makes it hard?" Then, make the difficulty disappear! You may not be able to do this *legally*, but who cares? Temporarily avoiding the hard part of a problem will allow you to make progress and may shed light on the difficulties. For example, if the problem involves big, ugly numbers, make them small and pretty. If a problem involves complicated algebraic fractions or radicals, try looking at a similar problem without such terms. At best, pretending that the difficulty isn't there will lead to a bold solution, as in Example 2.1.1 on page 17. At worst, you will be forced to focus on the key difficulty of your problem, and possibly formulate an intermediate question, whose answer will help you with the problem at hand. And eliminating the hard part of a problem, even temporarily, will allow you to have some fun and raise your confidence. If you cannot solve the problem as written, at least you can make progress with its easier cousin!

Here are a few examples that illustrate the use of these strategies. We will not concentrate on solving problems here, just making some *initial progress*. It is important to keep in mind that any progress is OK. Never be in a hurry to solve a problem! The process of investigation is just as important. You may not always believe this, but try:

> *Time spent thinking about a problem is always time worth spent. Even if you seem to make no progress at all.*

Example 2.2.1 (Russia, 1995) The sequence $a_0, a_1, a_2, \ldots$ satisfies

$$a_{m+n} + a_{m-n} = \frac{1}{2}(a_{2m} + a_{2n}) \tag{1}$$

for all nonnegative integers m and n with $m \geq n$. If $a_1 = 1$, determine a_{1995}.

  Partial Solution: Equation (1) is hard to understand without experimentation. Let's try to build up some values of a_n. First, we keep things simple and try $m = n = 0$ which yields $a_0 + a_0 = a_0$, so $a_0 = 0$. Now use the known indices 0, 1: plug $m = 1, n = 0$ into (1) and we have $2a_1 = \frac{1}{2}(a_2 + a_0) = a_2/2$, so $a_2 = 4a_1 = 4$. More generally, if we just fix $n = 0$, we have $2a_m = a_{2m}/2$, or $a_{2m} = 4a_m$. Now let's plug in $m = 2, n = 1$:

$$a_{2+1} + a_{2-1} = \frac{1}{2}(a_4 + a_2).$$

Since $a_4 = 4a_2 = 4 \cdot 4 = 16$, we have $a_3 + a_1 = \frac{1}{2}(16 + 4)$, so $a_3 = 9$. At this point, we are ready to venture the conjecture that $a_n = n^2$ for all $n = 1, 2, 3 \ldots$. If this conjecture is true, a likely method of proof will be **mathematical induction**. See page 50 below for an outline of this technique and page 52 for the continuation of this problem.

Example 2.2.2 (AIME 1985) The numbers in the sequence

$$101, 104, 109, 116, \ldots$$

are of the form $a_n = 100 + n^2$, where $n = 1, 2, 3, \ldots$. For each n, let d_n be the greatest common divisor[5] of a_n and a_{n+1}. Find the maximum value of d_n as n ranges through the positive integers.

Partial Solution: Just writing out the first few terms of a_n leads us to speculate that the maximum value of d_n is 1, since consecutive terms of the sequence seem to be relatively prime. But first, we should look at simpler cases. There is probably nothing special about the number 100 in the definition of a_n, except perhaps that $100 = 10^2$. Let's look at numbers defined by $a_n = u + n^2$, where u is fixed. Make a table.

u	a_1	a_2	a_3	a_4	a_5	a_6	a_7
1	2	**5**	**10**	17	26	37	50
2	3	6	11	**18**	**27**	38	51
3	4	7	12	19	28	**39**	**52**

We marked with boldface the pair of consecutive terms in each row which had the largest GCD (at least the largest for the first 7 terms of that row). Notice that when $u = 1$, then a_2 and a_3 had a GCD of 5. When $u = 2$, then a_4 and a_5 had a GCD of 9, and when $u = 3$, then a_6 and a_7 had a GCD of 13. There is a clear pattern: we conjecture that in general, for any fixed positive integer u, then a_{2u} and a_{2u+1} will have a GCD of $4u + 1$. We can explore this conjecture with a little algebra:

$$a_{2u} = u + (2u)^2 = 4u^2 + u = u(4u + 1),$$

while

$$a_{2u+1} = u + (2u + 1)^2 = 4u^2 + 4u + 1 + u = 4u^2 + 5u + 1 = (4u + 1)(u + 1),$$

and, sure enough, a_{2u} and a_{2u+1} share the common factor $4u + 1$.

This is encouraging progress, but we are not yet done. We have merely shown that $4u + 1$ is a *common* factor of a_{2u} and a_{2u+1}, but we want to show that it is the *greatest* common factor. And we also need to show that the value $4u + 1$ is the greatest possible GCD value for pairs of consecutive terms. Neither of these items is hard to prove if you know some simple number theory tools. We will continue this problem in Example 7.1.8 on page 248.

Example 2.2.3 Lockers in a row are numbered $1, 2, 3, \ldots, 1000$. At first, all the lockers are closed. A person walks by, and opens every other locker, starting with locker #2. Thus lockers $2, 4, 6, \ldots, 998, 1000$ are open. Another person walks by, and changes the "state" (i.e., closes a locker if it is open, opens a locker if it is closed) of every third locker, starting with locker #3. Then another person changes the state of every fourth locker, starting with #4, etc. This process continues until no more lockers can be altered. Which lockers will be closed?

---

[5] See 3.2.4, 3.2.15 and Section 7.1 for more information about the greatest common divisor.

Partial Solution: Most likely, there is nothing special about the number 1000 in this problem. Let us **make it easier**: Simplify the problem by assuming a much smaller number of lockers, say 10, to start. Now **get your hands dirty** by making a table, using the notation "o" for open and "x" for closed. Initially (step 1) all 10 are closed. Here is a table of the state of each locker at each pass. We stop at step 10, since further passes will not affect the lockers.

step	Locker #									
	1	2	3	4	5	6	7	8	9	10
1	x	x	x	x	x	x	x	x	x	x
2	x	o	x	o	x	o	x	o	x	o
3	x	o	o	o	x	x	x	o	o	o
4	x	o	o	x	x	x	x	x	o	o
5	x	o	o	x	o	x	x	x	o	x
6	x	o	o	x	o	o	x	x	o	x
7	x	o	o	x	o	o	o	x	o	x
8	x	o	o	x	o	o	o	o	o	x
9	x	o	o	x	o	o	o	o	x	x
10	x	o	o	x	o	o	o	o	x	o

We see that the closed lockers are numbered $1, 4, 9$; a reasonable conjecture is that only perfect square-numbered lockers will remain closed in general. We won't prove the conjecture right now, but we can make substantial progress by looking at the **penultimate step**. What determines if a locker is open or closed? After filling out the table, you know the answer: the **parity** (evenness or oddness) of the number of times the locker's state changed. A locker is closed or open according as the number of state changes was odd or even. Apply the penultimate step idea once more: what causes a state change? When does a locker get touched? Simplify things for a moment, and just focus on one locker, say #6. It was altered at steps $1, 2, 3$ and 6, a total of 4 times (an even number, hence the locker remains open). Look at locker #10. It was altered at steps $1, 2, 5$ and 10. Now it's clear:

Locker #n is altered at step k if and only if k divides n.

Hence we have restated our conjecture into an unrecognizably different, yet equivalent, form:

Prove that an integer has an odd number of divisors if and only if it is a perfect square.

There are many ways to think about this particular problem. See Example 3.1 on page 75 and Problem 6.1.21 on page 212 for two completely different approaches.

Example 2.2.4 (LMO 1988) There are 25 people sitting around a table, and each person has two cards. One of the numbers $1, 2, 3, \ldots, 25$ is written on each card, and each number occurs on exactly two cards. At a signal, each person passes one of her cards—the one with the smaller number—to her right-hand neighbor. Prove that, sooner or later, one of the players will have two cards with the same numbers.

Partial Solution: At first, this problem seems impenetrable. How can you prove something like this? Small doses of getting our hands dirty and making it easier go a long way. Let's help our understanding by making the problem easier. There is probably nothing special about the number 25, although we are immediately alerted to both squares and odd numbers. Let's try an example with 2 people. If each person has the cards numbered 1 and 2, we see that the pattern is periodic: each person just passes the number 1 to her neighbor, and always each person holds numbers 1 and 2. So the conclusion is *not* true for 2 people. Does **parity** (evenness or oddness) matter, then? Perhaps! Use the notation $\dfrac{a}{b}$ to indicate a person holding cards numbered a and b. Consider four people, initially holding

$$\begin{array}{|c|c|c|c|} \hline 1 & 1 & 2 & 2 \\ \hline 3 & 3 & 4 & 4 \\ \hline \end{array}.$$

We see that at each turn, everyone will hold a 1 or 2 paired with a 3 or 4. So again, the conclusion is not true, and for any even number of people, we can make sure that it never is true. For example, if there are $10 = 2 \cdot 5$ people, just start by giving each person one card chosen from $\{1, 2, \ldots, 5\}$ and one card chosen from $\{6, 7, \ldots, 10\}$. Then, at every turn, each person holds a card numbered in the range 1–5 paired with one in the range 6–10, so no one ever holds two cards with the same numbers.

Now we turn our attention to the case where there is an odd number of people. Here is an example involving 5 people.

$$\begin{array}{|c|c|c|c|c|} \hline 1 & 2 & 3 & 1 & 4 \\ \hline 4 & 3 & 5 & 2 & 5 \\ \hline \end{array}$$

We arranged the table so that the top row cards are smaller than the bottom row, so we know that these are the cards to be passed on the next turn. After that, we sort them again so that the top level contains the smaller numbers:

$$\begin{array}{|c|c|c|c|c|} \hline 1 & 2 & 3 & 1 & 4 \\ \hline 3 & 4 & 5 & 2 & 5 \\ \hline \end{array} \xrightarrow{\text{(pass)}} \begin{array}{|c|c|c|c|c|} \hline 4 & 1 & 2 & 3 & 1 \\ \hline 3 & 4 & 5 & 2 & 5 \\ \hline \end{array}$$

$$\xrightarrow{\text{(sort)}} \begin{array}{|c|c|c|c|c|} \hline 3 & 1 & 2 & 2 & 1 \\ \hline 4 & 4 & 5 & 3 & 5 \\ \hline \end{array}.$$

At this point, we can repeat the process, but before doing so, we observe that already the largest number in the top row is at most as big as the smallest number in the bottom row. So after we shift the top numbers, we no longer have to sort. We see that eventually the number 3 in the top will join the number 3 in the bottom:

$$\begin{array}{|c|c|c|c|c|} \hline 3 & 1 & 2 & 2 & 1 \\ \hline 4 & 4 & 5 & 3 & 5 \\ \hline \end{array} \xrightarrow{\text{(pass)}} \begin{array}{|c|c|c|c|c|} \hline 1 & 3 & 1 & 2 & 2 \\ \hline 4 & 4 & 5 & 3 & 5 \\ \hline \end{array}$$

$$\xrightarrow{\text{(pass)}} \begin{array}{|c|c|c|c|c|} \hline 2 & 1 & 3 & 1 & 2 \\ \hline 4 & 4 & 5 & 3 & 5 \\ \hline \end{array} \xrightarrow{\text{(pass)}} \begin{array}{|c|c|c|c|c|} \hline 2 & 2 & 1 & 3 & 1 \\ \hline 4 & 4 & 5 & 3 & 5 \\ \hline \end{array}.$$

So, what actually happened? We were able to get the two 3's to coincide, because the top and bottom rows stopped "mixing," and there was a 3 in the top and a 3 in the bottom. Will this always happen? Can you come up with a general argument? And what role was played by the odd parity of 5? A few more experiments, perhaps with 7 people, should help you finish this off.

In the next example, not only do we fail to solve the problem, we explore a conjecture that we know to be false! Nevertheless, we make partial progress: we develop an understanding of how the problem works, even if we do not attain a complete solution.

Example 2.2.5 (Putnam 83) Let $f(n) = n + \lfloor \sqrt{n} \rfloor$. Prove that, for every positive integer m, the sequence

$$m, f(m), f(f(m)), f(f(f(m))), \ldots$$

contains the square of an integer.[6]

Partial Solution: At first, it seems rather difficult. The function $f(n)$ has a strange definition and the desired conclusion is also hard to understand. Let's first get oriented: the problem asks us to show that something involving $f(m)$ and squares is true for *all* positive integers m. The only way to proceed is by getting our hands dirty. We need to understand how the function $f(m)$ works. So we start experimenting and making tables.

m	1	2	3	**4**	5	6	7	8	**9**
$f(m)$	2	3	4	**6**	7	8	9	10	**12**

m	10	11	12	13	14	15	**16**	17	18
$f(m)$	13	14	15	16	17	18	**20**	21	22

The pattern seems simple: $f(m)$ increases by 1 as m increases, until m is a perfect square, in which case $f(m)$ increases by 2 (boldface). Whenever you observe a pattern, you should try to see why it is true. In this case, it is not hard to see what is going on. For example, if $9 \le m < 16$, the quantity $\lfloor \sqrt{m} \rfloor$ has the constant value 3. Hence $f(m) = m + 3$ for $9 \le m < 16$. Likewise, $f(m) = m + 4$ for $16 \le m < 25$, which accounts for the "skip" that happens at perfect squares.

Now that we understand f pretty well, it is time to look at repeated iterations of this function. Again, experiment and make a table! We will use the notation

$$f^r(m) = f(f(\cdots f(m) \cdots)), \tag{2}$$

where the right-hand side contains r fs, and we will indicate squares with boldface. Notice that we don't bother trying out values of m which are squares, since then we

---

[6]Recall that $\lfloor x \rfloor$ is the greatest integer less than or equal to x. For more information, see page 160.

would trivially be done. Instead, we start with values of m that are one more than a perfect square, etc. Also, even when we achieve a perfect square, we continue to fill in the table. This an important work habit:

Don't skimp on experimentation! Keep messing around until you think you understand what is going on. Then mess around some more.

m	$f(m)$	$f^2(m)$	$f^3(m)$	$f^4(m)$	$f^5(m)$	$f^6(m)$
5	7	**9**	12	15	18	22
6	8	10	13	**16**	20	24
7	**9**	12	15	18	22	26
8	10	13	**16**	20	24	28
50	57	**64**	72	80	88	97
51	58	65	73	**81**	90	99
101	111	**121**	132	143	154	166
102	112	122	133	**144**	156	168
103	113	123	134	145	157	**169**

Now more patterns emerge. It seems that if m has the form $n^2 + 1$, where n is an integer, then $f^2(m)$ is a perfect square. Likewise, if m has the form $n^2 + 2$, then $f^4(m)$ appears to be a perfect square. For numbers of the form $n^2 + 3$, such as 7 and 103, it is less clear: $f^6(103)$ is a perfect square, yet $f(7)$ is a perfect square while $f^6(7)$ is not. So we know that the following "conjecture" is not quite correct:

If $m = n^2 + b$, then $f^{2b}(m)$ is a perfect square.

Nevertheless, the wishful thinking strategy demands that we at least examine this statement. After all, we wish to prove that for any m, there will be an r such that $f^r(m)$ is a perfect square. Before we dive into this, let's pause and consider: what is the chief difficulty with this problem? It is the perplexing $\lfloor \sqrt{m} \rfloor$ term. So we should first focus on this expression. Once we understand it, we will really understand how $f(m)$ works. Define $g(m) = \lfloor \sqrt{m} \rfloor$. Then what is $g(n^2 + b)$ equal to? If b is "small enough," then $g(n^2 + b) = n$. We can easily make this more precise, either by experimenting or just thinking clearly about the algebra. For what values of m is $g(m) = n$? The answer:

$$n^2 \le m < (n+1)^2 = n^2 + 2n + 1.$$

In other words,

$$g(n^2 + b) = n \text{ if and only if } 0 \le b < 2n + 1.$$

Now let us look at the "conjecture." For example, consider the case where $b = 1$. Then $m = n^2 + 1$ and $g(m) = n$ and

$$f(m) = m + g(m) = n^2 + 1 + n.$$

Iterating the function f once more, we have

$$f^2(m) = f(n^2 + n + 1) = n^2 + n + 1 + g(n^2 + n + 1)$$
$$= n^2 + n + 1 + n = n^2 + 2n + 1 = (n+1)^2,$$

so indeed the "conjecture" is true for $b = 1$.

Now let's look at $b = 2$. Then $m = n^2 + 2$ and $g(m) = n$ provided that $2 < 2n + 1$, which is true for all positive integers n. Consequently,

$$f(m) = m + g(m) = n^2 + 2 + n,$$

and

$$f^2(n^2 + 2) = n^2 + n + 2 + g(n^2 + n + 2).$$

Since $n + 2 < 2n + 1$ is true for $n > 1$, we have $g(n^2 + n + 2) = n$ and consequently, if $n > 1$,

$$f^2(n^2 + 2) = n^2 + n + 2 + n = (n + 1)^2 + 1.$$

Now we have discovered something fascinating: if m is 2 more than a perfect square, and $m > 1^2 + 2 = 3$, then two iterations of f yields a number which is 1 more than a perfect square. We know from our earlier work that two more iterations of f will then give us a perfect square. For example, let $m = 6^2 + 2 = 38$. Then $f(m) = 38 + 6 = 44$ and $f(44) = 44 + 6 = 50$ and $f^2(50) = 64$ (from the table). So $f^4(38) = 64$. The only number of the form $n^2 + 2$ that doesn't work is $1^2 + 2 = 3$, but in this case, $f(3) = 4$, so we are done.

We have made significant partial progress. We have shown that if $m = n^2 + 1$ or $m = n^2 + 2$, then finitely many iterations of f will yield a perfect square. And we have a nice direction in which to work. Our goal is getting perfect squares. The way we measure partial progress toward this goal is by writing our numbers in the form $n^2 + b$, where $0 \leq b < 2n + 1$. In other words, b is the "remainder." Now a more intriguing conjecture is

If m has remainder b, then $f^2(m)$ has remainder $b - 1$.

If we can establish this conjecture, then we are done, for eventually, the remainder will become zero. Unfortunately, this conjecture is not quite true. For example, if $m = 7 = 2^2 + 3$, then $f^2(7) = 12 = 3^2 + 3$. Even though this "wishful thinking conjecture" is false, careful analysis will uncover something very similar, which *is* true, and this will lead to a full solution. We leave this analysis to you.

Example 2.2.6 (Putnam 1991) Let $\mathbf{A}$ and $\mathbf{B}$ be different $n \times n$ matrices with real entries. If $\mathbf{A}^3 = \mathbf{B}^3$ and $\mathbf{A}^2\mathbf{B} = \mathbf{B}^2\mathbf{A}$, can $\mathbf{A}^2 + \mathbf{B}^2$ be invertible?

Solution: This is the sort of problem that most students shy away from, even those who excelled at linear algebra. But it is really not hard at all when approached with confidence. First of all, we note the given: $\mathbf{A} \neq \mathbf{B}$, $\mathbf{A}^3 = \mathbf{B}^3$ and $\mathbf{A}^2\mathbf{B} = \mathbf{B}^2\mathbf{A}$. The conclusion is to determine if $\mathbf{C} = \mathbf{A}^2 + \mathbf{B}^2$ is invertible. Either $\mathbf{C}$ is invertible or it is not. How do we show that a matrix is invertible? One way is to show that its determinant is non-zero. That seems difficult, since the matrices are $n \times n$, where n is arbitrary, and the formula for a determinant is very complicated once $n \geq 3$. Other ways to show that a matrix $\mathbf{C}$ is invertible is by showing that $\mathbf{C}b_i \neq 0$ for each basis vector $b_1, b_2, \ldots, b_n$. But that is also hard, since we need to find a basis.

The hunt for a penultimate step for invertibility has failed. That's OK; now we think about *non*-invertibility. That turns out to be a little easier: all we need to do is find a *single* non-zero vector v such that $\mathbf{C}v = 0$. Now that is a manageable penultimate step. Will it work? We have no way of knowing, but the strategic ideas of wishful thinking and make it easier demand that we investigate this path.

We need to use the given, start with $\mathbf{C}$, and somehow get zero. Once again, wishful thinking tells us to look at constructions such as $\mathbf{A}^3 - \mathbf{B}^3$, since $\mathbf{A}^3 = \mathbf{B}^3$. Starting with $\mathbf{C} = \mathbf{A}^2 + \mathbf{B}^2$, the most direct approach to getting the cubic terms seems fruitful (recall that matrix multiplication is not commutative, so that $\mathbf{B}^2\mathbf{A}$ is not necessarily equal to $\mathbf{A}\mathbf{B}^2$):

$$(\mathbf{A}^2 + \mathbf{B}^2)(\mathbf{A} - \mathbf{B}) = \mathbf{A}^3 - \mathbf{A}^2\mathbf{B} + \mathbf{B}^2\mathbf{A} - \mathbf{B}^3 = \mathbf{A}^3 - \mathbf{B}^3 + \mathbf{B}^2\mathbf{A} - \mathbf{A}^2\mathbf{B} = 0.$$

Now we are done! Since $\mathbf{A} \neq \mathbf{B}$, the matrix $\mathbf{A} - \mathbf{B} \neq 0$. Therefore there exists a vector u such that $(\mathbf{A} - \mathbf{B})u \neq 0$. Now just set $v = (\mathbf{A} - \mathbf{B})u$ and we have

$$\mathbf{C}v = \Big((\mathbf{A}^2 + \mathbf{B}^2)(\mathbf{A} - \mathbf{B})\Big)u = \mathbf{0}u = 0.$$

Thus $\mathbf{A}^2 + \mathbf{B}^2$ is always non-invertible.   ∎

Example 2.2.7 (Leningrad Mathematical Olympiad, 1988) Let $p(x)$ be a polynomial with real coefficients. Prove that if

$$p(x) - p'(x) - p''(x) + p'''(x) \geq 0$$

for every real x, then $p(x) \geq 0$ for every real x.

Partial Solution: If you have never seen a problem of this kind before, it is quite perplexing. What should derivatives have to do with whether a function is nonnegative or not? And why is it important that $p(x)$ be a polynomial?

We have to simplify the problem. What is the most difficult part? Obviously, the complicated expression $p(x) - p'(x) - p''(x) + p'''(x)$. **Factoring** is an important algebraic tactic (see page 163) Motivated by the factorization

$$1 - x - x^2 + x^3 = (1 - x)(1 - x^2),$$

we write

$$p(x) - p'(x) - p''(x) + p'''(x) = (p(x) - p''(x)) - (p(x) - p''(x))'.$$

In other words, if we let $q(x) = p(x) - p''(x)$, then

$$p(x) - p'(x) - p''(x) + p'''(x) = q(x) - q'(x).$$

So now we have a simpler problem to examine:

> *If $q(x)$ is a polynomial and $q(x) - q'(x) \geq 0$ for all real x, what can we say about $q(x)$?*

Is it possible as well that $q(x) \geq 0$ for every real x? This may or may not be true, nor may it solve the original problem, but it is certainly worth investigating. Wishful thinking demands that we look into this.

The inequality $q(x) - q'(x) \geq 0$ is equivalent to $q'(x) \leq q(x)$. Consequently, if $q(x) < 0$, then $q'(x)$ must also be negative. Thus, if the graph of $y = q(x)$ ever drops below the x-axis (going from left to right), then it must stay below the x-axis, for the function $q(x)$ will always be decreasing! We have three cases.

- The graph of $y = q(x)$ does cross the x-axis. By the above reasoning, it must only cross once, going from positive to negative (since once it is negative, it stays negative). Furthermore, since $q(x)$ *is a polynomial*, we know that

$$\lim_{x \to -\infty} q(x) = +\infty \quad \text{and} \quad \lim_{x \to +\infty} q(x) = -\infty,$$

because any polynomial $q(x) = a_n x^n + a_{n-1} x^{n-1} + \cdots + a_0$ is dominated by its highest-degree term $a_n x^x$ for large enough (positive or negative) x. Furthermore, $q(x)$ must have odd degree n and $a_n < 0$. For example, the graph of the polynomial $q(x) = -x^7 + x^2 + 3$ has the appropriate behavior.

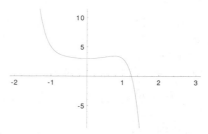

However, this function does not satisfy the inequality $q'(x) \leq q(x)$: We have $q(x) = -x^7 + x^2 + 3$ and $q'(x) = -7x^6 + 2x$. Both polynomials are dominated by their highest-degree term. When x is a large positive number, both $q(x)$ and $q'(x)$ will be negative, but $q(x)$ will be larger in absolute value, since its dominant term is degree 7 while the dominant term of $q'(x)$ is degree 6. In other words, for large enough positive x, we will have $q(x) < q'(x)$. Certainly this argument is a general one: if the graph of $y = q(x)$ crosses the x-axis, then the inequality $q'(x) \leq q(x)$ will not be true for all x. So this case is impossible.

- The graph of $y = q(x)$ stays below (or just touches) the x-axis. Then, since $q(x)$ is a polynomial, it must have even degree with a negative leading coefficient. For example, $q(x) = -5x^8 - 200$ would have the right kind of graph. However, the previous argument still applies; for large enough positive x, we will have $q(x) < q'(x)$. So this case is not possible, either.

- The graph of $y = q(x)$ stays above (or just touches) the x-axis, i.e., $q(x) \geq 0$. This case must be true, since we have eliminated the other possibilities! However, it is instructive to see why the previous argument doesn't lead to a contradiction. Now $q(x)$ must have even degree with leading coefficient *positive*, for example, $q(x) = x^2 + 10$ has the right kind of graph. But now, $q'(x) = 2x$. When x is a large positive number, $q(x) > q'(x)$ because the leading coefficients are positive. That's the key.

Anyway, we've managed to prove a very nice assertion:

If $q(x)$ is a polynomial with real coefficients satisfying $q(x) \geq q'(x)$ for all real x, then $q(x)$ has even degree with a positive leading coefficient, and is always nonnegative.

This fact should give us confidence for wrapping up the original problem. We know that $q(x) = p(x) - p''(x)$ has even degree with positive leading coefficient, hence the same is true of $p(x)$. So we have reduced the original problem to a seemingly easier one:

> *Prove that if $p(x)$ has even degree with positive leading coefficient, and $p(x) - p''(x) \geq 0$ for all real x, then $p(x) \geq 0$ for all real x.*

Example 2.2.8 (Putnam, 1990) Find all real-valued continuously differentiable functions f on the real line such that for all x,

$$(f(x))^2 = \int_0^x [(f(t))^2 + (f'(t))^2] \, dt + 1990.$$

Partial Solution: What is the worst thing about this problem? It contains both differentiation *and* integration. Differential equations are bad enough, but integral-differential equations are worse! So the strategy is obvious: **make it easier** by differentiating both sides of the equation with respect to x:

$$\frac{d}{dx}(f(x))^2 = \frac{d}{dx}\left(\int_0^x [(f(t))^2 + (f'(t))^2] \, dt + 1990\right).$$

The left-hand side is just $2f(x)f'(x)$ (by the Chain Rule), and the right-hand side becomes $(f(x))^2 + (f'(x))^2$ (the derivative of the constant 1990 vanishes).

Now we have reduced the problem to a differential equation,

$$2f(x)f'(x) = (f(x))^2 + (f'(x))^2.$$

This isn't pretty (yet), but is much nicer than what we started with. Do you see what to do next?

Problems and Exercises

For most of these, your task is just to experiment and get your hands dirty and come up with conjectures. Do not worry about proving your conjectures at this point. The idea is to stay loose and uninhibited, to get used to brainstorming. Some of the questions ask you to conjecture a formula or an **algorithm**. By the latter, we mean a computational procedure which is not a simple formula, but nevertheless is fairly easy to explain and carry out. For example, $f(n) = \sqrt{n!}$ is a formula but the following is an algorithm:

> Compute the sum of every third digit of the base-3 expansion of n. If the sum is even, that is $f(n)$. Otherwise, square it, and that is $f(n)$.

We will return to many of the problems later and develop more rigorous proofs. But it is important that you get your hands dirty *now* and start thinking about them. We reiterate: it is not important at this time to actually "solve" the problems. Having a bunch of partially solved, possibly true conjectures at the back of your mind is not only OK, it is ideal. "Backburner" problems ferment happily, intoxicating your brain with ideas. Some of this fermentation is conscious, some is not. Some ideas will work, others will fail. The more ideas, the better!

(By the way, a few of the problems were deliberately chosen to be similar to some of the examples. Remember, one of your orientation strategies is to ask, "Is there a similar problem?")

2.2.9 Define $f(x) = 1/(1-x)$ and denote r iterations of the function f by f^r [see equation (2) on page 35]. Compute $f^{1999}(2000)$.

2.2.10 (Putnam 1990) Let

$$T_0 = 2, \quad T_1 = 3, \quad T_2 = 6,$$

and for $n \geq 3$,

$$T_n = (n+4)T_{n-1} - 4nT_{n-2} + (4n-8)T_{n-3}.$$

The first few terms are

$$2, 3, 6, 14, 40, 152, 784, 5168, 40576, 363392.$$

Find a formula for T_n of the form $T_n = A_n + B_n$, where (A_n) and (B_n) are well-known sequences.

2.2.11 Let **N** denote the natural numbers $\{1, 2, 3, 4, \ldots\}$. Consider a function $f : \mathbf{N} \to \mathbf{N}$ which satisfies $f(1) = 1$, $f(2n) = f(n)$ and $f(2n+1) = f(2n) + 1$ for all $n \in \mathbf{N}$. Find a nice simple algorithm for $f(n)$. Your algorithm should be a single sentence long, at most.

2.2.12 Look at, draw, or build several (at least eight) **polyhedra** , i.e., solids with polygonal faces. Below are two examples: a box and a three- dimensional "ell" shape. For each polyhedron, count the number of vertices (corners), faces and edges. For example, the box has 8 vertices, 6 faces and 12 edges, while the ell has 12 vertices, 8 faces and 18 edges. Find a pattern and conjecture a rule which connects these three numbers.

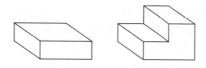

2.2.13 Into how many regions is the plane divided by n lines in **general position** (no two lines parallel; no three lines meet in a point)?

2.2.14 A **great circle** is a circle drawn on a sphere that is an "equator;" i.e., its center is also the center of the sphere. There are n great circles on a sphere, no three of which meet at any point. They divide the sphere into how many regions?

2.2.15 For each integer $n > 1$, find *distinct* positive integers x and y such that

$$\frac{1}{x} + \frac{1}{y} = \frac{1}{n}.$$

2.2.16 For each positive integer n, find positive integer solutions $x_1, x_2, \ldots, x_n$ to the equation

$$\frac{1}{x_1} + \frac{1}{x_2} + \cdots + \frac{1}{x_n} + \frac{1}{x_1 x_2 \cdots x_n} = 1.$$

2.2.17 Consider a triangle drawn on the coordinate plane, all of whose vertices are **lattice points** (points with integer coordinates). Let A, B and I respectively denote the area, number of boundary lattice points and number of interior lattice points of this triangle. For example, the triangle with vertices at $(0, 0)$, $(2, 0)$, $(1, 2)$ has (verify!) $A = 2$, $B = 4$, $I = 1$. Can you find a simple relationship between A, B and I that holds for any triangle with vertices at lattice points?

2.2.18 (British Mathematical Olympiad, 1996) Define

$$q(n) = \left\lfloor \frac{n}{\lfloor \sqrt{n} \rfloor} \right\rfloor \quad (n = 1, 2, \ldots).$$

Determine (with proof) all positive integers n for which $q(n) > q(n+1)$.

2.2.19 Bay Area Rapid Food sells chicken nuggets. You can buy packages of 7 or packages of 11. What is the largest integer n such that there is no way to buy exactly n nuggets? Can you generalize this?

2.2.20

(a) Find a nice simple formula for

$$1 + 2 + 3 + \cdots + n,$$

where n is any positive integer.

(b) Find a nice simple formula for

$$1^3 + 2^3 + 3^3 + \cdots + n^3,$$

where n is any positive integer.

(c) Experiment and conjecture the generalization to the above: For each positive integer k, is there a nice formula for

$$1^k + 2^k + 3^k + \cdots + n^k?$$

2.2.21 Define $s(n)$ to be the number of ways that the positive integer n can be written as an ordered sum of at least one positive integer. For example,

$$4 = 1 + 3 = 3 + 1 = 2 + 2 = 1 + 1 + 2 = 1 + 2 + 1 = 2 + 1 + 1 = 1 + 1 + 1 + 1,$$

so $s(4) = 8$. Conjecture a general formula.

2.2.22 Find infinitely many positive integer solutions to the equation

$$x^2 + y^2 + z^2 = w^2.$$

2.2.23 Note that $6 = 1^2 - 2^2 + 3^2$ and $7 = -1^2 + 2^2 + 3^2 - 4^2 - 5^2 + 6^2$. Investigate, generalize, conjecture.

2.2.24 (Turkey, 1996) Let

$$\prod_{n=1}^{1996} \left(1 + nx^{3^n}\right) = 1 + a_1 x^{k_1} + a_2 x^{k_2} + \ldots + a_m x^{k_m},$$

where $a_1, a_2, \ldots, a_m$ are nonzero and $k_1 < k_2 < \ldots < k_m$. Find a_{1996}.

2.2.25 *The Gallery Problem.* The walls of a museum gallery form a polygon with n sides, not necessarily regular or even convex. Guards are placed at fixed locations inside the gallery. Assuming that the guards can turn their heads, but do not walk around, what is the minimum number of guards needed to assure that every inch of wall can be observed? When $n = 3$ or $n = 4$, it is obvious that just one guard suffices. This also is true for $n = 5$, although it takes a few pictures to get convinced (the non-convex case is the interesting one). But for $n = 6$, one can create galleries that require 2 guards. Here are pictures of the $n = 5$ and $n = 6$ cases. The dots indicate positions of guards.

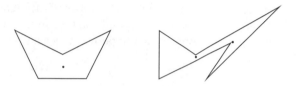

Can you discover a general formula for the number of guards, as a function of n?

2.2.26 *The Josephus Problem.* A group of n people are standing in a circle, numbered consecutively clockwise from 1 to n. Starting with person #2, we remove every other person, proceeding clockwise. For example, if $n = 6$, the people are removed in the order $2, 4, 6, 3, 1$, and the last person remaining is #5. Let $j(n)$ denote the last person remaining.

(a) Compute $j(n)$ for $n = 2, 3, \ldots 25$.

(b) Find a way to compute $j(n)$ for any positive integer $n > 1$. You may not get a "nice" formula, but try to find a convenient algorithm which is easy to compute by hand or machine.

2.2.27 Let $g(n)$ be the number of odd terms in the row of Pascal's Triangle which starts with $1, n \ldots$. For example, $g(6) = 4$, since the row

$$1, 6, 15, 20, 15, 6, 1$$

contains 4 odd numbers. Conjecture a formula (or an easy way of computing) $g(n)$.

2.2.28 (1991 Putnam) For each integer $n \geq 0$, let $S(n) = n - m^2$, where m is the greatest integer with $m^2 \leq n$. Define a sequence $(a_k)_{k=0}^{\infty}$ by $a_0 = A$ and $a_{k+1} = a_k + S(a_k)$ for $k \geq 0$. For what positive integers A is this sequence eventually constant?

2.2.29 Complete the solution started in Example 2.2.4.

2.2.30 Complete the solution started in Example 2.2.5.

2.2.31 Let $\{x\}$ denote the closest integer to the real number x. For example, $\{3.1\} = 3$ and $\{4.7\} = 5$. Now define $f(n) := n + \{\sqrt{n}\}$. Prove that, for every positive integer m, the sequence

$$m, \ f(m), \ f(f(m)), \ f(f(f(m))), \ldots$$

never contains the square of an integer. (Compare this with Example 2.2.5 on page 35.)

2.2.32 Complete the solution started in Example 2.2.7 on page 38.

2.2.33 Complete the solution started in Example 2.2.8 on page 40.

2.2.34 *Cautionary Tales.* It is easy to be seduced by the ease of experimentation-conjecture. But this is only part of mathematical investigation. Sometimes a relatively uninformed investigation leads us astray. Here are two examples. There are many other examples like this; see [10] for a wonderful discussion.

(a) Let $f(n) := n^2 + n + 41$. Is $f(n)$ a prime for all positive integers n?

(b) Let $t(n)$ be the maximum number of different areas that you can divide a circle into when you place n points on the circumference and draw all the possible line segments connecting the points. It is easy to check (verify!) that

$$t(1) = 1, \quad t(2) = 2, \quad t(3) = 4, \quad t(4) = 8, \quad t(5) = 16.$$

The conjecture that $t(n) = 2^{n-1}$ is practically inescapable. Yet $t(6)$ is equal to 31, not 32 (again, verify!), so something else is going on. Anyway, can you deduce the correct formula for $t(n)$?

2.3   Methods of Argument

As we said earlier, the solution to every problem involves two parts: the investigation, during which you discover what is going on, and the argument, in which you convince others (or maybe just yourself!) of your discoveries. Your initial investigation may suggest a tentative method of argument. Of course, sometimes a problem divides into cases or sub- problems, each of which may require completely different methods of argument.

Arguments should be rigorous and clear. However, "rigor" and "clarity" are both subjective terms. Certainly, you should avoid glaring logical flaws or gaps in your reasoning. This is easier said than done, of course. The more complicated the argument, the harder it is to decide if it is logically correct. Likewise, you should avoid deliberately vague statements, of course, but the ultimate clarity of your argument depends more on its intended audience than anything else. For example, many professional mathematicians would accept "Maximize balanced wires!" as a complete and clear solution to the Affirmative Action problem (Example 2.1.9 on page 23).

This book is much more concerned with the process of investigation and discovery than with polished mathematical argument. Nevertheless, a brilliant idea is useless if it cannot be communicated to anyone else. Furthermore, fluency in mathematical argument will help you to steer and modify your investigations.[7]

At the very least, you should be comfortable with three distinct styles of argument: straightforward **deduction** (also known as "direct proof"), argument by **contradiction**, and **mathematical induction**. We shall explore them below, but first, a few brief notes about style.

Common Abbreviations and Stylistic Conventions

1. Most good mathematical arguments start out with clear statements of the hypothesis and conclusion. The successful end of the argument is usually marked with a symbol. We use the Halmos symbol, but some other choices are the abbreviations

   QED for the Latin *quod erat demonstrandum* ("which was to be demonstrated") or the English "quite elegantly done";

   AWD for "and we're done";

   W[5] for "which was what we wanted."

2. Like ordinary exposition, mathematical arguments should be in complete sentences with nouns and verbs. Common mathematical verbs are

$$=, \quad \neq, \quad \leq, \quad \geq, \quad <, \quad >, \quad \in, \quad \subset, \quad \implies, \quad \iff.$$

   (The last four mean "is an element of," "is a subset of," "implies" and "is equivalent to," respectively.)

3. Complicated equations should always be displayed on a single line, and labeled, if referred to later. For example:

$$\int_{-\infty}^{\infty} e^{-x^2}\, dx = \sqrt{\pi}. \tag{3}$$

4. Often, as you explore the penultimate step of an argument (or sub-argument), you want to mark this off to your audience clearly. The abbreviations **TS** and **ISTS** ("to show" and "it is sufficient to show") are particularly useful for this purpose.

5. A nice bit of notation, borrowed from computer science, which is slowly becoming more common in mathematics is ":=" for "is defined to be." For example, $A := B + C$ introduces a new variable A and defines it to be the sum of the already defined variables B and C. Think of the colon as the point of an arrow; we always distinguish between left and right. The thing on the left side of the ":=" is the new definition (usually a simple variable) and the thing on the right is an expression using already defined variables. See 2.3.3 for an example of this notation.

---

[7]This section is deliberately brief. If you would like a more leisurely treatment of logical argument and methods of proof, including mathematical induction, we recommend Chapters 0 and 4.1 of [8].

6. A strictly formal argument may deal with many logically similar cases. Sometimes it is just as clear to single out one illustrative case or example. When this happens, we always alert the audience with **WLOG** ("without loss of generality"). Just make sure that you really can argue the specific and truly prove the general. For example, suppose you intended to prove that $1 + 2 + 3 + \cdots + n = n(n+1)/2$ for all positive integers n. It would be wrong to argue, "WLOG, let $n = 5$. Then $1 + 2 + 3 + 4 + 5 = 15 = 5 \cdot 6/2$. QED." This argument is certainly not general!

Deduction and Symbolic Logic

"Deduction" here has nothing to do with Sherlock Holmes. Also known as "direct proof," it is merely the simplest form of argument in terms of logic. A deductive argument takes the form "If P, then Q" or "$P \Longrightarrow Q$" or "P implies Q." Sometimes the overall structure of an argument is deductive, but the smaller parts use other styles. If you have isolated the penultimate step, then you have reduced the problem into the simple deductive statement

$$\text{The truth of the penultimate step} \quad \Longrightarrow \quad \text{The conclusion.}$$

Of course, establishing the penultimate step may involve other forms of argument.

Sometimes both $P \Longrightarrow Q$ and $Q \Longrightarrow P$ are true. In this case we say that P and Q are **logically equivalent**, or $P \Longleftrightarrow Q$. To prove equivalence, we first prove one direction (say, $P \Longrightarrow Q$) and then its **converse** $Q \Longrightarrow P$.

Keep track of the direction of the your implications. Recall that $P \Longrightarrow Q$ is *not* logically equivalent to its **converse** $Q \Longrightarrow P$. For example, consider the true statement "Dogs are mammals." This is equivalent to "If you are a dog, then you are a mammal." Certainly, the converse "If you are a mammal, then you are a dog" is not true!

However, the **contrapositive** of $P \Longrightarrow Q$ is the statement (not Q)$\Longrightarrow$(not P), and these two *are* logically equivalent. The contrapositive of "Dogs are mammals" is the true statement "Non-mammals are not dogs," clearly true.

Viewed "globally," most arguments have a simple deductive structure. But "locally," the individual pieces of the argument can take many forms. We turn now to the most common of these alternate forms, argument by contradiction.

Argument by Contradiction

Instead of directly trying to prove something, we start by assuming that it is false, and show that this assumption leads us to an absurd conclusion. A contradiction argument is usually helpful for proving directly that something *cannot* happen. Here is a simple number theory example.

Example 2.3.1 Show that

$$b^2 + b + 1 = a^2 \tag{4}$$

has no positive integer solutions.

Solution: We wish to show that the equality (4) cannot be true. So assume, to the contrary, that (4) is true. If $b^2 + b + 1 = a^2$, then $b < a$, and $a^2 - b^2 = b + 1$. As in Example 2.2.7 on page 38, we employ the useful tactic of factoring, which yields

$$(a - b)(a + b) = b + 1. \tag{5}$$

Since $a > b \geq 1$, we must have $a - b \geq 1$ and $a + b \geq 2 + b$, so the left-hand side of (5) is greater than or equal to $1 \cdot (b + 2)$, which is strictly greater than the right-hand side of (5). This is an impossibility, so the original assumption, that (4) is true, must in fact be false. ∎

Here is another impossibility proof, a classical argument from ancient Greece.

Example 2.3.2 Show that $\sqrt{2}$ is not rational.

Solution: Let us suppose that $\sqrt{2}$ is rational. Then $\sqrt{2}$ can be expressed as a quotient of two positive integers (without loss of generality we can assume both numerator and denominator are positive). Now we shall use the extreme principle: of all the possible ways of doing this, pick the quotient for which the denominator is *smallest*.

Thus we write $\sqrt{2} = a/b$ where $a, b \in \mathbf{N}$, where b is as small as possible. This means the fraction a/b is in "lowest terms," for if it were not, we could divide both a and b by a positive integer greater than 1, making both a and b smaller, contradicting the minimality of b. In particular, it is impossible that both a and b are even.

However, $\sqrt{2} = a/b$ implies that $2b^2 = a^2$, so a^2 is even (since it is equal to 2 times an integer). But this implies that a must be even as well (for if a were odd, a^2 would also be odd), and hence a is equal to 2 times an integer. So we can write $a = 2t$, where $t \in \mathbf{N}$. Substituting, we get

$$2b^2 = a^2 = (2t)^2 = 4t^2,$$

so $b^2 = 2t^2$. But now by exactly the same reasoning, we conclude that b is also even! This was impossible, so we have contradicted the original assumption, that $\sqrt{2}$ is rational. ∎

Argument by contradiction can be used to prove "positive" statements as well. Study the next example.

Example 2.3.3 (Greece, 1995) If the equation

$$ax^2 + (c + b)x + (e + d) = 0$$

has real roots greater than 1, show that the equation

$$ax^4 + bx^3 + cx^2 + dx + e = 0$$

has at least one real root.

Solution: The hypothesis is that $P(x) := ax^2 + (c + b)x + (e + d) = 0$ has real roots greater than 1, and the desired conclusion is that $Q(x) := ax^4 + bx^3 + cx^2 +$

$dx + e = 0$ has at least one real root. Let us assume that the conclusion is false, i.e., that $Q(x)$ has *no* real roots. Thus $Q(x)$ is always positive or always negative for all real x. Without loss of generality, assume that $Q(x) > 0$ for all real x, in which case $a > 0$.

Now our strategy is to use the inequality involving Q to produce a contradiction, presumably by using the hypothesis about P in some way. How are the two polynomials related? We can write

$$Q(x) = ax^4 + bx^3 + cx^2 + dx + e$$
$$= ax^4 + (c + b)x^2 + (e + d) + bx^3 - bx^2 + dx - d,$$

hence

$$Q(x) = P(x^2) + (x - 1)(bx^2 + d). \tag{6}$$

Now let y be a root of P. By hypothesis, $y > 1$. Consequently, if we set $u := +\sqrt{y}$, we have $u > 1$ and $P(u^2) = 0$. Substituting $x = u$ into (6) yields

$$Q(u) = P(u^2) + (u - 1)(bu^2 + d) = (u - 1)(bu^2 + d).$$

Recall that we assumed Q to always be positive, so $(u - 1)(bu^2 + d) > 0$. But we can also plug in $x = -u$ into (6) , and we get

$$Q(-u) = P(u^2) + (-u - 1)(bu^2 + d) = (-u - 1)(bu^2 + d).$$

So now we must have both $(u - 1)(bu^2 + d) > 0$ and $(-u - 1)(bu^2 + d) > 0$. But this is impossible, since $u - 1$ and $-u - 1$ are respectively positive and negative (remember, $u > 1$). We have achieved our contradiction, so our original assumption that Q was always positive had to be false. We conclude that Q must have at least one real root.   ∎

Why did contradiction work in this example? Certainly, there are other ways to prove that a polynomial has at least one real root. What helped us in this problem was the fact that the negation of the conclusion produced something that was easy to work with. Once we assumed that Q had no real roots, we had a nice inequality which we could play with fruitfully. When you begin thinking about a problem, it is always worth asking,

> *What happens if we negate the conclusion? Will we have something that we can easily work with?*

If the answer is "yes," then try arguing by contradiction. It won't always work, but that is the nature of investigation. To return to our old mountaineering analogy, we are trying to climb. Sometimes the conclusion seems like a vertical glass wall, but its negation has lots of footholds. Then the negation is easier to investigate than the conclusion. It's all part of the same underlying **opportunistic** strategic principle:

> ***Anything** that furthers your investigation is worth doing.*

The next example involves some basic number theory, a topic which we develop in more detail in chapter 7. However, it is important to learn at least a minimal amount of "basic survival" number theory as soon as possible. We will discuss several number theory problems in this and the next chapter. First let's introduce some extremely useful and important notation.

Let m be a positive integer. If $a - b$ is a multiple of m, we write

$$a \equiv b \pmod{m}$$

(read "a is congruent to b modulo m"). For example,

$$10 \equiv 1 \pmod{3}, \quad 17 \equiv 102 \pmod{5}, \quad 2 \equiv -1 \pmod{3}, \quad 32 \equiv 0 \pmod{8}.$$

This notation, invented by Gauss, is very convenient. Here are several facts that you should verify immediately.

- If you divide a by b and get a remainder of r, then $a \equiv r \pmod{b}$.
- The statement $a \equiv b \pmod{m}$ is equivalent to saying that there exists an integer k such that $a = b + mk$.
- If $a \equiv b \pmod{m}$ and $c \equiv d \pmod{m}$, then $a + c \equiv b + d \pmod{m}$ and $ac \equiv bd \pmod{m}$.

Example 2.3.4 Prove that if p is prime, then modulo p, every nonzero number has a *unique* multiplicative inverse; i.e., if x is not a multiple of p then there is a unique $y \in \{1, 2, 3, \ldots, p - 1\}$ such that $xy \equiv 1 \pmod{p}$. [For example, if $p = 7$, the multiplicative inverses $\pmod{7}$ of $1, 2, 3, 4, 5, 6$ are $1, 4, 5, 2, 3, 6$, respectively.]

Solution: Let $x \in \{1, 2, 3, \ldots, p - 1\}$ be nonzero modulo p; i.e., x is not a multiple of p. Now consider the $(p - 1)$ numbers

$$x, 2x, 3x, \ldots, (p - 1)x.$$

The crux idea: we claim that these numbers are all distinct modulo p. We will show this by contradiction. Assume, to the contrary, that they are not distinct. Then we must have

$$ux \equiv vx \pmod{p}, \tag{7}$$

for $u, v \in \{1, 2, 3, \ldots, p - 1\}$ with $u \neq v$. But (7) implies that

$$ux - vx = (u - v)x \equiv 0 \pmod{p};$$

in other words, $(u - v)x$ is a multiple of p. But x is not a multiple of p by hypothesis, and the value of $u - v$ is non-zero and at most $p - 2$ in absolute value (since the biggest difference would occur if one of x, y was 1 and the other was $p - 1$). Thus $u - v$ cannot be a multiple of p, either. Since p is a *prime* it is impossible, then, for the product $(u - v)x$ to be a multiple of p. That is the contradiction we wanted: we have proven that $x, 2x, 3x, \ldots, (p - 1)x$ are distinct modulo p.

Since those $p - 1$ distinct numbers are also non-zero modulo p (why?), exactly one of them has to equal 1 modulo p. Hence there exists a unique $y \in \{1, 2, 3, \ldots, p - 1\}$ such that $xy \equiv 1 \pmod{p}$. ∎

Mathematical Induction

This is a very powerful method for proving assertions that are "indexed" by integers; for example:

- The sum of the interior angles of any n-gon is $180(n-2)$ degrees.
- The inequality $n! > 2^n$ is true for all positive integers $n \geq 4$.

Each assertion can be put in the form,

P(n) is true for all integers $n \geq n_0$,

where $P(n)$ is a statement involving the integer n, and n_0 is the "starting point." There are two forms of induction, standard and strong.

Standard Induction

Here's how standard induction works:

1. Establish the truth of $P(n_0)$. This is called the "base case," and it is usually an easy exercise.
2. Assume that $P(n)$ is true for some arbitrary integer n. This is called the **inductive hypothesis**. Then show that the inductive hypothesis implies that $P(n+1)$ is also true.

This is sufficient to prove $P(n)$ for all integers $n \geq n_0$, since $P(n_0)$ is true by (1) and (2) implies that $P(n_0 + 1)$ is true, and now (2) implies that $P(n_0 + 1 + 1)$ is true, etc.

Here's an analogy. Suppose you arranged infinitely many dominos in a line, corresponding to statements $P(1), P(2), \ldots$. If you make the first domino fall to the right, then you can be sure that all of the dominos will fall, provided that whenever one domino falls, it knocks down its neighbor to the right.

Knocking the first one down is the same as establishing the base case. Showing that falling domino knocks down its neighbor is equivalent to showing that $P(n)$ implies $P(n+1)$ for all $n \geq 1$.

Let's use induction to prove the two examples above.

Example 2.3.5 Prove that the sum of the interior angles of any n-gon is $180(n-2)$ degrees.

Solution: The base case ($n_0 = 3$) is the well-known fact that the sum of the interior angles of any triangle is 180 degrees. Now assume that the theorem is true for n-gons. We will show that this implies truth for $(n+1)$-gons; i.e., the sum of the interior angles of any $(n+1)$-gon is $180(n+1-2) = 180(n-1)$ degrees.

Let S be an arbitrary $(n + 1)$-gon with vertices $v_1, v_2, \ldots, v_{n+1}$. Decompose S into the union of the triangle T with vertices v_1, v_2, v_3 and the n-gon U with vertices $v_1, v_3, \ldots, v_{n+1}$.

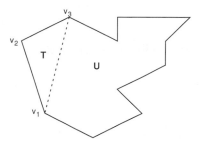

The sum of the interior angles of S is equal to the sum of the interior angles of T (which is 180 degrees), plus the sum of the interior angles of the U [which is $180(n - 2)$ by the inductive hypothesis]. Hence the sum is $180 + 180(n - 2) = 180(n - 1)$, just what we wanted. ∎

Example 2.3.6 Prove that if n is an integer greater than 3, then $n! > 2^n$.

Solution: The base case, $n_0 = 4$, is obviously true: $4! > 2^4$. Now assume that $n! > 2^n$ for some n. We wish to use this to prove the "next" case; i.e. we wish to prove that $(n + 1)! > 2^{n+1}$. Let's think strategically: the left-hand side of the inductive hypothesis is $n!$, and the left-hand side of the "goal" is $(n + 1)!$. How to get from one to the other? Multiply both sides of the inductive hypothesis by $n + 1$, of course. Multiplying an inequality by a positive number doesn't change its truth, so we get

$$(n + 1)! > 2^n(n + 1).$$

This is almost what we want, for the right-hand side of the "goal" is 2^{n+1}, and $n + 1$ is certainly greater than 2; i.e.,

$$(n + 1)! > 2^n(n + 1) > 2^n \cdot 2 = 2^{n+1}. \quad ∎$$

Induction arguments can be rather subtle. Sometimes it is not obvious what plays the role of the "index" n. Sometimes the crux move is a clever choice of this variable, and/or a neat method of traveling between $P(n)$ and $P(n + 1)$. Here is an example.

Example 2.3.7 The plane is divided into regions by straight lines. Show that it is always possible to color the regions with two colors so that adjacent regions are never the same color (like a checkerboard).

Solution: The statement of the problem does not involve integers directly. However, when we experiment, we naturally start out with one line, then two, etc., so the natural thing to "induct on" is the number of lines we draw. In other words, we shall prove the statement $P(n)$: "If we divide the plane with n lines, then we can color the regions with two colors so that adjacent regions are different colors." Let us call such a coloration "good." Obviously, $P(1)$ is true. Now assume that $P(n)$ is true. Draw in the $(n+1)$st line, and "invert" the coloration to the right of this line.

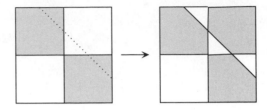

The regions to the left of the line still have a good coloration, and the regions to the right of the line also have a good coloration (just the inverse of their original colors). And the regions which share borders along the line will, by construction, be differently colored. Thus the new coloration is good, establishing $P(n+1)$.   ∎

Strong Induction

Strong induction gets its name because we use a "stronger" inductive hypothesis. After establishing the base case, we assume that for some n, *all* of the following are true:

$$P(n_0),\ P(n_0+1),\ P(n_0+2),\ \ldots,\ P(n),$$

and we use this assumption to prove that $P(n+1)$ is true. Sometimes strong induction will work when standard induction doesn't. With standard induction, we use only the truth of $P(n)$ to deduce the truth of $P(n+1)$. But sometimes, in order to prove $P(n+1)$, we need to know something about "earlier" cases. If you liked the domino analogy, consider a situation where the dominos have springs that keep them from falling, with the springs getting stiffer as n increases. Domino n requires not only the force of the nearest neighbor, but *all* of the falling neighbors to the left, in order to fall.

Our first example of strong induction continues the problem we began in Example 2.2.1 on page 31.

Example 2.3.8 Recall that the sequence $a_0, a_1, a_2, \ldots$ satisfied $a_1 = 1$ and

$$a_{m+n} + a_{m-n} = \frac{1}{2}(a_{2m} + a_{2n}) \tag{8}$$

for all nonnegative integers m and n with $m \geq n$. We deduced that $a_0 = 0$ and that $a_{2m} = 4a_m$ for all m, and we conjectured the proposition $P(n)$, which states that $a_n = n^2$.

Solution: Certainly the base case $P(0)$ is true. Now assume that $P(k)$ is true for $k = 0, 1, \ldots, u$. We have [let $m = u, n = 1$ in (8) above]

$$a_{u+1} + a_{u-1} = \frac{1}{2}(a_{2u} + a_2) = \frac{1}{2}(4a_u + a_2) = 2a_u + 2.$$

Now our stronger inductive hypothesis allows us to use the truth of both $P(u)$ and $P(u-1)$, so

$$a_{u+1} + (u-1)^2 = 2u^2 + 2,$$

and hence

$$a_{u+1} = 2u^2 + 2 - (u^2 - 2u + 1) = u^2 + 2u + 1 = (u+1)^2. \qquad \blacksquare$$

The previous example needed the truth of $P(u)$ and $P(u-1)$. The next example uses the truth of $P(k)$ for two arbitrary values.

Example 2.3.9 A **partition** of a set is a decomposition of the set into disjoint subsets. A **triangulation** of a polygon is a partition of the polygon into triangles, all of whose vertices are vertices of the original polygon. A given polygon can have many different triangulations. Here are two different triangulations of a 9-gon.

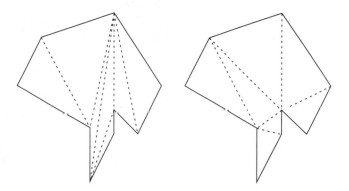

Given a triangulation, call two vertices **adjacent** if they are joined by the edge of a triangle. Suppose we decide to color the vertices of a triangulated polygon. How many colors do we need to use in order to guarantee that no two adjacent vertices have the same color? Certainly we need at least 3 colors, since that is required for just a single triangle. The surprising fact is that

Three colors always suffice for any triangulation of a polygon!

Proof: We shall induct on n, the number of vertices of the polygon. The statement $P(n)$ that we wish to prove for all integers $n \geq 3$ is

> *For any triangulation of an n-gon, it is possible to **3-color** the vertices (i.e., color the vertices using 3 colors) so that no two adjacent vertices are the same color.*

The base case $P(3)$ is obviously true. The inductive hypothesis is that

$$P(3), P(4), \ldots, P(n)$$

are all true. We will show that this implies $P(n+1)$. Given a triangulated $(n+1)$-gon, pick any edge and consider the triangle T with vertices x, y, z that contains this edge. This triangle cuts the $(n + 1)$-gon into two smaller triangulated polygons L and R (abbreviating left and right, respectively). It may turn out that one of L or R doesn't exist (this will happen if $n = 4$), but that doesn't matter for what follows. Color the vertices of L red, white and blue in such a way that none of its adjacent vertices are the same color. We know that this can be done by the inductive hypothesis!

Note that we have also colored two of the vertices (x and y) of T, and one of these vertices is also a vertex of R. Without loss of generality, let us assume that x is blue, and y is white.

By the inductive hypothesis, it is possible to 3-color R using red, white and blue so that no two of its vertices are the same color. But we have to be careful, since R shares two vertices (y and z) with T, one of which is already colored white. Consequently, vertex z must be red. If we were lucky, the coloring of R will coincide with the coloring of T. But what if it doesn't? No problem. Just rename the colors! In other words, interchange the roles of red, blue, and white in our coloring of R. For example, if the original coloring of R made y red and z blue, just recolor all of R's red vertices to be white and all blue vertices red.

We're done; we've successfully 3-colored an arbitrarily triangulated $(n + 1)$- gon so that no two adjacent vertices are the same color. So $P(n + 1)$ is true. ∎

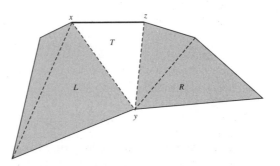

Notice that we really needed the stronger inductive hypothesis in this proof. Perhaps we could have arranged things as in Example 2.3.5 on page 50 so that we decomposed the $(n + 1)$-gon into a triangle and an n-gon. But it is not immediately obvious that among the triangles of the triangulation, there is a "boundary" triangle that we can pick. It is just easier to pick an arbitrary edge, and assume the truth of $P(k)$ for *all* $k \leq n$.

Behind the idea of strong induction is the notion that one should stay flexible when it comes to defining hypotheses and conclusions. In the following example, we need an unusually strong inductive hypothesis in order to make progress.

Example 2.3.10 Prove that

$$\left(\frac{1}{2}\right)\left(\frac{3}{4}\right)\cdots\left(\frac{2n-1}{2n}\right) \leq \frac{1}{\sqrt{3n}}.$$

Solution: We begin innocently by letting $P(n)$ denote the above assertion. $P(1)$ is true, since $1/2 \leq 1/\sqrt{3}$. Now we try to prove $P(n+1)$, given the inductive hypothesis that $P(n)$ is true. We would like to show that

$$\left\{ \left(\frac{1}{2}\right) \left(\frac{3}{4}\right) \cdots \left(\frac{2n-1}{2n}\right) \right\} \left(\frac{2n+1}{2n+2}\right) \leq \frac{1}{\sqrt{3n+3}}.$$

Extracting the quantity in the large brackets, and using the inductive hypothesis, it will suffice to prove

$$\frac{1}{\sqrt{3n}} \left(\frac{2n+1}{2n+2}\right) \leq \frac{1}{\sqrt{3n+3}}.$$

Unfortunately, this inequality is false! For example, if you plug $n = 1$ in, you get

$$\frac{1}{\sqrt{3}} \left(\frac{3}{4}\right) \leq \frac{1}{\sqrt{6}}$$

which would imply $\sqrt{2} \leq 4/3$, which is clearly absurd.

What happened? We employed wishful thinking, and got burned. It happens from time to time. The inequality that we wish to prove, while true, is very weak (i.e., asserts very little), especially for small n. The starting hypothesis of $P(1)$ is too weak to lead to $P(2)$, and we are doomed.

The solution: strengthen the hypothesis from the start. Let us replace the $3n$ with $3n + 1$. Denote the statement

$$\left(\frac{1}{2}\right) \left(\frac{3}{4}\right) \cdots \left(\frac{2n-1}{2n}\right) \leq \frac{1}{\sqrt{3n+1}}$$

by $Q(n)$. Certainly, $Q(1)$ is true; in fact, it is an *equality* ($1/2 = 1/\sqrt{4}$), which is about as sharp as an inequality can be! So let us try to prove $Q(n+1)$ using $Q(n)$ as the inductive hypothesis. As before, we try the obvious algebra, and hope that we can prove the inequality

$$\frac{1}{\sqrt{3n+1}} \left(\frac{2n+1}{2n+2}\right) \leq \frac{1}{\sqrt{3n+4}}.$$

Squaring and cross-multiplying reduce this to the alleged inequality

$$(3n+1)(4n^2 + 4n + 1) \leq 4(n^2 + 2n + 1)(3n + 1),$$

which reduces (after some tedious multiplying) to $19n \leq 20n$, and that is certainly true. So we are done.   ∎

Problems and Exercises

2.3.11 Let a, b, c be integers satisfying $a^2 + b^2 = c^2$. Give two different proofs that abc must be even,

   (a) by considering various parity cases;

(b) using argument by contradiction.

2.3.12

(a) Prove that $\sqrt{3}$ is irrational.

(b) Prove that $\sqrt{6}$ is irrational.

(c) If you attempt to prove $\sqrt{49}$ is irrational by using the same argument as before, where does the argument break down?

2.3.13 Prove that there is no smallest positive real number.

2.3.14 Prove that $\log_{10} 2$ is irrational.

2.3.15 Prove that $\sqrt{2} + \sqrt{3}$ is irrational.

2.3.16 Can the complex numbers be ordered? In other words, is it possible to define a notion of "inequality" so that any two complex numbers $a + bi$ and $c + di$ can be compared and it can be decided that one is "bigger" or one is "smaller" or they are both equal? (Using the norm function $|x + iy| = \sqrt{x^2 + y^2}$ is "cheating," for this converts each complex number into a real number and hence eludes the question of whether any two complex numbers can be compared.)

2.3.17 Prove that if a is rational and b is irrational, then $a + b$ is irrational.

2.3.18 True or false and why: If a and b are irrational, then a^b is irrational.

2.3.19 Prove the statements made in the discussion of congruence notation on page 49.

2.3.20 Prove the following generalization of Example 2.3.4 on page 49:

> Let m be a positive integer and let S denote the set of positive integers less than m which are relatively prime to m, i.e., share no common factor with m other than 1. Then for each $x \in S$, there is a unique $y \in m$ such that $xy \equiv 1 \pmod{m}$.

For example, if $m = 12$, then $S = \{1, 5, 7, 11\}$. The "multiplicative inverse" modulo 12 of each element $x \in S$ turns out to be x: $5 \cdot 5 \equiv 1 \pmod{12}, 7 \cdot 7 \equiv 1 \pmod{12}$, etc.

2.3.21 There are infinitely many primes. Of the many proofs of this important fact, perhaps the oldest was known to the ancient Greeks and written down by Euclid. It is a classic argument by contradiction. We start by assuming that there are only finitely many primes $p_1, p_2, p_3, \ldots, p_N$. Now (the ingenious crux move!) consider the number $Q := (p_1 p_2 p_3 \cdots p_N) + 1$.
  Complete the proof!

2.3.22 (Putnam 95) Let S be a set of real numbers which is closed under multiplication (that is, if a and b are in S, then so is ab). Let T and U be disjoint subsets of S whose union is S. Given that the product of any *three* (not necessarily distinct) elements of T is in T and that the product of any three elements of U is in U, show that at least one of the two subsets T, U is closed under multiplication.

2.3.23 (Russia 95) Is it possible to place 1995 different natural numbers along a circle so that for any two of these numbers, the ratio of the greatest to the least is a prime?

2.3.24 Complete the solution started in Example 1.1.2 on page 3.

2.3.25 If you haven't worked on it already, look at Problem 2.2.13 on page 41. The correct answer is $(n^2 + n + 2)/2$. Prove this by induction.

2.3.26 It is easy to prove that the product of any three consecutive integers is always divisible by 6, for at least one of the three integers is even, and at least one is divisible by 3. Done! This is a very easy proof, but as an *exercise*, prove the assertion using induction. It is less fun, but good practice.

2.3.27 Prove that a set with n elements has 2^n subsets, including the empty set and the set itself. For example, the set $\{a, b, c\}$ has the eight subsets

$$\emptyset, \{a\}, \{b\}, \{c\}, \{a, b\}, \{a, c\}, \{b, c\}, \{a, b, c\}.$$

2.3.28 Prove the formula for the sum of a geometric series:

$$a^{n-1} + a^{n-2} + \cdots + 1 = \left(\frac{a^n - 1}{a - 1} \right).$$

2.3.29 Prove that the absolute value of the sum of several real numbers or complex is at most equal to the sum of the absolute values of the numbers. Note: you will need to first verify the truth of the **triangle inequality**, which states that $|a + b| \leq |a| + |b|$ for any real or complex numbers a, b.

2.3.30 Prove that the magnitude of the sum of several vectors in the plane is at most equal to the sum of the magnitudes of the vectors.

2.3.31 Show that $7^n - 1$ is divisible by 6, for all positive integers n.

2.3.32 (Germany, 1995) Let x be a real number such that $x + 1/x$ is an integer. Prove that $x^n + 1/x^n$ is an integer, for all positive integers n.

2.3.33 Prove **Bernoulli's Inequality** , which states that if $x > -1$, $x \neq 0$ and n is a positive integer greater than 1, then

$$(1 + x)^n > 1 + nx.$$

2.3.34 After studying Problem 2.2.11 on page 41, you may have concluded that $f(n)$ is equal to the number of ones in the **binary** (base-2) representation of n. [For example, $f(13) = 3$, since 13 is 1101 in binary.] Prove this characterization of $f(n)$ using induction.

2.3.35 The Fibonacci sequence $f_1, f_2, f_3, \ldots$ was defined in Example 1.3.18. In the problems below, prove that each proposition holds for all positive integers n.

(a) $f_1 + f_3 + f_5 + \cdots + f_{2n-1} = f_{2n}$.

(b) $f_2 + f_4 + \cdots + f_{2n} = f_{2n+1} - 1$.

(c) $f_n < 2^n$.

(d) $f_n = \dfrac{1}{\sqrt{5}} \left\{ \left(\dfrac{1 + \sqrt{5}}{2} \right)^n - \left(\dfrac{1 - \sqrt{5}}{2} \right)^n \right\}$.

(e) If M is the matrix $\begin{bmatrix} 1 & 1 \\ 1 & 0 \end{bmatrix}$, then $M^n = \begin{bmatrix} f_{n+1} & f_n \\ f_n & f_{n-1} \end{bmatrix}$.

(f) $f_{n+1}f_{n-1} - f_n^2 = (-1)^n$.

(g) $f_{n+1}^2 + f_n^2 = f_{2n+1}$.

2.3.36 If you did Problem 2.2.17 on page 42, you probably discovered that if a triangle drawn on the coordinate plane has vertices at lattice points, then

$$A = \frac{1}{2}B + I - 1,$$

where A, B and I respectively denote the area, number of boundary lattice points and number of interior lattice points of this triangle. This is a special case of **Pick's Theorem**, which holds for *any* polygon with vertices at lattice points (including non-convex polygons). Prove Pick's Theorem with induction. Easy version: assume that it is true for triangles (i.e., assume the base case is true). Harder version: prove that it is true for triangles first!

2.3.37 Here are a few questions about Example 2.3.9 on page 53, where we proved that the vertices of any triangulation can be 3-colored so that no two adjacent vertices are the same color.

(a) Our proof was "nonconstructive," in that it showed that the coloring existed, but did not show how to achieve it. Can you come up with a "constructive" proof; i.e., can you outline a coloring *algorithm* that will work on an arbitrary polygon?

(b) On page 54, we raised the question of whether any arbitrary triangulation of a polygon has a "boundary triangle," i.e., a triangle that cuts the original polygon into two, rather than three, pieces. It sure seems obvious. Prove it.

(c) In the diagram for the triangulation proof, the "central" triangle split the polygon into two parts, called L and R. What if the central triangle had split the polygon into *three* parts?

2.3.38 Consider a $2^{1999} \times 2^{1999}$ square, with a single 1×1 square removed. Show that no matter where the small square was removed, it is possible to tile this "giant square minus tiny square" with ells (see Example 1.3.19 on page 12 for another problem involving tiling by ells).

2.3.39 (IMO 97) An $n \times n$ matrix (square array) whose entries come from the set $S = \{1, 2, \ldots, 2n-1\}$ is called a *silver* matrix if, for each $i = 1, \ldots, n$, the ith row and the ith column together contain all the members of S. Show that silver matrices exist for infinitely many values of n.

2.4   Other Important Strategies

Many strategies can be applied at different stages of a problem, not just the beginning. Here we focus on just a few powerful ideas. Learn to keep them in the back of your mind during any investigation. We will also discuss more advanced strategies in later chapters.

Draw a Picture!

Central to the open-minded attitude of a "creative" problem solver is an awareness that problems can and should be reformulated in different ways. Often, just translating something into pictorial form does wonders. For example, the monk problem (Example 2.1.2 on page 19) had a stunningly creative solution. But what if we just interpreted the situation with a simple distance-time graph?

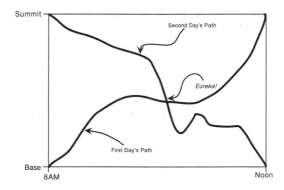

It's obvious that no matter how the two paths are drawn, they must intersect somewhere!

Whenever a problem involves several algebraic variables, it is worth pondering whether some of them can be interpreted as coordinates. The next example uses both vectors and lattice points. (See Problem 2.2.17 on page 42 and Problem 2.3.36 on page 58 for practice with lattice points.)

Example 2.4.1 How many ordered pairs of real numbers (s, t) with $0 < s, t < 1$ are there such that both $3s + 7t$ and $5s + t$ are integers?

Solution: One may be tempted to interpret (s, t) as a point in the plane, but that doesn't help much. Another approach is to view $(3s + 7t, 5s + t)$ as a point. For any s, t we have

$$(3s + 7t, 5s + t) = (3, 5)s + (7, 1)t.$$

The condition $0 < s, t < 1$ means that $(3s + 7t, 5s + t)$ is the endpoint of a vector which lies *inside* the parallelogram with vertices $(0, 0)$, $(3, 5)$, $(7, 1)$ and $(3, 5) + (7, 1) = (10, 6)$. The picture below illustrates the situation when $s = 0.4, t = 0.7$.

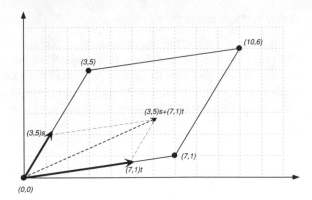

Since both $3s + 7t$ and $5s + t$ are to be integers, $(3s + 7t, 5s + t)$ is a lattice point. Consequently, counting the ordered pairs (s, t) with $0 < s, t < 1$ such that both $3s + 7t$ and $5s + t$ are integers is equivalent to counting the lattice points inside the parallelogram. This is easy to do by hand; the answer is 31.   ∎

This problem can be generalized nicely using Pick's Theorem (Problem 2.3.36 on page 58). See Problem 2.4.17 below.

Pictures Don't Help? Recast the Problem in Other Ways!

The powerful idea of converting a problem from words to pictures is just one aspect of the fundamental peripheral vision strategy. Open your mind to other ways of reinterpreting problems. One example that you have already encountered was Example 2.1.7 on page 22, where what appeared to be a sequence of numbers was actually a sequence of *descriptions* of numbers. Another example was the locker problem (Example 2.2.3 on page 32) in which a combinatorial question metamorphosed into a number theory lemma. "Combinatorics ⟷ Number Theory" is one of the most popular and productive such "crossovers," but there are many other possibilities. Some of the most spectacular advances in mathematics occur when someone discovers a new reformulation for the first time. For example, Descartes's idea of recasting geometric questions in a numeric/algebraic form led to the development of analytic geometry, which then led to calculus.

Our first example is a classic problem.

Example 2.4.2 Remove the two diagonally opposite corner squares of a chessboard. Is it possible to **tile** this shape with thirty-one 2×1 "dominos"? (In other words, every square is covered and no dominos overlap.)

Solution: At first, it seems like a geometric/combinatorial problem with many cases and subcases. But it is really just a question about counting colors. The two corners that were removed were both (without loss of generality) white, so the shape we are interested in contains 32 black and 30 white squares. Yet any domino, once it is placed, will occupy exactly one black and one white square. The 31 dominos thus require 31 black and 31 white squares, so tiling is impossible.   ∎

The idea of introducing coloring to reformulate a problem is quite old. Nevertheless, it took several years before anyone thought to use this method on the Gallery Problem (Problem 2.2.25 on page 43). This problem was first proposed, by Victor Klee, in 1973, and solved shortly thereafter by Václav Chvátal. However, his proof was rather complicated. In 1978, S. Fisk discovered the elegant coloring argument which we present below.[8] If you haven't thought about this problem yet, please do so before reading the solution below.

Example 2.4.3 *Solution to the Gallery Problem:* If we let $g(n)$ denote the minimum number of guards required for an n-sided gallery, we get $g(3) = g(4) = g(5) = 1$ and $g(6) = 2$ by easy experimentation. Trying as hard as we can to draw galleries with "hidden" rooms, it seems impossible to get a 7-sided or 8-sided gallery needing more than 2 guards, yet we can use the idea of the 6-sided, 2-guard gallery on page 43 to create a 9-sided gallery, which seems to need 3 guards. Here are examples of an 8-sided and 9-sided gallery, with dots indicating guards.

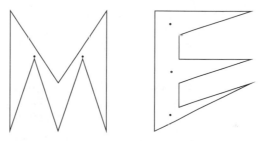

If this pattern persists, we have the tentative conjecture that $g(n) = \lfloor n/3 \rfloor$. A key difficulty with this problem, though, is that even when we draw a gallery, it is hard to be sure how many guards are needed. And as n becomes large, the galleries can become pretty complex.

A coloring reformulation comes to the rescue: Triangulate the gallery polygon. Recall that we proved, in Example 2.3.9 on page 53, that we can color the vertices of this triangulation with 3 colors in such a way that no two adjacent vertices are the same color. Now, pick a color, and station guards at all the vertices with that color. These guards will be able to view the entire gallery, since every triangle in the triangulation is guaranteed to have a guard at one of its vertices! Here is an example of a triangulation of a 15-sided gallery.

---

[8]See [14] for a nice discussion of Chvátal's proof and [19] for an exhaustive treatment of this and related problems.

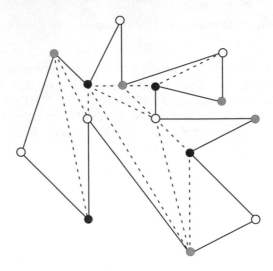

This procedure works for all three colors. One color must be used at $\lfloor n/3 \rfloor$ vertices at most (since otherwise, each color is used on more than $\lfloor n/3 \rfloor$ vertices, which would add up to more than n vertices). Choose that color, and we have shown that at most $\lfloor n/3 \rfloor$ guards are needed.

Thus, $g(n) \leq \lfloor n/3 \rfloor$. To see that $g(n) = \lfloor n/3 \rfloor$, we need only produce an example, for each n, that requires $\lfloor n/3 \rfloor$ guards. That is easy to do; just adapt the construction used for the 9-gon on page 61. If $\lfloor n/3 \rfloor = r$, just make r "spikes," etc.   ■

The next example (used for training the 1996 USA team for the IMO) is rather contrived. At first glance it appears to be an ugly algebra problem. But it is actually something else, crudely disguised (at least to those who remember trigonometry well).

Example 2.4.4 Find a value of x satisfying the equation

$$5(\sqrt{1-x} + \sqrt{1+x}) = 6x + 8\sqrt{1-x^2}.$$

Solution: Notice the constants $5, 6, 8$ and the $\sqrt{1-x^2}$ and $\sqrt{1 \pm x}$ terms. It all looks like 3-4-5 triangles (maybe 6-8-10 triangles) and trig. Recall the basic formulas:

$$\sin^2 \theta + \cos^2 \theta = 1, \quad \sin \frac{\theta}{2} = \sqrt{\frac{1 - \cos \theta}{2}}, \quad \cos \frac{\theta}{2} = \sqrt{\frac{1 + \cos \theta}{2}}.$$

Make the **trig substitution** $x = \cos \theta$. We choose cosine rather than sine, because $\sqrt{1 \pm \cos \theta}$ is involved in the half-angle formulas, but $\sqrt{1 \pm \sin \theta}$ is not. This substitution looks pretty good, since we immediately get $\sqrt{1 - x^2} = \sin \theta$, and also

$$\sqrt{1 \pm x} = \sqrt{1 \pm \cos \theta} = \sqrt{2}\sqrt{\frac{1 \pm \cos \theta}{2}}.$$

Thus the original equation becomes

$$5\sqrt{2}\left(\sin \frac{\theta}{2} + \cos \frac{\theta}{2}\right) = 6\cos \theta + 8\sin \theta. \tag{9}$$

Now we introduce a simple trig **tool**: Given an expression of the form $a\cos\theta + b\sin\theta$, write

$$a\cos\theta + b\sin\theta = \sqrt{a^2+b^2}\left(\frac{a}{\sqrt{a^2+b^2}}\cos\theta + \frac{b}{\sqrt{a^2+b^2}}\sin\theta\right).$$

This is useful, because $\dfrac{a}{\sqrt{a^2+b^2}}$ and $\dfrac{b}{\sqrt{a^2+b^2}}$ are respectively the sine and cosine of the angle $\alpha := \arctan\dfrac{a}{b}$.

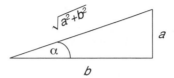

Consequently,

$$a\cos\theta + b\sin\theta = \sqrt{a^2+b^2}(\sin\alpha\cos\theta + \cos\alpha\sin\theta)$$
$$= \sqrt{a^2+b^2}\sin(\alpha+\theta).$$

In particular, we have

$$\sin x + \cos x = \sqrt{2}\sin\left(x + \frac{\pi}{4}\right).$$

Applying this, equation (9) becomes (note that $\sqrt{6^2+8^2} = 10$)

$$5\sqrt{2}\sqrt{2}\sin\left(\frac{\theta}{2} + \frac{\pi}{4}\right) = 10\left(\frac{3}{5}\cos\theta + \frac{4}{5}\sin\theta\right).$$

Hence, if $\alpha = \arctan(3/4)$, we have

$$\sin\left(\frac{\theta}{2} + \frac{\pi}{4}\right) = \sin(\alpha + \theta).$$

Equating angles yields $\theta = \dfrac{\pi}{2} - 2\alpha$. Thus

$$x = \cos\theta = \sin(2\alpha) = 2\sin\alpha\cos\alpha = 2\left(\frac{3}{5}\right)\left(\frac{4}{5}\right) = \frac{24}{25}. \qquad\blacksquare$$

Converting a problem to geometric or pictorial forms usually helps, but in some cases the reverse is true. The classic example, of course, is analytic geometry, which converts pictures into algebra. Here is a more exotic example: a problem that is geometric on the surface, but not at its core.

Example 2.4.5 We are given n planets in space, where n is a positive integer. Each planet is a perfect sphere and all planets have the same radius R. Call a point on the surface of a planet *private* if it cannot be seen from any other planet. (Ignore things such as the height of people on the planet, clouds, perspective, etc. Also, assume that the planets are not touching each other.) It is easy to check that if $n = 2$, the total private area is $4\pi R^2$, which is just the total area of one planet. What can you say about $n = 3$? Other values of n?

Partial Solution: A bit of experimentation convinces us that if $n = 3$, the total private area is also equal to the total area of one planet. Playing around with larger n suggests the same result. We conjecture that the total private area is always equal exactly to the area of one planet, no matter how the planets are situated. It appears to be a nasty problem in solid geometry, but must it be? The notions of "private" and "public" seem to be linked with a sort of duality; perhaps the problem is really not geometric, but *logical.* We need some "notation." Let us assume that there is a universal coordinate system, such as longitude and latitude, so that we can refer to the "same" location on any planet. For example, if the planets were little balls floating in a room, the location "north pole" would mean the point on a planet which was closest to the ceiling.

Given such a universal coordinate system, what can we say about a planet P which has a private point at location x? Without loss of generality, let x be at the "north pole." Clearly, the centers of all the other planets must lie on the south side of the P's "equatorial" plane. But that renders the north poles of these planets public, for their north poles are visible from a point in the southern hemisphere of P (or from the southern hemisphere of an planet that lies between). In other words, we have shown pretty easily that

If location x is private on one planet, it is public on all the other planets.

After this nice discovery, the penultimate step is clear: to prove that

Given any location x, it must be private on some planet.

We leave this as an exercise (problem?) for the reader.

The above examples just scratched the surface of the vast body of crossover ideas. While the concept of reformulating a problem is strategic, its implementation is tactical, frequently requiring specialized knowledge. We will discuss several other crossover ideas in detail in Chapter 4.

Change Your Point of View

Changing the point of view is just another manifestation of peripheral vision. Sometimes a problem is hard only because we choose the "wrong" point of view. Spending a few minutes searching for the "natural" point of view can pay big dividends. Here is a classic example.

Example 2.4.6 A person dives from a bridge into a river and swims upstream through the water for 1 hour at constant speed. She then turns around and swims downstream through the water at the same rate of speed. As the swimmer passes under the bridge, a bystander tells her that her hat fell into the river as she originally dived. The swimmer continues downstream at the same rate of speed, catching up with the hat at another bridge exactly 1 mile downstream from the first one. What is the speed of the current in miles per hour?

Solution: It is certainly possible to solve this in the ordinary way, by letting x equal the current and y equal the speed of the swimmer, etc. But what if we look at things from the *the hat's* point of view? The hat does not think that it moves. From its point of view, the swimmer abandoned it, and then swam away for an hour at a certain speed (namely, the speed of the current plus the speed of the swimmer). Then the swimmer turned around and headed back, *going at exactly the same speed*. Therefore, the swimmer retrieves the hat in exactly one hour after turning around. The whole adventure thus took two hours, during which the hat traveled one mile downstream. So the speed of the current is $\frac{1}{2}$ miles per hour. ∎

For another example of the power of a "natural" point of view, see the "Four Bugs" problem (Example 3.1.6 on page 72). This classic problem combines a clever point of view with the fundamental tactic of **symmetry**.

Problems and Exercises

2.4.7 Pat works in the city and lives in the suburbs with Sal. Every afternoon, Pat gets on a train which arrives at the suburban station at exactly 5pm. Sal leaves the house before 5 and drives at a constant speed so as to arrive at the train station at exactly 5pm to pick up Pat. The route that Sal drives never changes.

One day, this routine is interrupted, because there is a power failure at work. Pat gets to leave early, and catches a train which arrives at the suburban station at 4pm. Instead of phoning Sal to ask for an earlier pickup, Pat decides to get a little exercise, and begins walking home along the route that Sal drives, knowing that eventually Sal will intercept Pat, and then will make a U-turn, and they will head home together in the car. This is indeed what happens, and Pat ends up arriving at home 10 minutes earlier than on a normal day. Assuming that Pat's walking speed is constant, and that the U-turn takes no time, and that Sal's driving speed is constant, for how many minutes did Pat walk?

2.4.8 Two towns, A and B, are connected by a road. At sunrise, Pat begins biking from A to B along this road, while simultaneously Dana begins biking from B to A. Each person bikes at a constant speed, and they cross paths at noon. Pat reaches B at 5pm while Dana reaches A at 11:15pm. When was sunrise?

2.4.9 A bug is crawling on the coordinate plane from $(7, 11)$ to $(-17, -3)$. The bug travels at constant speed one unit per second everywhere but quadrant II (negative x- and positive y-coordinates), where it travels at $\frac{1}{2}$ units per second. What path should the bug take to complete its journey in minimal time? Generalize!

2.4.10 What is the first time after 12 o'clock that the hour and minute hands meet? This is an amusing and moderately hard algebra exercise, well worth doing if you never did it before. However, this problem can be solved in a few seconds *in your head* if you avoid messy algebra and just consider the "natural" point of view. Go for it!

2.4.11 Complete the solution of the Planets problem, started in Example 2.4.5 on page 63.

2.4.12 (Putnam 1984) Find the minimum value of

$$(u - v)^2 + \left(\sqrt{2 - v^2} - \frac{9}{v}\right)^2$$

for $0 < u < \sqrt{2}$ and $v > 0$.

2.4.13 A bug sits on one corner of a unit cube, and wishes to crawl to the diagonally opposite corner. If the bug could crawl through the cube, the distance would of course be $\sqrt{3}$. But the bug has to stay on the surface of the cube. What is the length of the shortest path?

2.4.14 Let a and b be integers greater than one which have no common divisors. Prove that

$$\sum_{i=1}^{b-1} \left\lfloor \frac{ai}{b} \right\rfloor = \sum_{j=1}^{a-1} \left\lfloor \frac{bj}{a} \right\rfloor,$$

and find the value of this common sum.

2.4.15 Let a_0 be any real number greater than 0 and less than 1. Then define the sequence $a_1, a_2, a_3, \ldots$ by $a_{n+1} = \sqrt{1 - a_n}$ for $n = 0, 1, 2, \ldots$. Show that

$$\lim_{n \to \infty} a_n = \frac{\sqrt{5} - 1}{2},$$

no matter what value is chosen for a_0.

2.4.16 For positive integers n, define S_n to be the minimum value of the sum

$$\sum_{k=1}^{n} \sqrt{(2k - 1)^2 + a_k^2},$$

as the $a_1, a_2, \ldots, a_n$ range through all positive values such that

$$a_1 + a_2 + \cdots + a_n = 17.$$

Find S_{10}.

2.4.17 (Taiwan, 1995) Let a, b, c, d be integers such that $ad - bc = k > 0$ and

$$\mathrm{GCD}(a, b) = \mathrm{GCD}(c, d) = 1.$$

Prove that there are exactly k ordered pairs of real numbers (x_1, x_2) satisfying $0 \le x_1, x_2 < 1$ and both $ax_1 + bx_2$ and $cx_1 + dx_2$ are integers.

2.4.18 Several marbles are placed on a circular track of circumference one meter. The width of the track and the radii of the marbles are negligible. Each marble is randomly given an orientation, clockwise or counterclockwise. At time zero, each marble begins to travel with speed one meter per minute, where the direction of travel depends on the orientation. Whenever two marbles collide, they bounce back with no change in speed, obeying the laws of inelastic collision.

What can you say about the possible locations of the marbles after one minute, with respect to their original positions? There are three factors to consider: the number of marbles, their initial locations, and their initial orientations.

Chapter 3

Fundamental Tactics for Solving Problems

Now we turn to the tactical level of problem solving. Recall that **tactics** are broadly applicable mathematical methods that often simplify problems. Strategy alone rarely solves problems; we need the more focused power of tactics (and often highly specialized **tools** as well) to finish the job. Of the many different tactics, this chapter will explore some of the most important ones that can be used in many different mathematical settings.

Most of the strategic ideas in Chapter 2 were plain common sense. In contrast, the tactical ideas in this chapter, while easy to use, are less "natural," as few people would think of them. Let's return to our mountaineering analogy for a moment. An important climbing tactic is the rather non-obvious idea (meant to be taken literally):

> *Stick your butt out!*

The typical novice climber sensibly hugs the rock face that he is attempting to climb, for it is not intuitive to push *away* from the rock. Yet once he grits his teeth and pushes out his rear end, a miracle happens: the component of gravity which is perpendicular to the rock rises, which increases the frictional force on his feet and immediately produces a more secure stance.

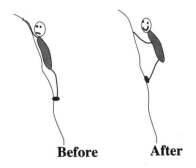

Before   **After**

Likewise, you may find that some of the tactical ideas below are peculiar. But once you master them, you will notice a dramatic improvement in your problem solving.

Many fundamental problem-solving tactics involve the *search for order*. Often problems are hard because they seem "chaotic" or disorderly; they appear to be missing parts (facts, variables, patterns) or the parts do not seem connected. Finding (and using) order can quickly simplify such problems. Consequently, we will begin by studying problem-solving tactics that help us find or impose order where there seemingly is none. The most dramatic form of order is our first topic, **symmetry**.

Symmetry involves finding or imposing order in a concrete way, for example, by reflection. Other tactics find or exploit order, but in more abstract, almost "metaphorical" ways. We shall discuss three such methods of "pseudosymmetry." All of them rely on very simple observations that sometimes yield amazingly useful information. The first tactic, the **extreme principle**, which we encountered with the Affirmative Action problem (Example 2.1.9 on page 23), works by focusing on the largest and smallest entities within a problem. Next, the **pigeonhole principle** stems from the nearly vacuous observation that if you have more guests than spare rooms, some of the guests will have to share rooms. Our final tactic, the concept of **invariants**, shows how much information comes your way when you restrict your attention to a narrow aspect of your problem which does not change (such as parity). This is an extremely powerful idea, fundamental to mathematics, that underlies many seemingly different tactics and tools.

3.1   Symmetry

We all have an intuitive idea of symmetry; for example, everyone understands that circles are symmetrical. It is helpful, however, to define symmetry in a formal way, if only because this will expand our notion of it. We call an object **symmetric** if there are one or more non-trivial "actions" which leave the object unchanged. We call the actions that do this the **symmetries** of the object.[1]

We promised something formal, but the above definition seems pretty vague. What do we mean by an "action"? Almost anything! We are being deliberately vague, because our aim is to let you see symmetry in as many situations as possible. Here are a few examples.

Example 3.1.1 A square is symmetric with respect to reflection about a diagonal. The reflection is one of the several symmetries of the square. Other symmetries include rotation clockwise by 90 degrees and reflection about a line joining the midpoints of two opposite sides.

Example 3.1.2 A circle has infinitely many symmetries, for example, rotation clockwise by α degrees for any α.

Example 3.1.3 The doubly infinite sequence

$$\ldots, 3, 1, 4, 3, 1, 4, 3, 1, 4, \ldots$$

---

[1] We are deliberately avoiding the language of transformations and automorphisms that would be demanded by a mathematically precise definition.

is symmetrical with respect to the action "shift everything three places to the right (or left)."

Why is symmetry important? Because it gives you "free" information. If you know that something is, say, symmetric with respect to 90-degree rotation about some point, then you only need to look at one-quarter of the object. And you also know that the center of rotation is a "special" point, worthy of close investigation. You will see these ideas in action below, but before we begin, let us mention two things to keep in mind as you ponder symmetry:

- The strategic principles of peripheral vision and rule-breaking tell us to look for symmetry in unlikely places, and not to worry if something is almost, but not quite symmetrical. In these cases, it is wise to proceed as if symmetry is present, since we will probably learn something useful.

- An informal alternate definition of symmetry is "harmony." This is even vaguer than our "formal" definition, but it is not without value. Look for harmony, and beauty, whenever you investigate a problem. If you can do something that makes things more harmonious or more beautiful, *even if you have no idea how to define these two terms*, then you are often on the right track.

Geometric Symmetry

Most geometric investigations profit by a look at symmetry. Ask these questions about symmetry:

- Is it present?
- If not, can it be imposed?
- How can it then be exploited?

Here is a simple but striking example.

Example 3.1.4 A square is inscribed in a circle which is inscribed in a square. Find the ratio of the areas of the two squares.

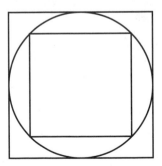

Solution: The problem can certainly be solved algebraically (let x equal the length of the small square, then use Pythagorean theorem, etc.) but there is a nicer approach.

The original diagram is full of symmetries. We are free to rotate and/or reflect many shapes and still preserve the areas of the two squares. How do we choose from all these possibilities? We need to use the hypothesis that the objects are inscribed in one another. If we rotate the small square by 45 degrees, its vertices now line up with the points of tangency between the circle and the large square, and instantly the solution emerges.

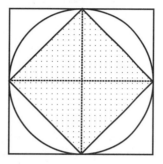

The small square obviously has half the area of the larger! ∎

The simplest geometric symmetries are rotational and reflectional. Always check to see if rotations or reflections will impose order on your problem. The next example shows the power of imposing reflectional symmetry.

Example 3.1.5 Your cabin is 2 miles due north of a stream which runs east-west. Your grandmother's cabin is located 12 miles west and 1 mile north of your cabin. Every day, you go from your cabin to Grandma's, but first visit the stream (to get fresh water for Grandma). What is the length of the route with minimum distance?

Solution: First, **draw a picture**! Label your location by Y and Grandma's by G. Certainly, this problem can be done with calculus, but it is very ugly (you need to differentiate the sum of two radicals, for starters). The problem appears to have no symmetry in it, but the stream is practically begging you to reflect across it! Draw a sample path, (shown below as YA followed by AG) and look at its reflection. We call the reflections of your house and Grandma's house Y' and G', respectively.

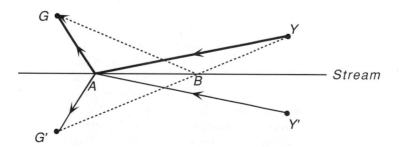

While you are carrying water to Grandma, your duplicate in an alternate universe is doing the same, only south of the stream. Notice that $AG = AG'$, so the length of your path is *unchanged* if you visited the reflected Grandma instead of the real one. Since the shortest distance between two points is the straight line YBG', the optimal path will be YB followed by BG. Its length is the same as the length of YBG', which is just the hypotenuse of a right triangle with legs 5 and 12 miles. Hence our answer is 13 miles.   ■

When pondering a symmetrical situation, you should always focus briefly on the "fixed" objects which are unchanged by the symmetries. For example, if something is symmetric with respect to reflection about an axis, that axis is fixed and worthy of study (the stream in the previous problem played that role). Here is another example, a classic problem which exploits rotational symmetry along with a crucial fixed point.

Example 3.1.6 Four bugs are situated at each vertex of a unit square. Suddenly, each bug begins to chase its counterclockwise neighbor. If the bugs travel at 1 unit per minute, how long will it take for the four bugs to crash into one another?

Solution: The situation is rotationally symmetric in that there is no one "distinguished" bug. If their starting configuration is that of a square, then they will always maintain that configuration. This is the key insight, believe it or not, and it is a very profitable one!

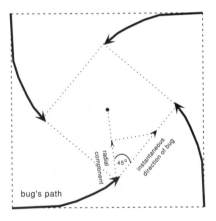

As time progresses, the bugs form a shrinking square which rotates counterclockwise. The center of the square does not move. This center, then, is the only "distinguished" point, so we focus our analysis on it. Many otherwise intractable problems become easy once we shift our focus to the natural frame of reference; in this case we should consider a *radial* frame of reference, one that rotates with the square. For example, pick one of the bugs (it doesn't matter which one!), and look at the line segment from the center of the square to the bug. This segment will rotate counterclockwise, and (more importantly) shrink. When it has shrunk to zero, the bugs will have crashed into one another. How fast is it shrinking? Forget about the fact that the line is rotating. *From the point of view of this radial line,* the bug is always traveling at a 45° angle. Since the bug travels at unit speed, its radial velocity component is just $1 \cdot \cos 45° = \sqrt{2}/2$ units per minute, i.e., the radial line shrinks at this speed. Since the original length of the radial line was $\sqrt{2}/2$, it will take just 1 minute for the bugs to crash. ∎

Here is a simple calculus problem that can certainly be solved easily in a more conventional way. However, our method below illustrates the power of the Draw a Picture strategy coupled with symmetry, and can be applied in many harder situations.

Example 3.1.7 Compute $\displaystyle\int_0^{\frac{1}{2}\pi} \cos^2 x\, dx$ in your head.

Solution: Mentally draw the sine and cosine graphs from 0 to $\frac{1}{2}\pi$, and you will notice that they are symmetric with respect to reflection about the vertical line $x = \frac{1}{4}\pi$. Thus

$$\int_0^{\frac{1}{2}\pi} \cos^2 x\, dx = \int_0^{\frac{1}{2}\pi} \sin^2 x\, dx.$$

Consequently, the value that we seek is

$$\frac{1}{2}\int_0^{\frac{1}{2}\pi} (\cos^2 x + \sin^2 x)\, dx = \frac{1}{2}\int_0^{\frac{1}{2}\pi} 1 \cdot dx = \frac{\pi}{4}. \qquad ∎$$

The next problem, from the 1995 IMO, is harder than the others, but only in a "technical" way. In order to solve it, you need to be familiar with **Ptolemy's Theorem**, which states

> Let $ABCD$ be a cyclic quadrilateral, i.e., a quadrilateral whose vertices lie on a circle. Then

$$AB \cdot CD + AD \cdot BC = AC \cdot BD.$$

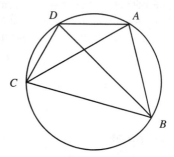

See [3], p. 42, for a proof. A key feature of cyclic quadrilaterals is the easily verified fact that

A quadrilateral is cyclic if and only if its opposite angles are supplementary (add up to 180 degrees).

Example 3.1.8 Let $ABCDEF$ be a convex hexagon with $AB = BC = CD$ and $DE = EF = FA$, such that $\angle BCD = \angle EFA = \pi/3$. Suppose G and H are points in the interior of the hexagon such that $\angle AGB = \angle DHE = 2\pi/3$. Prove that $AG + GB + GH + DH + HE \geq CF$.

Solution: First, as with all geometry problems, draw an accurate diagram, using pencil, compass and ruler. Look for symmetry. Note that BCD and EFA are equilateral triangles, so that $BD = BA$ and $DE = AE$. By the symmetry of the figure, it is seems profitable to reflect about BE. Let C' and F' be the reflections of C and F respectively.

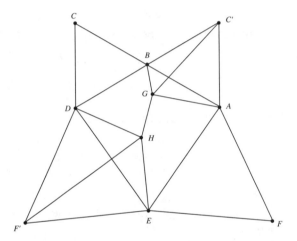

A tactic that often works for geometric inequalities is to look for a way to compare the sum of several lengths with a single length, since the shortest distance between two points is a straight line.

Quadrilaterals $AGBC'$ and $HEF'D$ are cyclic, since the opposite angles add up to 180 degrees. Ptolemy's Theorem then implies that $AG \cdot BC' + GB \cdot AC' = C'G \cdot AB$. Since ABC' is equilateral, this implies that $AG + GB = C'G$. Similarly, $DH + HE = HF'$. The shortest path between two points is a straight line. It follows that

$$CF = C'F' \leq C'G + GH + HF' = AG + GB + GH + DH + HE,$$

with equality if and only if G and H both lie on $C'F'$.   ∎

Algebraic Symmetry

Don't restrict your notions of symmetry to physical or geometric objects. For example, sequences can have symmetry, like this row of Pascal's Triangle:

$$1, 6, 15, 20, 20, 15, 6, 1.$$

That's only the beginning. In just about any situation where you can imagine "pairing" things up, you can think about symmetry. And thinking about symmetry almost always pays off.

The Gaussian Pairing Tool

Carl Friedrich Gauss (1777-1855) was certainly one of the greatest mathematicians of all times. Many stories celebrate his precocity and prodigious mental power. No one knows how true these stories are, because many of them are attributable only to Gauss himself. The following anecdote has many variants. We choose one of the simplest.

When Gauss was 10, his teacher punished his class with a seemingly tedious sum:

$$1 + 2 + 3 + \cdots + 98 + 99 + 100.$$

While the other students slowly added the numbers, little Carl discovered a short-cut and immediately arrived at the answer of 5,050. He was the only student to find the correct sum. His insight was to notice that 1 could be paired with 100, 2 with 99, 3 with 98, etc. to produce 50 identical sums of 101. Hence the answer of $101 \cdot 50 = 5050$. Another, more formal way of doing this is to write the sum in question twice, first forward, then backward:

$$\begin{aligned} S &= 1 &+ 2 &+ \cdots + 99 &+ 100, \\ S &= 100 &+ 99 &+ \cdots + 2 &+ 1. \end{aligned}$$

Then, of course, $2S = 100 \cdot 101$. The advantage of this method is that it does not matter whether the number of terms was even or odd (notice that this is an issue with the original pairing method).

This is a pretty good trick, especially for a 10-year-old, and it has many applications. Don't be restricted to sums. Look for any kind of *symmetry* in a problem and then investigate whether a clever pairing of items can simplify things. First, let's tackle the "Locker Problem," which we first encountered in Example 2.2.3 on page 32. Recall that we reduced this problem to

Prove that $d(n)$ is odd if and only if n is a perfect square,

where $d(n)$ denotes the number of divisors of n, including 1 and n. Now we can use the Gaussian pairing tool. The symmetry here is that one can always pair a divisor d of n with n/d. For example, if $n = 28$, it is "natural" to pair the divisor 2 with the divisor 14. Thus, as we go through the list of divisors of n, each divisor will have a unique "mate" *unless n is a perfect square*, in which case $\sqrt{n}$ is paired with itself. For example, the divisors of 28 are $1, 2, 4, 7, 14, 28$, which can be rearranged into the pairs $(1, 28), (2, 14), (4, 7)$, so clearly $d(28)$ is even. On the other hand, the divisors of the perfect square 36 are $1, 2, 3, 4, 6, 9, 12, 18, 36$, which pair into $(1, 36), (2, 18), (3, 12), (4, 9), (6, 6)$. Notice that 6 is paired with itself, so the actual count of divisors is odd (several true pairs plus the 6 which can only be counted once). We can conclude that $d(n)$ is odd if and only if n is a perfect square.[2]   ■

---

[2]There is a completely different argument, using basic counting methods and parity. See problem 6.1.21 on page 212.

The argument used can be repeated almost identically, yet in a radically different context, to prove a well-known theorem in number theory.

Example 3.1.9 *Wilson's Theorem.* Prove that for all primes p,

$$(p-1)! \equiv -1 \pmod{p}.$$

Solution: First try an example. Let $p = 13$. Then the product in question is $1 \cdot 2 \cdot 3 \cdot 4 \cdot 5 \cdot 6 \cdot 7 \cdot 8 \cdot 9 \cdot 10 \cdot 11 \cdot 12$. One way to evaluate this product modulo 13 would be to multiply it all out, but that is not the problem solver's way! We should look for smaller subproducts that are easy to compute modulo p. The easiest numbers to compute with are 0, 1 and -1. Since p is a prime, none of the factors $1, 2, 3, \ldots p - 1$ are congruent to 0 modulo p. Likewise, no subproducts can be congruent to 0. On the other hand, the primality of p implies that every nonzero number has a *unique* multiplicative inverse modulo p; i.e., if x is not a multiple of p then there is a unique $y \in \{1, 2, 3, \ldots, p - 1\}$ such that $xy \equiv 1 \pmod{p}$. (Recall the proof of this assertion in Example 2.3.4 on page 49.)

Armed with this information, we proceed as in the locker problem: Pair each element $x \in \{1, 2, 3, \ldots, p - 1\}$ with its "natural" mate $y \in \{1, 2, 3, \ldots, p - 1\}$ such that $xy \equiv 1 \pmod{p}$. For example, if $p = 13$, the pairs are

$$(1, 1), (2, 7), (3, 9), (4, 10), (5, 8), (6, 11), (12, 12),$$

and we can rewrite

$$12! = 1 \cdot 2 \cdot 3 \cdot 4 \cdot 5 \cdot 6 \cdot 7 \cdot 8 \cdot 9 \cdot 10 \cdot 11 \cdot 12$$

as

$$1 \cdot (2 \cdot 7)(3 \cdot 9)(4 \cdot 10)(5 \cdot 8)(6 \cdot 11) \cdot 12 \equiv 1 \cdot 1 \cdot 1 \cdot 1 \cdot 1 \cdot 1 \cdot 12 = -1 \pmod{13}.$$

Notice that 1 and 12 were the only elements that paired with themselves. Notice also that $12 \equiv -1 \pmod{13}$. We will be done in general if we can show that x is its own multiplicative inverse modulo p if and only if $x = \pm 1$. But this is easy: If $x^2 \equiv 1 \pmod{p}$, then

$$x^2 - 1 = (x - 1)(x + 1) \equiv 0 \pmod{p}.$$

Because p is prime, this implies that either $x - 1$ or $x + 1$ is a multiple of p; i.e., $x \equiv \pm 1 \pmod{p}$ are the *only* possibilities.   ∎

Symmetry in Polynomials and Inequalities

Algebra problems with many variables or of high degree are often intractable unless there is some underlying symmetry to exploit. Here is a lovely example.

Example 3.1.10 Solve $x^4 + x^3 + x^2 + x + 1 = 0$.

Solution: While there are other ways of approaching this problem (see page 137), we will use the symmetry of the coefficients as a starting point to impose yet more symmetry, on the degrees of the terms. Simply divide by x^2:

$$x^2 + x + 1 + \frac{1}{x} + \frac{1}{x^2} = 0.$$

This looks no simpler, but note that now there is more symmetry, for we can collect "like" terms as follows:

$$x^2 + \frac{1}{x^2} + x + \frac{1}{x} + 1 = 0. \tag{1}$$

Now make the **substitution** $u := x + \frac{1}{x}$. Note that

$$u^2 = x^2 + 2 + \frac{1}{x^2},$$

so (1) becomes $u^2 - 2 + u + 1 = 0$, or $u^2 + u - 1 = 0$, which has solutions

$$u = \frac{-1 \pm \sqrt{5}}{2}.$$

Solving $x + \frac{1}{x} = u$, we get $x^2 - ux + 1 = 0$, or

$$x = \frac{u \pm \sqrt{u^2 - 4}}{2}.$$

Putting these together, we have

$$x = \frac{\frac{-1 \pm \sqrt{5}}{2} \pm \sqrt{\left(\frac{-1 \pm \sqrt{5}}{2}\right)^2 - 4}}{2} = \frac{-1 \pm \sqrt{5} \pm i\sqrt{2\sqrt{5} \pm 10}}{4}. \qquad \blacksquare$$

The last few steps are mere "technical details." The two crux moves were to increase the symmetry of the problem and then make the symmetrical substitution $u = x + x^{-1}$.

In the next example, we use symmetry to reduce the complexity of an inequality.

Example 3.1.11 Prove that

$$(a+b)(b+c)(c+a) \geq 8abc$$

is true for all positive numbers a, b and c, with equality only if $a = b = c$.

Solution: Observe that the alleged inequality is symmetric, in that it is unchanged if we permute any of the variables. This suggests that we not multiply out the left side (rarely a wise idea!) but instead look at the factored parts, for the sequence

$$a+b, \quad b+c, \quad c+a$$

can be derived by just looking at the term $a + b$ and then performing the **cyclic permutation** $a \mapsto b, b \mapsto c, c \mapsto a$ once and then twice.

The simple 2-variable version of the **Arithmetic-Geometric-Mean** inequality (see section 5.5 for more details) implies

$$a + b \geq 2\sqrt{ab}.$$

Now, just perform the cyclic permutations

$$b + c \geq 2\sqrt{bc}$$

and

$$c + a \geq 2\sqrt{ca}.$$

The desired inequality follows by multiplying these three inequalities. ∎

It is worth exploring the concept of cyclic permutation in more detail. Given an n-variable expression $f(x_1, x_2, \ldots, x_n)$ we will denote the **cyclic sum** by

$$\sum_\sigma f(x_1, x_2, \ldots, x_n) :=$$

$$f(x_1, x_2, \ldots, x_n) + f(x_2, x_3, \ldots, x_n, x_1) + \cdots + f(x_n, x_1, \ldots, x_{n-1}).$$

For example, if our variables are x, y and z, then

$$\sum_\sigma x^3 = x^3 + y^3 + z^3 \quad \text{and} \quad \sum_\sigma xz^2 = xz^2 + yx^2 + zy^2.$$

Let us use this notation to factor a symmetric cubic in three variables.

Example 3.1.12 Factor $a^3 + b^3 + c^3 - 3abc$.

Solution: We hope for the best, and proceed naively, making sure that our guesses stay symmetric. The simplest guess for a factor would be $a + b + c$, so let us try it. Multiplying $a + b + c$ by $a^2 + b^2 + c^2$ would give us the cubic terms, with some error terms. Specifically, we have

$$a^3 + b^3 + c^3 - 3abc = (a + b + c)(a^2 + b^2 + c^2) - \sum_\sigma (a^2 b + b^2 a) - 3abc$$

$$= (a + b + c)(a^2 + b^2 + c^2) - \sum_\sigma (a^2 b + b^2 a + abc)$$

$$= (a + b + c)(a^2 + b^2 + c^2) - \sum_\sigma (ab(a + b + c))$$

$$= (a + b + c)\left(a^2 + b^2 + c^2 - \sum_\sigma ab\right)$$

$$= (a + b + c)(a^2 + b^2 + c^2 - ab - bc - ac).$$

Notice how the $\sum_\sigma$ notation saves time and, once you get used to it, reduces the chance for an error.

Problems and Exercises

3.1.13 Find the length of the shortest path from the point (3,5) to the point (8,2) that touches the x-axis and also touches the y-axis.

3.1.14 In Example 1.2.1 on page 6, we saw that the product of four consecutive integers is always one less than a square. What can you say about the product of four consecutive terms of an arbitrary arithmetical progression, e.g., $3 \cdot 8 \cdot 13 \cdot 18$?

3.1.15 Find (and prove) a nice formula for the product of the divisors of any integer. For example, if $n = 12$, the product of its divisors is

$$1 \cdot 2 \cdot 3 \cdot 4 \cdot 6 \cdot 12 = 1728.$$

You may want use the $d(n)$ function (defined in the solution to the locker problem on page 75) in your formula.

3.1.16 A polynomial in several variables is called **symmetric** if it is unchanged when the variables are permuted. For example, if $f(x, y, z) := x^2 + y^2 + z^2 + xyz$ is symmetric, since

$$f(x, y, z) = f(x, z, y) = f(y, x, z) = f(y, z, x) = f(z, x, y) = f(z, y, x).$$

A polynomial in several variables is called **homogeneous** of degree r if all of the terms are degree r. For example, $g(x, y) := x^2 + 5xy$ is homogeneous of degree 2. (The $5xy$ term is considered to have degree 2, since it is the product of two degree-1 terms. In general, the k-variable polynomial $g(x_1, x_2, \ldots, x_k)$ is homogeneous of degree r if for all t, we have

$$g(tx_1, tx_2, \ldots, tx_k) = t^r g(x_1, x_2, \ldots, x_k).$$

Given three variables x, y, z, we define the **elementary symmetric functions**

$$s_1 := x + y + z,$$
$$s_2 := xy + yz + zx,$$
$$s_3 := xyz.$$

The elementary symmetric function s_k is symmetric, homogeneous of degree k, with all coefficients equal to 1. The elementary symmetric functions can be defined for any number of variables. For example, for four variables x, y, z, w, they are

$$s_1 := x + y + z + w,$$
$$s_2 := xy + xz + xw + yz + yw + zw,$$
$$s_3 := xyz + xyw + xzw + yzw,$$
$$s_4 := xyzw.$$

(a) Verify that

$$x^2 + y^2 + z^2 = (x + y + z)^2 - 2(xy + yz + zx) = s_1^2 - 2s_2,$$

where the s_i are the elementary symmetric functions in 3 variables.

(b) Likewise, express $x^3 + y^3 + z^3$ as a polynomial in the elementary symmetric functions.

(c) Do the same for $(x + y)(x + z)(y + z)$.

(d) Do the same for $xy^4 + yz^4 + zx^4 + xz^4 + yx^4 + zy^4$.

(e) Can any symmetric polynomial in 3 variables be written as a polynomial in the elementary symmetric functions?

(f) Can any polynomial (not necessarily symmetric) in 3 variables be written as a polynomial in the elementary symmetric functions?

(g) Generalize to more variables. If you are confused, look at the 2-variable case ($s_1 := x + y$, $s_2 := xy$).

(h) What is the relationship, if any, between cyclic sums and elementary symmetric functions?

3.1.17 Consider Example 3.1.6, the Four Bugs Problem. As the bugs travel, they "turn." For example, if one bug starts out facing due north, but then gradually comes to face due west, it will have turned 90°. It may even be that the bugs turn more than 360°. How much does each bug turn (in degrees) before they crash into each other?

3.1.18 Consider the following two-player game. Each player takes turns placing a penny on the surface of a rectangular table. No penny can touch a penny which is already on the table. The table starts out with no pennies. The last player who makes a legal move wins. Does the first player have a winning strategy?

3.1.19 A billiard ball strikes ray $\overrightarrow{BC}$ at point C, with angle of incidence α as shown. The billiard ball continues its path, bouncing off line segments $\overline{AB}$ and $\overline{BC}$ according to the rule: angle of incidence equals angle of reflection. If $AB = BC$, determine the number of times the ball will bounce off the two line segments (including the first bounce, at C). Your answer will be a function of α and β.

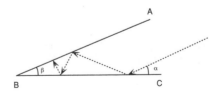

3.1.20 A projectile is launched from the very center of the floor of a rectangular room which is 40 feet wide with a very high ceiling. The projectile hits the wall at a height exactly 10 feet above the floor, reflects off this wall (obeying the "angle of incidence equals angle of reflection" rule), hits the opposite wall, and reflects again, finally landing back exactly where it was launched, without hitting the ceiling. This is possible because the projectile does not travel along straight lines, but instead travels along parabolic segments due to gravity. When the projectile is at its highest point, how high above the floor is it?

3.1.21 Recall that the ellipse is defined to be the locus of all points in the plane, the sum of whose distance to two fixed points (the foci) is a constant. Prove the *reflection property of the ellipse:* if a pool table is built with an elliptical wall and you shoot a ball from one focus to *any* point on the wall, the ball will reflect off the wall and travel straight to the other focus.

3.1.22 Recall that the parabola is defined to be the locus of all points in the plane, such that the distance to a fixed points (the foci) is equal to the distance to a fixed line (the directrix). Prove the *reflection property of the parabola:* if a beam of light, traveling perpendicular to the directrix, strikes any point on the concave side of a parabolic mirror, the beam will reflect off the mirror and travel straight to the focus.

3.1.23 A spherical, 3-dimensional planet has center at $(0, 0, 0)$ and radius 20. At any point (x, y, z) on the surface of this planet, the temperature is $T(x, y, z) := (x + y)^2 + (y - z)^2$ degrees. What is the average temperature of the surface of this planet?

3.1.24 (Putnam 1980) Evaluate

$$\int_0^{\pi/2} \frac{dx}{1 + (\tan x)^{\sqrt{2}}}.$$

3.1.25 (Hungary, 1906) Let K, L, M, N designate the centers of the squares erected on the four sides (outside) of a rhombus. Prove that the polygon $KLMN$ is a square.

3.1.26 Sharpen the problem above by showing that the conclusion still holds if the rhombus is merely an arbitrary parallelogram.

3.1.27 (Putnam 1992) Four points are chosen at random on the surface of a sphere. What is the probability that the center of the sphere lies inside the tetrahedron whose vertices are at the four points? (It is understood that each point is independently chosen relative to a uniform distribution on the sphere.)

3.1.28 *Symmetry in Probability.* Imagine dropping three pins at random on the unit interval $[0, 1]$. They separate the interval into 4 pieces. What is the average length of each piece? It seems obvious that the answer "should" be 1/4, and this would be true if the probability distributions (mean, standard deviation, etc.) for each of the four lengths are identical. And indeed, this is true. One way to see this is to imagine that we are not actually dropping 3 pins on a line segment, but instead dropping 4 pins on a circle with circumference 1 unit. Wherever the fourth point lands, cut the circle there and "unwrap" it to form the unit interval. Ponder this argument until it makes sense. Then try the next few problems!

(a) An ordinary deck of 52 cards with 4 aces is shuffled, and then the cards are drawn one by one until the first ace appears. On the average, how many cards are drawn?

(b) (Jim Propp) Given a deck of 52 cards, extract 26 of the cards at random in one of the $\binom{52}{26}$ possible ways and place them on the top of the deck in the same relative order as they were before being selected. What is the expected number of cards that now occupy the same position in the deck as before?

(c) Given any sequence of n distinct integers, we compute its "swap number" in the following way: Reading from left to right, whenever we reach a number which is less than the first number in the sequence, we swap its position with the first number in the sequence. We continue in this way until we get to the end of the sequence. The swap number of the sequence is the total number of swaps. For example, the sequence $3, 4, 2, 1$ has a swap number of 2, for we swap 3 with 2 to get $2, 4, 3, 1$ and then we swap 2 with 1 to get $1, 4, 3, 2$. Find the average value of the swap numbers of the $7! = 5040$ different permutations of the integers $1, 2, 3, 4, 5, 6, 7$.

3.2   The Extreme Principle

When you begin grappling with a problem, one of the difficulties is that there are just so many things to keep track of and understand. A problem may involve a sequence with many (perhaps infinitely many) elements. A geometry problem may use many different lines and other shapes. A good problem solver *always* tries to organize this mass of stuff. A fundamental tactic is the **extreme principle:**

> *If possible, assume that the elements of your problem are "in order."*
> *Focus on the "largest" and "smallest" elements, as they may be constrained in interesting ways.*

This seems almost trite, yet it works wonders in some situations (for example, the Affirmative Action problem of page 23). Here is a simple example.

Example 3.2.1 Let B and W be finite sets of black and white points, respectively, in the plane, with the property that every line segment which joins two points of the same color contains a point of the other color. Prove that both sets must lie on a single line segment.

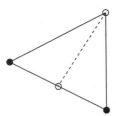

    Solution: After experimenting, it seems that if the points do not all lie on a single line, that there cannot be finitely many of them. It may be possible to prove this with many complicated diagrams showing that "you can always draw a new point," but this isn't easy. The extreme principle comes to the rescue: Assume that the points do not all lie on a line. Then they form at least one triangle. *Consider the triangle of smallest area.* Two of its vertices are the same color, so between them is a point of the other color, but this forms a smaller triangle—a contradiction!   ∎

Note how the extreme principle immediately cracked the problem. It was so easy that it almost seemed like cheating. The structure of the argument used is a typical one: Work by contradiction; assume that whatever you wish to prove is not true. Then look at the minimal (or maximal) element and develop an argument that creates a smaller (or larger) element, which is the desired contradiction. As long as a set of real numbers is finite, it will have a minimum and a maximum element. For infinite sets, there may be no extreme values (for example, consider the infinite set $\{1, 2, 1/2, 2^2, 1/2^2, 2^3, 1/2^3, \ldots\}$, which has neither a minimum nor a maximum element), but if the set consists of positive integers, we can use the **Well-Ordering Principle**:

Every non-empty set of positive integers has a least element.

Here is another simple example. The tactic used is the extreme principle; the crux move is a clever geometric construction.

Example 3.2.2 (Korea, 1995) Consider finitely many points in the plane such that, if we choose any three points A, B, C among them, the area of triangle ABC is always less than 1. Show that all of these points lie within the interior or on the boundary of a triangle with area less than 4.

Solution: Let triangle ABC have the largest area among all triangles whose vertices are taken from the given set of points. Let $[ABC]$ denote the area of triangle ABC. Then $[ABC] \leq 1$. Let triangle LMN be the triangle whose **medial triangle** is ABC. (In other words, A, B, C are the midpoints of the sides of triangle LMN. See figure.)

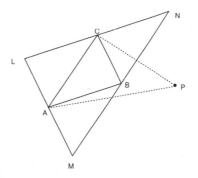

Then $[LMN] = 4[ABC] \leq 4$. We claim that the set of points must lie on the boundary or in the interior of LMN. Suppose a point P lies outside LMN. Then we can connect P with two of the vertices of ABC forming a triangle with larger area than ABC, contradicting the maximality of $[ABC]$. ∎

Always be aware of order and maximum/minimum in a problem, and always assume, if possible, that the elements are arranged in order (we call this **monotonizing**). Think of this as "free information." The next example illustrates the principle once again that a close look at the maximum (and minimum) elements often pays off. You first encountered this as Problem 1.1.4 on page 4. We break down the solution into two parts, the investigation, followed by a formal write-up.

Example 3.2.3 I invite 10 couples to a party at my house. I ask everyone present, including my wife, how many people they shook hands with. It turns out that everyone shook hands with a different number of people. If we assume that no one shook hands with his or her partner, how many people did my wife shake hands with? (I did not ask myself any questions.)

Investigation: This problem seems intractable. There doesn't appear to be enough information. Nevertheless, we can **make it easier** by looking at a simpler case, one where there are are, say, 2 couples in addition to the host and hostess.

The host discovers that of the 5 people he interrogated, there are 5 different "handshake numbers." Since these numbers range from 0 to 4 inclusive (no one shakes with their partner), the 5 handshake numbers discovered are 0, 1, 2, 3 and 4. Let's call these people $P_0, P_1, \ldots, P_4$, respectively, and let's **draw a picture**, including the host in our diagram (with the label H).

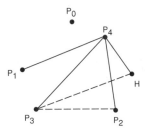

It is *interesting* to look at the two people with the *extreme* handshake numbers, i.e., P_0 and P_4. Consider P_0, the "least gregarious" member of the party. He or she shook hands with no one. What about the "most gregarious" person, P_4? P_4 shakes hands with everyone possible, which means that everyone *except for P_4's partner* shook P_4's hand! So everyone who is not P_4's partner has a non-zero handshake number. So P_4 and P_0 must be partners!

At this point, we can relax, for we have achieved the crux move. Our instinct tells us that probably P_3 and P_1 are also partners. How can we prove this? Let's try to adapt the argument we just used. P_3 shook hands with all but one person other than his or her partner. P_1 only shook hands with one person. Do we have any more information? Yes: P_4 was the one person that P_1 shook hands with and one of the people that P_3 shook hands with. In other words, *if we exclude P_4 and his or her partner, P_0,* then P_1 and P_3 play the role that P_0 and P_4 played; i.e., they are respectively the least and most gregarious, and by the same reasoning, must be partners (P_3 shakes hands with 2 out of the 3 people P_1, P_2, H, and we know that P_1 doesn't shake hands with P_3, so the only possible person P_3 can be partnered with is P_1).

Now we are done, for the only people left are P_2 and H. They must be partners, so P_2 is the hostess, and she shakes hands with 2 people. It is easy to adapt this argument to the general case of n couples.

Formal Solution: We shall use induction. For each positive integer n, define the statement $P(n)$ to be

> *If the host invites n couples, and if no one shakes hands with his or her partner, and if each of the $2n + 1$ people interrogated by the host shook a different number of hands, then the hostess shook n hands.*

It is easy to check that $P(1)$ is true by drawing a diagram and working out the only logical possibility. We need to show that $P(n)$ implies $P(n+1)$, and we'll be done. So assume that $P(n)$ is true. Now consider a party with $n+1$ couples (other than host and hostess) satisfying the hypotheses (no one shakes with partner; all handshake numbers are different). Then all handshake numbers from 0 to $2(n+1) = 2n + 2$ inclusive will occur among the $2n + 3$ people interrogated by the host. Consider the person X, who shook the maximum number of $2n + 2$ hands. This person shook hands with all but two of the $2n + 4$ people at the party. Since no one shakes hands with themselves or their partner, X shook hands with everyone possible. So the only person who could be partnered with X had to have been the person Y, who shook zero hands, for everyone else shook hands with X and is hence ineligible as a partner.

Now, let us remove X and Y from the party. If we no longer count handshakes involving these two people, we are reduced to a party with n invited couples, and everyone's handshake number (other than the host, about whom we have no information) has dropped by exactly 1, since everyone shook hands with X and no one shook hands with Y. But the inductive hypothesis $P(n)$ tells us that the hostess at this "abridged" party shook n hands. But in reality, the hostess shook one more hand—that of the X whom we just removed. So in the party with $n + 1$ invited guests, the hostess shook $n + 1$ hands, establishing $P(n+1)$. ∎

Often, problems involve the extreme principle plus a particular argument style, such as contradiction or induction. A common tactic is to use the extremes to "strip down" a problem into a smaller version, as above. The formal argument then uses induction. Here is another example. As above, we break down our solution into an informal investigation, followed by a formal write-up.

Example 3.2.4 (St. Petersberg City Olympiad, 1996) Several positive integers are written on a blackboard. One can erase any two distinct integers and write their greatest common divisor and least common multiple instead. Prove that eventually the numbers will stop changing.

Investigation: Before we begin, we mention a few simple number theory definitions. See Chapter 7 for more details.

- For integers a and b, the notation $a|b$ means "a divides b"; i.e., there exists an integer m such that $am = b$.
- The **greatest common divisor** of a and b is defined to be the largest integer g satisfying both $g|a$ and $g|b$. The notation used is $\text{GCD}(a, b)$. Sometimes (a, b) is also used.

- The **least common multiple** of a and b is defined to be the smallest *positive* integer u satisfying both $a|u$ and $b|u$. The notation used is $\mathrm{LCM}(a, b)$. Sometimes $[a, b]$ is also used. We have to specify that the least common multiple is positive, otherwise it would be $-\infty$!

Start with a simple example of two numbers such as $10, 15$. This is immediately transformed into $5, 30$, which thereafter never changes, since $5|30$. Here is a more complicated example. We will write in boldface the two elements which get erased and replaced,

$$
\begin{array}{cccc}
11 & \mathbf{16} & \mathbf{30} & 72 \\
\mathbf{11} & 2 & 240 & \mathbf{72} \\
1 & 2 & \mathbf{240} & \mathbf{792} \\
1 & 2 & 24 & 7920,
\end{array}
$$

and once again, the sequence will not change after this, since

$$1|2, \quad 2|24, \quad 24|7920.$$

A few more experiments (do them!) leads easily to the following conjecture:

> *Eventually, the sequence will form a chain where each element will divide the next (when arranged in order). Moreover, the least element and the greatest element of this chain are respectively the greatest common divisor and least common multiple of all the original numbers.*

How do we use the extreme principle to prove this? Focus on the least element of the sequence at each stage. In the example above, the least elements were $11, 2, 1, 1$. Likewise, look at the greatest elements. In our example, these elements were $72, 240, 792, 7290$. We observe that the sequence of least elements is non-increasing, while the sequence of greatest elements is non-decreasing. Why is this true? Let $a < b$ be two elements which get erased and replaced by

$$\mathrm{GCD}(a, b), \mathrm{LCM}(a, b).$$

Notice that if $a|b$, then $a = \mathrm{GCD}(a, b)$ and $b = \mathrm{LCM}(a, b)$, so the sequence is unchanged. Otherwise we have $\mathrm{GCD}(a, b) < a$ and $\mathrm{LCM}(a, b) > b$. Consequently, if x is the current least element, on the next turn, either

- We erase x and another element y (of course we assume that x does not divide y), replacing x with $\mathrm{GCD}(x, y) < x$, producing a new, smaller least element.
- We erase two elements, neither of which was equal to x. In this case, the smaller of the two new elements created is either smaller than x, producing a new least element, or greater than or equal to x, in which case x is still the least element.

A completely analogous argument works for the greatest element, but for now, let's just concentrate on the least element. We know that the least element either stays the same or decreases each time we erase-and-replace. Eventually, it will hit rock bottom. Why? The maximum possible value that we can ever encounter is the least common multiple of the original elements, so there are only finitely many possible values for our sequence at each stage. Either eventually the sequence cannot change, in which case we're done—this is what what we wanted to prove—or eventually the sequence will repeat. Thus we can distinguish a *smallest* least element, i.e., the smallest possible least element ℓ that ever occurs as our sequence evolves. We claim that ℓ must divide all other elements in the sequence in which it appears. For if not, then we would have $\ell \nmid x$ for some element x, and then if we erase-and-replace the pair ℓ, x, the replaced element $\mathrm{GCD}(\ell, x) < \ell$, contradicting the minimality of ℓ.

Why is this good? Because once the minimal element ℓ is attained, it can be ignored, since ℓ divides all the other elements and hence the sequence will not change if we try to erase-and-replace two elements, one of which is ℓ. Hence the "active" part of the sequence has been shortened by one element.

This is just enough leverage to push through an induction proof! Let us conclude with a formal argument.

Formal Solution: We will prove the assertion by induction on the number of elements n in the sequence. The base case for $n = 2$ is obvious: a, b becomes

$$\mathrm{GCD}(a, b), \mathrm{LCM}(a, b)$$

which then no longer changes.

Now assume the inductive hypothesis that all sequences of length n will eventually become unchanging. Consider an $(n + 1)$-element sequence

$$u_1, u_2, \ldots, u_n, u_{n+1}.$$

We will perform the erase-and-replace operation repeatedly on this sequence, with the understanding that we only perform operations which produce change, and at each stage, we will arrange the terms in increasing order. We make some simple observations:

1. The least element in the sequence at any possible stage is at least equal to the greatest common divisor of all of the original elements.
2. The greatest element in the sequence at any possible stage is at most equal to the least common multiple of all of the original elements.
3. The least element at any given stage is always less than or equal to the least element at the previous stage.
4. Since we arrange the terms in increasing order, observations (1) and (2) imply that there are only finitely many possible sequences possible.

As we perform erase-and-replace operations to the sequence, let ℓ_i be the least element of the sequence at stage i. There are two possible scenarios.

- Eventually we will no longer be able to change the sequence, in which case the least-element sequence $\ell_1, \ell_2, \ell_3, \ldots$ terminates.
- At some point, say stage k, we will return to a sequence that previously occurred (because of observation #4 above). Then the least-element sequence $\ell_1, \ell_2, \ell_3, \ldots$ may be infinite, but $\ell_k = \ell_{k+1} = \ell_{k+2} = \cdots$, due to observation #3 above.

Consequently, for each initial sequence $u_1, u_2, \ldots, u_n, u_{n+1}$, depending on how it evolves, there will be a stage k and a number $\ell := \ell_k$ which is *smallest* least element that ever occurs at *any* stage.

This number ℓ must divide all the other elements of the sequence at stage k and all later stages, for otherwise, we could erase-and-replace ℓ with another element to produce a smaller least element, which contradicts the minimality of ℓ. Therefore, once we reach stage k, the least element ℓ is out of the picture—the only possibility for erasing-and-replacing to produce change is to use the remaining n elements. But the inductive hypothesis says that eventually, these n elements will no longer change. Thus, eventually, our original $(n + 1)$-element sequence will no longer change.   ∎

In more complicated problems, it is not always obvious what entities should be monotonized, and the Well-Ordering Principle is not always true for infinite sets. In situations involving infinite sets, sometimes extremal arguments work, but you need to be careful.

Example 3.2.5 Let $f(x)$ be a polynomial with real coefficients of degree n such that $f(x) \geq 0$ for all $x \in \mathbf{R}$. Define $g(x) := f(x) + f'(x) + f''(x) \cdots + f^{(n)}(x)$. Show that $g(x) \geq 0$ for all $x \in \mathbf{R}$.

Solution: There are several things that we can apply extreme arguments to. Since $f(x) \geq 0$, we might want to look at the value of x at which $f(x)$ is minimal. First we need to prove that this value actually exists, i.e., that there is an $x_0 \in \mathbf{R}$ such that $f(x_0)$ is minimal. [This is not true for all functions, such as $f(x) = 1/x$.] Write

$$f(x) = a_n x^n + a_{n-1} x^{n-1} + \cdots + a_1 x + a_0,$$

where each $a_i \in \mathbf{R}$. Since $f(x)$ is always non-negative, the leading coefficient a_n must be positive, since the leading term $a_n x^n$ dominates the value of $f(x)$ when x is a large positive or negative number. Moreover, we know that n is even. So we know that

$$\lim_{x \to -\infty} f(x) = \lim_{x \to +\infty} f(x) = +\infty,$$

and consequently, $f(x)$ has a minimum value. Notice that $g(x)$ has the same leading term as $f(x)$ so $g(x)$ also has a minimum value. Indeed, $g(x)$ is what we will focus our attention on. We wish to show that this polynomial is always non-negative, so a promising strategy is contradiction. Assume that $g(x) < 0$ for some values of x. Consider its *minimum* value, achieved when $x = x_0$. Then $g(x_0) < 0$. Now, what is the relationship between $g(x)$ and $f(x)$? Since $f(x)$ is nth degree, $f^{n+1}(x) = 0$ and

$$g'(x) = f'(x) + f''(x) + \cdots + f^{(n)}(x) = g(x) - f(x).$$

Consequently,

$$g'(x_0) = g(x_0) - f(x_0) < 0,$$

since $g(x_0)$ is strictly negative and $f(x_0)$ is non-negative by hypothesis. But this contradicts the fact that $g(x)$ achieves its minimum value at x_0, for then $g'(x_0)$ should equal zero!   ∎

In the next example, Gaussian Pairing plus monotonization help solve a rather difficult problem, from the 1994 IMO in Hong Kong.

Example 3.2.6 Let m and n be positive integers. Let $a_1, a_2, \ldots, a_m$ be distinct elements of $\{1, 2, \ldots, n\}$ such that whenever $a_i + a_j \leq n$ for some i, j, $1 \leq i \leq j \leq m$, there exists k, $1 \leq k \leq m$, with $a_i + a_j = a_k$. Prove that

$$\frac{a_1 + a_2 + \cdots + a_m}{m} \geq \frac{n+1}{2}.$$

Solution: This is not an easy problem. Part of the difficulty is figuring out the statement of the problem! Let us call those sequences which have the property described in the problem "good" sequences. In other words, a good sequence is a sequence $a_1, a_2, \ldots, a_m$ of distinct elements of $\{1, 2, \ldots, n\}$ such that whenever $a_i + a_j \leq n$ for some i, j, $1 \leq i \leq j \leq m$, there exists k, $1 \leq k \leq m$, with $a_i + a_j = a_k$. Begin by plugging in easy values for m and n, say $m = 4, n = 100$. We have the sequence a_1, a_2, a_3, a_4 of distinct positive integers which may range between 1 and 100, inclusive. One possible such sequence is $5, 93, 14, 99$. Is this a good sequence? Note that $a_1 + a_2 = 98 \leq 100$, so we need there to be a k such that $a_k = 98$. Alas, there is not. Improving the original sequence, we try $5, 93, 98, 99$. Now check for pairs that add up to 100 or less. There is only one $(5, 93)$, and sure enough, their sum is an element of the sequence. So our sequence is good. Now compute the average, and indeed, we have

$$\frac{5 + 93 + 98 + 99}{4} \geq \frac{100 + 1}{2}.$$

But why? Let's try to construct another good sequence, with $m = 6, n = 100$. Start with $11, 78$. Since $11 + 78 = 89 \leq 100$, there must be a term in the sequence which equals 89. Now we have $11, 78, 89$. Notice that $89 + 11 = 100 \leq 100$, so 100 also must be in our sequence. Now we have $11, 78, 89, 100$, and if we include any more small terms, we need to be careful. For example, if we included 5, that would force the sequence to contain $11 + 5 = 16$ and $78 + 5 = 83$ and $89 + 5 = 94$, which is impossible, since we have only 6 terms. On the other hand, we could append two large terms without difficulty. One possible sequence is $11, 78, 89, 100, 99, 90$. And once more, we have

$$\frac{11 + 78 + 89 + 100 + 99 + 90}{6} \geq \frac{101}{2}.$$

Again, we need to figure out just why this is happening. Multiplying by 6, we get

$$11 + 78 + 89 + 100 + 99 + 90 \geq 6 \cdot \frac{101}{2} = 3 \cdot 101,$$

which strongly suggests that we try to *pair* the 6 terms to form 3 sums, each greater than 101. And indeed, that is easy to do:

$$11 + 78 + 89 + 100 + 99 + 90 = (11 + 100) + (78 + 89) + (99 + 90).$$

Can we do this in general? We can hope so. Notice that by hoping this, we are actually attempting to prove something slightly stronger. After all, it could be that the terms don't always pair nicely, as they did in this case, but nevertheless, the sum of all the terms is always big enough. However, there is no harm in trying to prove a stronger statement. Sometimes stronger statements are easier to prove.

Let's try to work out an argument that handles all sequences with $m = 6, n = 100$. First **monotonize**! This is an important simplification. Assume without loss of generality that our sequence is good and that

$$a_1 < a_2 < a_3 < a_4 < a_5 < a_6.$$

We have strict inequalities above because each term in a good sequence is *distinct*, that being one of the features of a good sequence.

We want to see if we can pair the terms so that the sum of each pair is greater than or equal to $n + 1$. Let's pause for a moment and think about strategy. The hypothesis is that the sequence is good (and monotonized), and the conclusion that we desire is a set of three inequalities. It is hard to prove 3 different inequalities directly (and if n were bigger, then there would be even more). A more promising approach is contradiction. For then, we need only assume that one of the inequalities fails, and if that gets a contradiction, we are done. And not only that, if we assume that two things sum to less than $n + 1$, that means the sum is $\leq n$, which is involved in the definition of a good sequence—very promising, indeed! So, appealing to the symmetry of the monotonized sequence, we will assume that at least one of the sums

$$a_1 + a_6, \quad a_2 + a_5, \quad a_3 + a_4$$

is less than or equal to n. Let's pick $a_1 + a_6$. If this sum is less than or equal to n, goodness implies that one of the a_k is equal to $a_1 + a_6$. But this is impossible, because the a_i are positive and *monotone*. A contradiction!

Now try assuming that $a_2 + a_5 \leq n$. Then this sum equals a_k for some $k, 1 \leq k \leq 6$. This sum is strictly greater than a_5, so we must have $a_2 + a_5 = a_6$. So far, no contradiction. But we have not exhausted the hypotheses of goodness. The sum $a_1 + a_5$ is strictly less than $a_2 + a_5$ (by monotonicity); hence $a_1 + a_5 = a_j$ for some $j, 1 \leq j \leq 6$. But $a_j > a_5$, which forces $a_j = a_6$. But this can't be, since $a_1 + a_5$ is strictly less than $a_2 + a_5 = a_6$, since all terms are *distinct*. Another contradiction.

Likewise, if we assume that $a_3 + a_4 \leq n$, goodness will imply that $a_3 + a_4$ equals either a_5 or a_6. But the sums $a_1 + a_4$ and $a_2 + a_4$ are also $\leq n$, and greater than a_4, and *distinct*. But this is a contradiction; we have three distinct sums ($a_3 + a_4, a_2 + a_4$ and $a_1 + a_4$) with only two possible values (a_5 and a_6).

Finally, we can produce a general argument. Before you read it, try to write it up on your own. The following argument assumes that n is even. You will need to alter it slightly for the case where n is odd.

Let $a_1, a_2, \ldots, a_m$ be good, and without loss of generality assume that $a_1 < a_2 < \cdots < a_m$. We will prove that $a_1 + a_2 + \cdots + a_m \geq m(n+1)/2$ by showing the stronger result that each of the pairs $a_1 + a_n, a_2 + a_{n-1}, \ldots a_{\frac{n}{2}}, a_{\frac{n}{2}+1}$ is greater than or equal to $n + 1$.

Assume that this is not the case. Then for some $j \leq \frac{n}{2}$, we must have $a_j + a_{n-j} \leq n$. Goodness implies that for some $k, 1 \leq k \leq m$, we have $a_j + a_{n-j} = a_k$. In fact, $k > n - j$, because the terms are positive and the sequence is monotone increasing.

Likewise, each of the j sums

$$a_1 + a_{n-j}, a_2 + a_{n-j}, \ldots, a_j, a_{n-j}$$

is less than or equal to n and distinct, and each greater than a_{n-j}. Goodness implies that each of these sums is equal to a_l for some $l > n - j$. There are only $j - 1$ choices for l, but there are j different sums. This is impossible, so we have achieved our contradiction.   ■

Problems and Exercises

3.2.7 Imagine an infinite chessboard that contains a positive integer in each square. If the value in each square is equal to the average of its four neighbors to the north, south, west, and east, prove the values in all the squares are equal.

3.2.8 There are 2000 points on a circle, and each point is given a number which is equal to the average of the numbers of its two nearest neighbors. Show that all the numbers must be equal.

3.2.9 Given a set of coins in the plane, all with different diameters. Show that one of them is tangent to at most 5 of the others.

3.2.10 (Canada, 1987) On a large, flat field, n people ($n > 1$) are positioned so that for each person the distances to all the other people are different. Each person holds a water pistol and at a given signal fires and hits the person who is closest. When n is odd, show that there is at least one person left dry. Is this always true when n is even?

3.2.11 Our solution to the GCD-LCM problem (Example 3.2.4 on page 85) omitted a proof that the sequence eventually forms a chain where each element divides the next (when arranged in increasing order). Modify the argument to do this (if you understand the original argument, you should be able to do this with just a tiny bit of work).

3.2.12 (Russia, 1996) A **palindrome** is a number or word that is the same when read forward and backward, for example, "176671" and "civic." Can the number obtained by writing the numbers from 1 to n in order ($n > 1$) be a palindrome?

3.2.13 Place the integers $1, 2, 3, \ldots, n^2$ (without duplication) in any order onto an $n \times n$ chessboard, with one integer per square. Show that there exist two adjacent entries whose difference is at least $n + 1$. (Adjacent means horizontally or vertically or diagonally adjacent.)

3.2.14 After experimenting with Problem 2.1.26 on page 28, you probably have come to the conclusion that the only possible configurations are those where the degree of each crossing point (i.e., number of lines emanating from the crossing point) is even or those where exactly two crossing points have odd degree, and in this latter case, the path must begin at one of the odd-degree points and end at the other one. Now you are ready to prove this conjecture. Use an extremal argument. Consider the longest "legal" path and try to argue (by contradiction) that this path includes *all* the lines in your figure.

3.2.15 *The extreme principle in number theory.* In Example 3.2.4 on page 85, we encountered the notions of greatest common divisor and least common multiple. Here we mention another simple number theory idea, the **division algorithm**:

> *Let a and b be positive integers, $b \geq a$. Then there exist integers q, r satisfying $q \geq 1$ and $0 \leq r < a$ such that*
>
> $$b = qa + r.$$

In other words, if you divide b by a, you get a positive integer quotient q with a remainder r that is at least zero (equals zero if $a|b$) but smaller than a. The division algorithm is an "obvious" idea, one that you have seen since grade school. It is actually a consequence of the well-ordering principle.

(a) As a warm-up, prove the division algorithm rigorously, by considering the minimum nonnegative value of $b - at$ as t ranges through the positive integers.

(b) Another warm-up: prove (this should take two seconds!) that if $a|b$ and $a|c$, then $a|(bx + cy)$ for any integers x, y (positive or negative).

(c) Now show that if $a|m$ and $b|m$, then $\text{LCM}(a, b)|m$.

(d) Finally, show that for any integers a, b, the greatest common divisor of a and b is equal to the minimum value of $ax + by$, as x and y range through all integers (positive and negative). For example, if $a = 7, b = 11$, we have $\text{GCD}(7, 11) = 1$ (since 7 and 11 are primes, they share no common divisors except 1), but also $1 = (-3) \cdot 7 + 2 \cdot 11$.

3.2.16 Let $P(x) = a_n x^n + a_{n-1} x^{n-1} + \cdots + a_0$ be a polynomial with integer coefficients and let q be a prime. If q is a factor of each of $a_{n-1}, a_{n-2}, \ldots, a_1, a_0$, but q is not a factor of a_n, and q^2 is not a factor of a_0, then $P(x)$ is irreducible over the rationals; i.e., $P(x)$ cannot be factored into two non-constant polynomials with rational coefficients.

3.3   The Pigeonhole Principle

Basic Pigeonhole

The **pigeonhole principle**,[3] in its simplest incarnation, states the following:

> *If you have more pigeons than pigeonholes, and you try to stuff them into the holes, then at least one hole must contain at least two pigeons.*

Amazingly, this trivial idea is revered by most mathematicians, for it is at least as powerful as the Gaussian pairing trick. For example, the pigeonhole principle played a crucial role in the solution of at least a third of the 1994 Putnam exam problems. A few examples will convince you of its power.

---

[3]The pigeonhole principle is sometimes also called the **Dirichlet Principle** in honor of the 19th-century mathematician Peter Dirichlet.

Example 3.3.1 Every point on the plane is colored either red or blue. Prove that now matter how the coloring is done, there must exist two points, exactly a mile apart, which are the same color.

Solution: Just messing around quickly yields a solution. Pick a point, any point. Without loss of generality it is red. Draw a circle of radius one mile with this point as center. If any point on the circumference of this circle is red, we are done. If all the points are blue, we are still done, for we can find two points on the circumference which are one mile apart (why?).

That wasn't hard, but that wasn't the pigeonhole solution. Consider this: just imagine the vertices of an equilateral triangle with side length one mile. There are three vertices, but only two colors available. The pigeonhole principle tells us that two vertices must be the same color! ∎

Here is another simple example.

Example 3.3.2 Given a unit square, show that if five points are placed anywhere inside or on this square, then two of them must be at most $\frac{\sqrt{2}}{2}$ units apart.

Solution: Partition the unit square into four $\frac{1}{2} \times \frac{1}{2}$ squares. By pigeonhole, one of these smaller squares must contain at least 2 points. Since the diagonal of each small square is $\frac{\sqrt{2}}{2}$, that is the maximum distance between the 2 points. ∎

Quick and beautiful solutions are characteristic of pigeonhole problems. The above example was quite simple. Solving most pigeonhole problems is often a three-part process:

1. Recognize that the problem might require the pigeonhole principle.
2. Figure out what the pigeons will be and what the holes must be. This is frequently the crux move.
3. After applying the pigeonhole principle, there is often more work to be done. Sometimes the pigeonhole principle yields only the "penultimate step," and sometimes just an intermediate result. The skilled problem solver plans for this when thinking of a strategy.

Here's a simple problem which illustrates the coordination of a good penultimate step with the pigeonhole principle. You will use this problem as a building-block in many other problems later.

Example 3.3.3 Show that among any $n + 1$ positive integers, there must be two whose difference is a multiple of n.

Solution: The penultimate step, of course, is realizing that the two desired numbers must have same remainder upon division by n. Since there are only n possible remainders, we are done. ∎

This next problem is a bit more intricate. Also note the strategic element of confidence present in the solution. This is not a terribly hard problem, for those who are brave and tenacious.

Example 3.3.4 (IMO 1972) Prove that from a set of ten distinct two-digit numbers (in the decimal system), it is possible to select two disjoint subsets whose members have the same sum.

Solution: We want to have two subsets have the same sum, so it is reasonable to make the subsets be the pigeons, and the sums be the holes. How do we do this precisely? First, let's look at the sums. The smallest possible sum is 10 and the largest possible sum is $99 + 98 + 97 + \cdots + 90$. Using the Gaussian pairing trick (for practice) this is $189 \cdot 5 = 945$. Consequently, there are $945 - 10 + 1 = 936$ different sums.

Now we need to count the number of pigeons. The number of subsets (see Section 6.1) is just $2^{10} = 1024$. Since $1024 > 936$, there are more pigeons than holes, so we are done: two of the subsets must have the same sum.

But are we done? Not quite yet. The problem specifies that the two subsets are disjoint! Don't panic. It would of course be bad problem solving form to abandon our solution at this point, for we have already achieved a *partial solution*: we have shown that there are two different subsets (perhaps not disjoint) that have the same sum. Can we use this to find two disjoint subsets with the same sum? Of course. Let A and B be the two sets. Divide into cases:

- A and B are disjoint. Done!
- A and B are not disjoint. From each set, remove the elements that they have in common. Now we have disjoint sets, but the sums are still the same (why?), so we are done! ∎

You might wonder, "What if, after removing the common elements, we had nothing left from one of the sets?" That is impossible. Why?

Intermediate Pigeonhole

Here is a more elaborate version of the pigeonhole principle, one that is used in practice more often than the basic pigeonhole principle described above. [The notation $\lceil x \rceil$ (the **ceiling** of x) means the smallest integer greater than or equal to x. For example, $\lceil \pi \rceil = 4$. See page 160 for more information.]

> *If you have p pigeons and h holes, then at least one of the holes contains at least $\lceil p/h \rceil$ pigeons.*

Notice that the basic pigeonhole principle is a corollary of this statement: we have $p > h$, so the quantity $\lceil p/h \rceil$ is at least 2.

Make sure you understand the statement of this "intermediate pigeonhole principle" by working through several examples. Make sure you understand why the statement is true.

If you are lucky, a single, beautiful application of a technique will solve your problem. But generally, we are not so lucky. It is not surprising that some problems require multiple applications of the pigeonhole principle. Each time the pigeonhole principle is used, information is gained. Here is an example which uses repeated application of the intermediate pigeonhole principle.

Example 3.3.5 (A. Soifer, S. Slobodnik) Forty-one rooks are placed on a 10×10 chessboard. Prove that there must exist 5 rooks, none of which attack each other. (Recall that rooks attack each piece located on its row or column.)

Solution: When you see the number 41 juxtaposed with 10, you know that the pigeonhole principle is lurking about, since 41 is just one more than $4 \cdot 10$. Once attuned to the pigeonhole principle, we cannot help but notice that $\lceil 41/10 \rceil = 5$, which is encouraging, for this is the number of rooks we seek. This of course is not a solution, but it does suggest that we probe carefully for one using the pigeonhole principle.

Let us do so. We seek 5 rooks that do not attack one another. Two rooks do not attack if they are located on different rows and different columns. So we need to find five rooks, each living in a different row, and each living in a different column. A vague strategy: we need 5 different rows, each with "lots" of rooks. Then we could pick a rook from one row, find another rook *in a different column* in the other row, etc. There are 10 rows, and 41 rooks, so the pigeonhole principle tells us that one row must contain at least $\lceil 41/10 \rceil = 5$ rooks. This is a start. Can we apply the pigeonhole principle again to get more information? Yes! What is happening with the *other* rows? We want to find other rows that have lots of rooks. We have isolated one row with at least 5 rooks. Now remove it! At most, we will remove 10 rooks. That leaves 31 rooks on the remaining 9 rows. The pigeonhole principle tells us that one of these 9 rows must contain at least $\lceil 31/9 \rceil = 4$ rooks.

Now we are on a roll. Removing this row, and "pigeonholing" once more, we deduce that there is another row containing at least $\lceil 21/8 \rceil = 3$ rooks. Continuing (verify!), we see that another row must have at least 2 rooks, and one more row must contain at least 1 rook.

Therefore there are five special rows on the chessboard, containing at least 5, 4, 3, 2 and 1 rooks, respectively. Now we can construct the "pacifistic quintuplet": Start by picking the rook in the row that has at least 1 rook. Then go to the row with at least 2 rooks. At least one of these rooks will not be in the same column as our first rook, so select it as the second rook. Next, look at the row that has at least 3 rooks. One of these 3 rooks lives neither in the same column as the first or the second rook. Select this as our third rook, etc., and we are done!   ∎

This rather elaborate problem was a good illustration of both the pigeonhole principle and the wishful thinking strategy, i.e., not giving up. When you think that a problem can be attacked with the pigeonhole principle, first try to do the job neatly. Look for a way to define pigeons and holes that yields a quick solution. If that doesn't work, don't give up! Use the pigeonhole principle once to gain a foothold, then try to use it again. Keep extracting information!

Advanced Pigeonhole

The examples here are harder problems. Some are hard because they require other specialized mathematical ideas together with the pigeonhole principle. Other problems require only basic pigeonhole, but the pigeons and/or holes are far from obvious.

The number theory problem below uses only basic pigeonhole, but with very cleverly chosen pigeons.

Example 3.3.6 (Colorado Springs Mathematical Olympiad, 1986) Let n be a positive integer. Show that if you have n integers, then either one of them is a multiple of n or a sum of several of them is a multiple of n.

Solution: We need to show that something is a multiple of n. How can we do this with the pigeonhole principle? By Example 3.3.3, you already know the answer: let the pigeonholes be the n different possible remainders one can have upon division by n. Then if two numbers lie in the same pigeonhole, their *difference* will be a multiple of n. (Why?)

But in Example 3.3.3, we had $n + 1$ pigeons to place into the n holes. In this problem, we have only n numbers. How do we create $n + 1$ or more pigeons? And also, how can we choose pigeons so that the thing that ends up being a multiple of n is either one of the original numbers or a sum of several of them?

If we can answer both of these questions, we will be done. The first question is rather mysterious, since we have so little to work with, but the second question has an straightforward possible answer: let the pigeons be sums themselves; if we choose them carefully, the differences of the pigeons will still be sums. How can that be done? Let the numbers in our set be $a_1, a_2, \ldots, a_n$. Consider the sequence

$$
\begin{aligned}
p_1 &= a_1, \\
p_2 &= a_1 + a_2, \\
&\vdots \\
p_n &= a_1 + a_2 + \cdots + a_n.
\end{aligned}
$$

We are using the letter p for pigeons; i.e., p_k denotes the kth pigeon. Notice that for any two distinct indices $i < j$, the difference $p_j - p_i$ will equal the sum $a_{i+1} + a_{i+2} + \cdots + a_j$. So our pigeons have the right behavior, but unfortunately, there are only n of them. But that is not as bad as it seems. Sometimes (as in Example 3.3.4) you can reduce the number of pigeonholes needed by *dividing into cases*.

We have n pigeons $p_1, p_2, \ldots, p_n$. There are two cases.

- One of the pigeons has a remainder of 0 upon division by n, in which case we are done! (Why?)

- None of the pigeons has a remainder of 0, so now we have just $n - 1$ pigeonholes to consider. With n pigeons, two of them must have the same remainder, so their difference, a sum of several of the original numbers, will be a multiple of n, and we are done. ∎

The next example is a famous problem, due to the great Paul Erdős,[4] that is remarkable because the difficulty is in the pigeonholes, not the pigeons.

Example 3.3.7 Let n be a positive integer. Choose any $(n + 1)$-element subset of $\{1, 2, \ldots, 2n\}$. Show that this subset must contain two integers, one of which divides the other.

Investigation: The language of the problem leads us to try the pigeonhole principle, fixing the $n + 1$ numbers as the pigeons. We need to create at most n pigeonholes, since there are $n + 1$ pigeons. And we want our pigeonholes chosen so that if two numbers inhabit the same hole, then one of them must divide the other. Each pigeonhole, then, is a set of integers with the property that if a and b are any two elements of the set, either a is a multiple of b or b is a multiple of a.

Let's try to construct such a set. If the set contains 7, then all the other numbers must be either factors of 7 or multiples of 7. Let's say the next number in the set is 21. Then the other numbers now must be either multiples of 21 or factors of 7, etc. So if 7 is the smallest number in the set, the set would be a list of numbers of the following form: $7, 7a, 7ab, 7abc, 7abcd, \ldots$, where a, b, c, d, etc. are positive integers.

Our task, then, is to partition the set $\{1, 2, \ldots, 2n\}$ into at most n disjoint subsets with the above property. This is not easy; the best thing to do at this point is experiment with small values of n. For example, let $n = 5$. Let's try to partition $\{1, 2, 3, 4, 5, 6, 7, 8, 9, 10\}$ into 5 disjoint subsets with the special property. Each set has a smallest element, so we need to pick 5 such "seeds." In striving for a general method—one that can be used for other values of n—the only "natural" collection of 5 seeds is $1, 3, 5, 7, 9$. (The list $2, 4, 6, 8, 10$ doesn't include 1, and 1 has to be the minimum element of one of the sets, so this list is not a "natural" candidate for our seeds.)

Notice that each seed is odd. To get the remaining numbers, we just have to multiply the seeds by 2. But that won't quite work, as we will not get all the numbers. If we keep on multiplying, though, we get the partition

$$\{1, 2, 4, 8\}; \quad \{3, 6\}; \quad \{5, 10\}; \quad \{7\}; \quad \{9\}.$$

This set of pigeonholes does the trick. If we choose any six numbers from

$$\{1, 2, \ldots, 10\},$$

then two of them must be contained in one of the above five sets. Some of the sets (in this case, just two) contain just one element, so the two numbers that "cohabit" a set cannot lie in these sets. Therefore the two cohabitors must live in either $\{1, 2, 4, 8\}$ or $\{3, 6\}$ or $\{5, 10\}$, and then we are done, for then one of the two cohabitors is a multiple of the other.

It is easy now to solve the problem in general.

Formal Solution: Each element of $\{1, 2, \ldots, 2n\}$ can be *uniquely* written in the form $2^r q$, where q is an odd integer and r is a non-negative integer. Each different odd

---

[4]Erdős, who died in 1996 at the age of 83, was the most prolific mathematician of modern times, having authored or co-authored more than 1,000 research papers.

number q defines a pigeonhole, namely all the elements of $\{1, 2, \ldots, 2n\}$ which have the form $2^r q$ for some positive integer r. (For example, if $n = 100$, the value $q = 11$ would define the pigeonhole $\{11, 22, 44, 88, 176\}$.) Since there are exactly n odd numbers between 1 and $2n$, we have defined n sets, and these sets are disjoint (they need to be disjoint; otherwise they cannot be "pigeonholes".) So we are done, for by the pigeonhole principle, two of the $n + 1$ numbers will lie in one of our n sets, which will force one of the two numbers to be a multiple of the other.   ■

The next problem, from the 1994 Putnam exam, involves some linear algebra, which makes it already pretty difficult. But the fun parts are the two crux moves: defining a function, and using the pigeonhole principle with the roots of a polynomial. Both ideas have many applications.

Example 3.3.8 Let A and B be 2×2 matrices with integer entries such that A, $A + B$, $A + 2B$, $A + 3B$, and $A + 4B$ are all invertible matrices whose inverses have integer entries. Show that $A + 5B$ is invertible and that its inverse has integer entries.

Solution: If X is an invertible matrix with integer entries and its inverse also has integer entries, then $\det X = \pm 1$. This follows from the fact that $\det X$ and $\det X^{-1}$ are both integers and that $\det X^{-1} = 1/\det X$. And conversely, if the determinant of a matrix with integer entries is ± 1, the inverse also will have integer entries (why?).

Now define the function $f(t) := \det(A + tB)$. Since A and B are 2×2 matrices with integer entries, $f(t)$ is a quadratic polynomial in t with integer coefficients (verify!). Since A, $A + B$, $A + 2B$, $A + 3B$, and $A + 4B$ are all invertible matrices whose inverses have integer entries, we know that the five numbers

$$f(0), \, f(1), \, f(2), \, f(3), \, f(4)$$

take on only the values 1 or -1. By the pigeonhole principle, three of these numbers must have the same value; without loss of generality, assume that it is 1. Then the quadratic polynomial $f(t)$ is equal to 1 when three different numbers are plugged in for t. This means that $f(t)$ must be the constant polynomial; i.e., $f(t) \equiv 1$. Therefore $\det(A + 5B) = f(5) = 1$, so $A + 5B$ is invertible and its inverse has integer entries.   ■

The above problem was mathematically very sophisticated, not just because it used linear algebra. Two other new ideas used were

- A quadratic polynomial $P(x)$ cannot assume the same value for three different values of x. This is really an application of the Fundamental Theorem of Algebra (see chapter 5).

- The **define a function** tool, which is part of a larger idea, the strategy of **generalizing** the scope of a problem before attacking it.

We conclude with a lovely argument that includes, among other neat ideas, an application of the pigeonhole principle to an infinite set. Please take some time to think about the problem before reading its solution.

Example 3.3.9 Let S be a region in the plane (not necessarily convex) with area greater than the positive integer n. Show that it is possible to translate S (i.e., slide without turning or distorting) so that S covers at least $n + 1$ lattice points.

Solution: Here is an example. The region S has area 1.36 units. At first, S covers just one lattice point, but it is possible to translate it down and to the right so that it covers two lattice points.

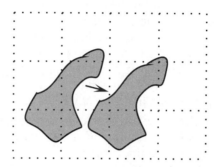

How do we develop a general argument? We will invent an algorithm that we will apply to our example region S which will work for any region. Our algorithm has several steps.

1. First, take S and decompose it into finitely many subregions, each lying on its own lattice square. In our example, S breaks up into the 5 regions a, b, c, d, e.

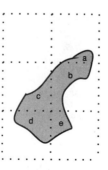

2. Next, "lift" each region up, with its lattice square. Think of each region as a picture drawn on a unit square. Stack these squares in a pile, as shown. Now imagine that we stick a pin through the squares, so that the pin pierces each square in the same place. Because the area of S is greater than 1, it must be possible to find a point such that the pin through that point pierces at least 2 subregions. In our example, the pin pierces both a and c.

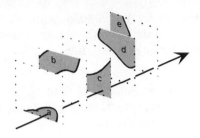

Why must this be true? It is really the pigeonhole principle! For if there were no such point, then the total area of S would be less than or equal to the area of a single unit square. Another way to think of it: imagine cutting out the subregions with a scissor and trying to stack them all in a unit square, without any overlapping. It would be impossible, since the total area is greater than 1. (By the same reasoning, if the total area were greater than the integer n, it would be possible to find a point where the pin pierces at least $n + 1$ subregions.)

3. Finally, we keep track of the point where our pin pierced the subregions, and rebuild S. The points are indicated with white circles. Since the two points are located in *exactly the same place* with respect to the lattice lines, it now is possible to translate S (in the direction of the arrow), so that both white points lie on lattice points, and we're done!   ■

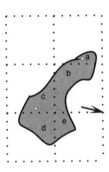

Problems and Exercises

3.3.10 Prove the intermediate pigeonhole principle, using argument by contradiction.

3.3.11 Show that in any finite gathering of people, there are at least two people who know the same number of people at the gathering (assume that "knowing" is a mutual relationship).

3.3.12 Use argument by contradiction to prove the following useful variant of the pigeonhole principle: Let $a_1, a_2, \ldots, a_n$ be positive integers. If

$$(a_1 + a_2 + \cdots + a_n) - n + 1$$

pigeons are put into n pigeonholes, then for some i, the statement "The ith pigeonhole contains at least a_i pigeons" must be true.

3.3.13 Notice how simple the pigeonhole argument was in Example 3.3.2 on page 93. What is wrong with the following "solution"?

> *Place 4 of the points on the vertices of the square; that way they are maximally separated from one another. Then the 5th point must be within $\sqrt{2}/2$ of one of the other 4 points, for the furthest from the corners it could be would be the center, and that is exactly $\sqrt{2}/2$ units from each corner.*

3.3.14 Show that the decimal expansion of a rational number must eventually become periodic.

3.3.15 7 points are placed inside a regular hexagon with side length 1. Show that at least two points are at most 1 unit apart.

3.3.16 Inside a 1×1 square, 101 points are placed. Show that some three of them form a triangle with area no more than 0.01.

3.3.17 Show that among any $n + 2$ integers, either there are two whose difference is a multiple of $2n$, or there are two whose sum is divisible by $2n$.

3.3.18 Chose any $(n + 1)$-element subset from $\{1, 2, \ldots, 2n\}$. Show that this subset must contain two integers that are relatively prime.

3.3.19 People are seated around a circular table at a restaurant. The food is placed on a circular platform in the center of the table, and this circular platform can rotate (this is commonly found in Chinese restaurants that specialize in banquets). Each person ordered a different entrée, and it turns out that no one has the correct entrée in front of him. Show that it is possible to rotate the platform so that at least *two* people will have the correct entrée.

3.3.20 Consider a sequence of N positive integers containing n distinct integers. If $N \geq 2^n$, show that there is a consecutive block of integers whose product is a perfect square. Is this inequality the best possible?

3.3.21 (Korea, 1995) For any positive integer m, show that there exist integers a, b satisfying

$$|a| \leq m, \quad |b| \leq m, \quad 0 < a + b\sqrt{2} \leq \frac{1 + \sqrt{2}}{m + 2}.$$

3.3.22 Show that for any positive integer n, there exists a positive multiple of n that contains only the digits 7 and 0.

3.3.23 To make sure that you understand example 3.3.7, explicitly construct the pigeonholes when $n = 25$. Verify that the solution works in this case.

3.3.24 A chess player prepares for a tournament by playing some practice games over a period of 8 weeks. She plays at least 1 game per day, but no more than 11 games per week. Show that there must be a period of consecutive days during which she plays exactly 23 games.

3.3.25 (Putnam 1994) Prove that the points of an isosceles triangle of side length 1 cannot be colored in four colors such that no two points at distance at least $2 - \sqrt{2}$ from each other receive the same color.

3.3.26 Example 3.3.5 on page 95 employed the pigeonhole principle several times in order to get a solution. There exists a fantastically simple solution which uses just one application of the (intermediate) pigeonhole principle. Can you find it?

3.3.27 The following problems are inspired by the very easy Example 3.3.1. However, not all of the variations below are easy. Have fun!

(a) Color the plane in 3 colors. Prove that there are two points of the same color 1 unit apart.

(b) Color the plane in 2 colors. Prove that one of these colors contains pairs of points at *every* mutual distance.

(c) Color the plane in 2 colors. Prove that there will always exist an equilateral triangle with all its vertices of the same color.

(d) Show that it is possible to color the plane in 2 colors in such a way that there cannot exist an equilateral triangle of side 1 with all vertices the same color.

(e) Color the plane in 2 colors. Show that there exists a rectangle, all of whose vertices are the same color.

3.4   Invariants

Our discussion of the extreme principle mentioned the importance of extracting key information from the chaos that most problems initially present. The *strategy* is to somehow "reduce" the problem so as to focus on certain essential entities. In Section 3.2, the *tactic* to implement this strategy was to focus on extreme values. There are other tactics with the same underlying strategy. Here we shall introduce the very rich topic of **invariants**.

An invariant, as the name suggests, is merely some aspect of a problem—usually a numerical quantity—that does not change, even if many other properties do change. Here are a few examples.

Example 3.4.1 *The Motel Room Paradox.* Recall Example 2.1.8 on page 22, the problem involving three women who check into a motel room. Denote g, p, d to be the amount the guests spent, the amount the porter pocketed for himself and the amount the motel desk received, respectively. Then the quantity

$$g - p - d$$

is an invariant, always equal to zero.

Example 3.4.2 *The Power of a Point Theorem.* Given a fixed point P and a fixed circle, draw a line through P which intersects the circle at X and Y. The **power of the point** P with respect to this circle is defined to be the quantity $PX \cdot PY$.

The Power of a Point Theorem (also known as **POP**) states that this quantity is invariant; i.e., does not depend on the line that is drawn. For example, in the picture below,

$$PX \cdot PY = PX' \cdot PY'.$$

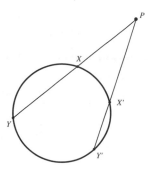

You undoubtedly learned this theorem in elementary geometry, at least for the case where the point P lies inside the circle. (The proof uses similar triangles, and considers the three cases: P inside, outside or on the circle.)

Example 3.4.3 *Euler's Formula.* You encountered this originally as Problem 2.2.12 on page 41, which asked you to conjecture a relationship between the number of vertices, edges and faces of any polyhedron. It turns out that if v, e, and f denote the number of vertices, edges and faces of a polyhedron without "holes," then

$$v - e + f = 2$$

always holds; i.e., the quantity $v - e + f$ is an invariant. This is known as **Euler's Formula**. See problem 3.4.37 for some hints on how to prove this formula.

Example 3.4.4 *Symmetry.* Even though we devoted the first section of this chapter to symmetry, this topic logically is contained within the concept of invariants. If a particular object (geometric or otherwise) contains symmetry, that is just another way of saying that the object itself is an invariant with respect to some transformation or set of transformations. For example, a square is an invariant with respect to rotations about its center of 0, 90, 180 and 270 degrees.

On a deeper level, the substitution $u := x + 1/x$, which helped solve

$$x^4 + x^3 + x^2 + x + 1 = 0$$

in Example 3.1.10 worked, because u is invariant with respect to a permutation of some of the roots.[5]

Example 3.4.5 *Divisibility by Nine.* Let $s(n)$ be the sum of the digits of the base-ten representation of the positive integer n. Then

$$n - s(n) \text{ is always divisible by 9.}$$

---

[5]This idea is the germ of one of the greatest achievements of 19th- century mathematics, **Galois theory**, which among other things develops a systematic way of determining which polynomials can be solved with radicals, and which cannot. For more information, consult Herstein's wonderful book [12].

For example, if $n = 136$, then

$$n - s(n) = 136 - (1 + 3 + 6) = 126 = 9 \cdot 14.$$

This was an example of a non-numeric invariant, although we could have recast things in a numeric form by saying, for example, that the remainder upon division by 9 of $n - s(n)$ is the invariant quantity zero.

Here is another example of a divisibility invariant.

Example 3.4.6 At first, a room is empty. Each minute, either one person enters or two people leave. After exactly 3^{1999} minutes, could the room contain $3^{1000} + 2$ people?

Solution: If there are n people in the room, after one minute, there will be either $n + 1$ or $n - 2$ people. The difference between these two possible outcomes is 3. Continuing for longer times, we see that

> *At any fixed time t, all the possible values for the population of the room differ from one another by multiples of 3.*

In 3^{1999} minutes, then, one possible population of the room is just 3^{1999} people (assuming that one person entered each time). This is a multiple of 3, so *all* the possible populations for the room have to also be multiples of 3. Therefore $3^{1000} + 2$ will not be a valid population. ∎

The simplest divisibility invariant is **parity**, which we used in Example 2.2.3 on page 32 and Example 2.2.4 on page 33. Let us explore this concept in more detail.

Parity

Integers are divided into two parity classes: even and odd. The even integers are divisible by 2, the odds are not. Note that zero is even. The following facts are easy to prove rigorously (do it!) but should have been known to you since you were little.

- The parity of a sum of a set of integers is odd if and only if the number of odd elements is odd.
- The parity of a product of a set of integers is odd if and only if there are no even elements in the set.

You may think that parity is a rather crude thing—after all, there are infinitely many integers yet only two kinds of parity—yet knowledge of parity is sometimes all that is needed, especially if parity is involved in the statement of the problem. Here is a fundamental example, a problem which has appeared in many forms. This version appeared in the 1986 Colorado Springs Mathematical Olympiad.

Example 3.4.7 If 127 people play in a singles tennis tournament, prove that at the end of tournament, the number of people who have played an odd number of games is even.

Solution: Many people unsuccessfully approached this problem by assuming that the tournament had a particular structure, such as double-elimination or round-robin, etc. But the problem doesn't specify this! The implication is that *no matter how the tournament is structured* the number of people who have played an odd number of games must be even. For example, one possible tournament would be one where no one plays any games. Then the number of people who have played an odd number of games is zero, which is even, as required.

There seem to be very few restrictions. Are there any? Yes, each game has exactly two people playing in it. In other words, if A plays with B then that game is counted *twice*: once as part of A's count, and once as part of B's count. More precisely, if we let g_i denote the number of games that player i has played at the end of the tournament, then the sum

$$g_1 + g_2 + g_3 + \cdots + g_{127}$$

must be even, since this sum counts every game that has been played exactly twice! Notice that this sum is always even, not just at the end of the tournament, but at any time!

Now we are done; the sum above is even, and is a sum of an odd number (127) of elements. If an odd number of them were odd, the sum would not be even. So the number of g_i which are odd is even. ∎

The next problem came from a 1906 Hungarian contest. We shall present two solutions; the first uses parity in a straightforward way and the second cleverly constructs another invariant first.

Example 3.4.8 Let $a_1, a_2, \ldots, a_n$ represent an arbitrary arrangement of the numbers $1, 2, 3, \ldots, n$. Prove that, if n is odd, the product

$$(a_1 - 1)(a_2 - 2)(a_3 - 3) \cdots (a_n - n)$$

is an even number.

Solution 1: It is helpful, of course, to look at a concrete case, say, $n = 11$. Employing the penultimate step strategy, we ask ourselves, what will force the product

$$(a_1 - 1)(a_2 - 2)(a_3 - 3) \cdots (a_{11} - 11)$$

to be even? Clearly, it is sufficient to show that one of the numbers

$$a_1 - 1, a_2 - 2, a_3 - 3, \ldots, a_{11} - 11$$

is even. How to do this? A good strategy is to try a proof by contradiction, since we need to show that just one of the numbers above is even, and we don't know which one. But if we start with the assumption that all of the numbers are odd, we have nice specific information to work with. So assume that each of

$$a_1 - 1, a_2 - 2, a_3 - 3, \ldots, a_{11} - 11$$

is odd. Now we can recover the parity of the original a_i. We see that 6 of them,

$$a_1, a_3, a_5, a_7, a_9, a_{11}$$

are even, while the remaining 5,

$$a_2, a_4, a_6, a_8, a_{10}$$

are all odd. This is a contradiction, because the a_i are a permutation of $1, 2, 3, \ldots, 11$, and this set contains 5 even and 6 odd elements. Clearly this argument extends to the general case, for the only special property of 11 that we used was the fact that it was odd.   ∎

Solution 2: The crux move: consider the *sum* of the terms. We have

$$(a_1 - 1) + (a_2 - 2) + \cdots + (a_n - n)$$
$$= (a_1 + a_2 + \cdots + a_n) - (1 + 2 + \cdots + n)$$
$$= (1 + 2 + \cdots + n) - (1 + 2 + \cdots + n)$$
$$= 0,$$

so the sum is an invariant; it is equal to zero no matter what the arrangement. A sum of an odd number of integers which equals zero (an even number) must contain at least one even number!   ∎

Both solutions were nice, but the second was especially clever. Try to incorporate the new idea here in future problems.

> *Be on the lookout for "easy" invariants. Check to see if you can rearrange your problem to get simple numbers such as zero or one.*

Example 3.4.9 Let $P_1, P_2, \ldots, P_{1997}$ be distinct points in the plane. Connect the points with the line segments $P_1 P_2, P_2 P_3, P_3 P_4, \ldots, P_{1996} P_{1997}, P_{1997} P_1$. Can one draw a line which passes through the interior of every one of these segments?

Solution: It is not obvious that parity is an issue here, but one should always keep parity in mind.

> *Whenever a problem involves integers, ask yourself if there are any parity restrictions. Experiment with different values than the given if necessary.*

This problem involves 1997 points. A few experiments with much smaller numbers of points, for example the range from 2 to 5, convinces you easily (do it!) that it is possible to draw the line if and only if the number of points is even. So parity seems to be important. Let us find a rigorous argument for a specific case, say 7 points. Once again, we will argue by contradiction because assuming that we *can* draw the line gives us lots of specific information that we can work with. So, assume there is a line L which passes through the interior of each segment. This line cuts the plane into two regions, which we will call the "left" and "right" sides of L. Without loss of generality, P_1 lies on the left side of L. This forces P_2 to lie on the right side of L, which in turn forces P_3 to lie in the left, P_4 in the right, etc. The important thing about this sequence is that P_7 ends up on the left side, along with P_1. Therefore L cannot pass through the interior of segment $P_1 P_7$, a contradiction.

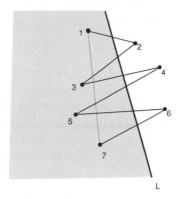

This argument certainly generalizes. As long as n is odd, then P_1 and P_n lie on the same side of L. ∎

The following is a rather sophisticated problem, one that requires a parity analysis combined with very resourceful and repeated use of the pigeonhole principle.

Example 3.4.10 (IMO 1985) Consider a set of 1,985 positive integers, not necessarily distinct, and none with prime factors bigger than 23. Prove that there must exist 4 integers in this set whose product is equal to the fourth power of an integer.

Solution: We shall present the solution in a terse, sketchy form. It will be up to you to fill in the details.

Every number in this set can be written in the form

$$2^{f_1}3^{f_2}5^{f_3}7^{f_4}11^{f_5}13^{f_6}17^{f_7}19^{f_8}23^{f_9},$$

where the exponents $f_1, f_2, \ldots, f_9$ are non-negative integers. The product of two such numbers, $2^{f_1}\cdots23^{f_9}$ and $2^{g_1}\cdots23^{g_9}$, will be a perfect square (square of an integer) if and only if the corresponding exponents have the same parity (evenness or oddness). In other words,

$$(2^{f_1}\cdots23^{f_9})\cdot(2^{g_1}\cdots23^{g_9})$$

will be a perfect square if and only if f_1 and g_1 have the same parity, f_2 and g_2 have the same parity, $\ldots$, f_9 and g_9 have the same parity. (Recall that zero is even.) To each number in this set there corresponds an ordered 9-tuple of the parities of the exponents. For example, the number $2^{10}3^57^117^123^{111}$ has the 9-tuple (even, odd, even, even, even, even, odd, even, odd).

There are 512 different 9-tuples possible. By repeated use of the pigeonhole principle, we conclude that 1472 of the integers in our set can be arranged into 736 pairs

$$a_1, b_1; \quad a_2, b_2; \quad \ldots; \quad a_{736}, b_{736}$$

such that each pair contains two numbers with identical 9-tuples of exponent parity. Thus the product of the numbers in each pair is a perfect square. In other words, if we let $c_i := a_i b_i$, then every number in the list

$$c_1, c_2, \ldots, c_{736}$$

is a perfect square. Thus each of

$$\sqrt{c_1}, \sqrt{c_2}, \ldots, \sqrt{c_{736}}$$

is an integer with no prime factor greater than 23. Employing the pigeonhole principle once more, we conclude that at least two numbers in the above list share the same 9-tuple of exponent parity. Without loss of generality, call these $\sqrt{c_k}$ and $\sqrt{c_j}$. Then $\sqrt{c_k}\sqrt{c_j}$ is an perfect square; i.e., $\sqrt{c_k}\sqrt{c_j} = n^2$, for some integer n. Thus $c_k c_j = n^4$. But $c_j c_k = a_j b_j a_k b_k$, so we have found four numbers from our original set of 1,985 integers whose product is the fourth power of an integer.   ∎

We conclude our discussion of parity with a famous problem, originally due (in a different form) to de Bruijn [4]. At least fourteen different solutions have been discovered, several of which use invariants in different ways (see [32] for a very readable account of these solutions). The solution we present, using parity, is perhaps the simplest, and is due to Andrei Gnepp. It is a beautiful solution; we urge you to first think about the problem before reading it. That way you will appreciate just how hard the problem is and how clever Gnepp's solution was!

Example 3.4.11 A rectangle is tiled with smaller rectangles, each of which has at least one side of integral length. Prove that the tiled rectangle also must have at least one side of integral length.

Solution: Here is an example. The large rectangle has been placed on the plane with grid lines marked at each unit, so that its lower-left corner is a lattice point. Notice that each of the small rectangles has at least one integer-length side, and the large 4×2.5 rectangle also has this property. (The circles in the diagram are used in our argument; we will explain them below.)

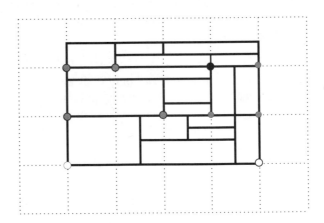

The first insight is to discover an appropriate penultimate step. Let us call the property of having at least one integral side "good." How do we show that the large rectangle is good? Orienting it so that the lower-left corner is a lattice point is the key:

If the rectangle weren't good, then it would have only one lattice point corner. But if the rectangle is good, then it will have either two lattice point corners (if one dimension was an integer), or four lattice point corners (if both length and width are integers).

In other words, parity plays a role: a rectangle with lower-left corner at a lattice point is good if and only if the number of corner lattice points is even! Let us count lattice point corners, hoping to show that the number of lattice point corners for the big rectangle must be even.

Of course we must use the hypothesis that each of the smaller rectangles is good. Consider the corners of a small rectangle. It may have no lattice point corners, or it may have two or four, but can never have just one or three, because of goodness. Consequently, if we count the number of corner lattice points on each small rectangle, and add them up, the sum, which we will call S, will be even.

But we overcounted some of the lattice points. For example, in the picture above, all the corner lattice points are indicated by circles. There are ten rectangles with corner lattice points, and each has exactly two corner lattice points, so $S = 20$. The white circles were counted exactly once, the grey ones were counted twice, and the black one was counted four times. So another way to count the sum S is

$$S = 1 \cdot (\text{\# white circles}) + 2 \cdot (\text{\# grey circles}) + 4 \cdot (\text{\# black circles})$$
$$= 1 \cdot 2 + 2 \cdot 7 + 4 \cdot 1 = 20.$$

In general, when we compute the sum S, we will overcount at some corners. Here are two simple observations that are easy to check.

- We will only count a corner point once, twice or four times—never three times.
- The only corner points that are counted exactly once are the corners of the large rectangle.

Coloring the lattice point corners as in our example, we see that if w, g, b denote the number of white, grey, and black circles, then

$$S = w + 2g + 4b.$$

Thus $w = S - 2g - 4b$ is the number of corner lattice points of the large rectangle. And since S is even, $S - 2g - 4b$ is also even: the large rectangle must be good! ∎

This solution is quite instructive. A terse outline that an experienced problem solver could understand might be, "Orient the large rectangle so that one corner is a lattice point, then consider the parity of the number of corner lattice points." There were two crux moves to this brilliant solution: first to anchor the lower-left corner on a lattice point (yielding "free" information) and then deducing the parity rule for goodness. The rest was a fairly standard argument (as you will see in Chapter 6, the tactic of counting something in two or more different ways is a pervasive one).

Modular Arithmetic and Coloring

Parity works amazingly well, but it is rather crude. After all, we are reducing the infinite universe of integers into a tiny world inhabited by just two entities, "even" and "odd." Sometimes we need to explore a more sophisticated world. Examples 3.4.5 and 3.4.6 used invariants with 9 and 3 possible values, respectively. These are both examples of **modular arithmetic**, that is, the reduction of our point of view from the infinite set of integers to the finite set of possible remainders modulo m, where m is chosen cleverly.

For practice, here's a quick proof of the assertion in the Divisibility by Nine example above. You may wish to recall the basic properties of congruence from page 49. Without loss of generality, let n be a 4-digit number with decimal representation $abcd$. Then

$$n = 10^3 a + 10^2 b + 10c + d.$$

Since $10 \equiv 1 \pmod 9$, we have $10^k \equiv 1^k = 1 \pmod 9$ for any nonnegative integer k. Consequently,

$$n = 10^3 a + 10^2 b + 10c + d \equiv 1 \cdot a + 1 \cdot b + 1 \cdot c + d \pmod 9.$$

You don't need to use congruence notation, but it is a convenient shorthand, and it may help you to systematize your thinking. The important thing is to be aware of the possibility that an invariant may be a quantity modulo m for a properly chosen m.

Example 3.4.12 A bubble-chamber contains three types of sub-atomic particles: 10 particles of type X, 11 of type Y, 111 of type Z. Whenever an X- and Y-particle collide, they both become Z-particles. Likewise, Y- and Z-particles collide and become X-particles and X- and Z-particles become Y-particles upon collision. Can the particles in the bubble chamber evolve so that only one type is present?

Solution: Let us indicate the population at any time by an ordered triple (x, y, z). Let's experiment a bit. We start with the population $(10, 11, 111)$. If an X- and Y- particle then collide, the new population will be $(9, 10, 113)$. Is there anything invariant? Notice that there is still 1 more Y-particle than X-particles, as before. However, where there were originally 100 more Z's than Y's, now there are 103 more. Doesn't look good, but let's stay loose and experiment some more. Another X-Y collision yields $(8, 9, 115)$. The population gap between X and Y is still 1, but the gap between Y and Z has grown to 106. Now let's have an X-Z collision. Our new population is $(7, 11, 114)$. The X-Y gap has grown from 1 to 4, while the Y-Z gap dropped back to 103.

If you are not attuned to the possibility of modular arithmetic, the evolution of population gaps from 1 to 4 or from 100 to 103 to 106 back to 103 may seem chaotic. But by now you have guessed that

The population gaps are invariant modulo 3.

To prove this formally, let (x, y, z) be the population at a certain time. Without loss of generality, consider an X-Z collision. The new population becomes $(x - 1, y + 2, z - 1)$ and you can easily verify that the difference between the X- and Z-populations is unchanged, while the difference between the X- and Y-populations has changed by 3 (likewise for the difference between the Z- and Y-populations).

We conclude the solution by noting that the initial population was $(10, 11, 111)$. The X-Y population gap is 1, and hence it must always be congruent to 1 modulo 3. Thus there is no way that the X and Y populations can ever be the same. ■

The use of coloring is related to parity and modular arithmetic, except that we are not constrained by the algebraic properties of integers. An example of coloring was the domino problem (Example 2.4.2 on page 60), which could have been recast as a parity problem. Here is another example, using 12 colors.

Example 3.4.13 Is is possible to tile a 66×62 rectangle with 12×1 rectangles?

Solution: Obviously, any large rectangle that is tiled by 12×1 rectangles must have an area which is divisible by 12. Indeed, $66 \cdot 62 = 12 \cdot 341$. So impossibility has not been ruled out. Nevertheless, experimenting with smaller configurations with the same property (i.e., $m \times n$ where neither m nor n is a multiple of 12, yet mn is) leads us to conjecture that the 66×62 rectangle *cannot* be tiled with 12×1 rectangles.

So let's assume that there is a tiling, and we shall look for a contradiction. Color the squares of 66×62 rectangle with 12 colors in a cyclic "diagonal" pattern as follows (we are assuming that the height is 66 and the width is 62):

1	12	11	$\cdots$	1	12
2	1	12	$\cdots$	2	1
3	2	1	$\cdots$	3	2
$\vdots$	$\vdots$	$\vdots$	$\ddots$	$\vdots$	$\vdots$
5	4	3	$\cdots$	5	4
6	5	4	$\cdots$	6	5

This coloring has the nice property that any 12×1 rectangle in the tiling consists of 12 differently colored squares. If the large rectangle can be tiled, it will be tiled by $66 \cdot 62/12 = 341$ 12×1 rectangles, and hence the large rectangle must contain 341 squares in each of the 12 colors. The important thing is not the number 341, but the fact that each color occurs in the same number of squares. We will call such a coloration "homogeneous."

Let's look more closely at the colored 66×62 rectangle. We can break it up into 4 sub-rectangles:

60×60	60×2
6×60	6×2

It is easy to check that the $60 \times 60, 60 \times 2$ and 6×60 sub- rectangles are all homogeneous, since each of these sub-rectangles has a dimension which is a multiple of 12. But the 6×2 sub-rectangle is colored as follows:

1	12
2	1
3	2
4	3
5	4
6	5

Consequently, the entire large rectangle is not homogeneous, contradicting the assumption that a tiling existed. So the tiling is impossible.   ■

Monovariants

A **monovariant** is a quantity which may or may not change at each step of a problem, but when it does change, it does so **monotonically** (in only one direction). Another term used for monovariant is **semi-invariant**. Monovariants are often used in the analysis of evolving systems, to show that certain final configurations must occur, and/or to determine the duration of the system. Many monovariant arguments also make use of the extreme principle (at least the well-ordering principle). Here is a very simple example.

Example 3.4.14 In an elimination-style tournament of a two-person game (for example, chess or judo), once you lose, you are out, and the tournament proceeds until only one person is left. Find a formula for the number of games that must be played in an elimination-style tournament starting with n contestants.

Solution: The number of people who are left in the tournament is clearly a monovariant over time. This number decreases by one each time a game is concluded. Hence if we start with n people, the tournament must end after exactly $n - 1$ games! ■

Here is a more subtle problem, one that shows the importance of combining the extreme principle with a monovariant.

Example 3.4.15 The n cards of a deck (where n is an arbitrary positive integer) are labeled $1, 2, \ldots, n$. Starting with the deck in any order, repeat the following operation: if the card on top is labeled k, reverse the order of the first k cards. Prove that eventually the first card will be 1 (so no further changes occur).

Investigation: For example, if $n = 6$ and the starting sequence was 362154, the cards evolve as follows:

$$362154 \to 263154 \to 623154 \to 451326 \to 315426$$
$$\to 513426 \to 243156 \to 423156 \to 132456.$$

It would be nice if the number of the card in the 1st place decreased monotonically, but it didn't (the sequence was $3, 2, 6, 4, 3, 5, 2, 4, 1$). Nevertheless, it is worth thinking about this sequence. We shall make use of a very simple, but important, general principle:

> *If there are only finitely many states as something evolves, either a state will repeat, or the evolution will eventually halt.*

In our case, either the sequence of 1st-place numbers repeats (since there are only finitely many), or eventually the 1st-place number will be 1 (and then the evolution halts). We would like to prove the latter. How do we exclude the possibility of repeats? After all, in our example, there were plenty of repeats!

Once again, the extreme principle saves the day. Since there are only finitely many possibilities as a sequence evolves, there exists a *largest* 1st-place value that ever occurs, which we will call L_1 (in the example above, $L_1 = 6$). So, at some point in the evolution of the sequence, the 1st-place number is L_1, and thereafter, no 1st-place number is ever larger than L_1. What happens immediately after L_1 occurs in the 1st place? We reverse the first L_1 cards, so L_1 appears in the L_1th place. We know that the 1st-place card can never be larger than L_1, but can it ever again equal L_1? The answer is no; as long as the 1st place value is less than L_1, the reversals will never touch the card in the L_1th place. We will never reverse more than the first L_1 cards (by the maximality of L_1), so the only way to get the card numbered L_1 to move at all would be if we reversed exactly L_1 places. But that would mean that the 1st-place and L_1th-place cards both had the value L_1, which is impossible.

That was the crux move. We now look at all the 1st-place values that occur after L_1 appeared in the 1st place. These must be *strictly* less than L_1. Call the maximum of these values L_2. After L_2 appears in 1st place, all subsequent 1st-place values will be strictly less than L_2 by exactly the same argument as before.

Thus we can define a *strictly decreasing* sequence of maximum 1st-place values. Eventually, this sequence must hit 1, and we are done!

The above was a little vague. It is instructive to give a more formal argument, especially to illustrate careful use of subscripts and notation.

Formal Solution: Let f_i be the value of the 1st card after i steps, $i = 1, 2, \ldots$. We wish to show that $f_m = 1$ for some m (and consequently, $f_n = 1$ for all $n \geq m$). Since $1 \leq f_i \leq n$, the number

$$L_1 := \max\{f_i : i \geq 1\}$$

exists. Also define

$$t_1 := \min\{t : f_t = L_1\};$$

i.e., t_1 is the first step at which the 1st card is equal to L_1. In the example above, $L_1 = 6$ and $t_1 = 3$.

We claim that if $t > t_1$, then $f_t < L_1$. To see this, notice that f_t cannot be greater than L_1, by the definition of L_1. To see that f_t cannot equal L_1, we argue by contradiction. Let t be the *first* step after t_1 such that $f_t = L_1$. At step t_1, the first L_1 cards were reversed, placing the value L_1 at the L_1th place. For all steps s after t_1 and before t, we have $f_s < L_1$, which means that the card with the value L_1 at the L_1th place was not moved. Hence at step t, it is impossible for $f_t = L_1$, since that means that the 1st and L_1th are the same (unless $L_1 = 1$, in which case we are done). This contradiction establishes the claim.

Now we define two sequences. For $r = 2, 3, \ldots$, define

$$L_r := \max\{f_i : i > t_{r-1}\} \quad \text{and} \quad t_r := \min\{t > t_{r-1} : f_t = L_r\}.$$

(In our example, $L_2 = 5, t_2 = 6$ and $L_3 = 4, t_3 = 8$.) As above, we assert that as long as $L_r > 1$, if $t > t_r$, then $f_t < L_r$ (for each $r \geq 1$).

Therefore, the sequence $L_1, L_2, \ldots$ is strictly decreasing, hence one of the L_r will equal 1, so eventually, $f_m = 1$ for some m. ∎

We will conclude the chapter with a wonderfully imaginative use of monovariants, John Conway's famous "Checker Problem."

Example 3.4.16 Put a checker on every lattice point (point with integer coordinates) in the plane with y-coordinate less than or equal to zero. The only legal moves are horizontal or vertical "jumping." By this we mean that a checker can leap over a neighbor, ending up 2 units up, down, right, or left of its original position, provided that the destination point is unoccupied. After the jump is complete, the checker that was jumped over is removed from the board. Here is an example.

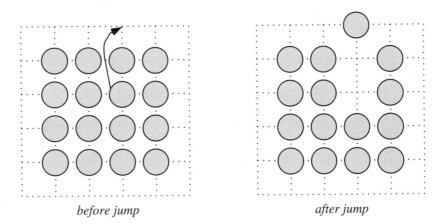

before jump   *after jump*

Is it possible to make a finite number of legal moves and get a checker to reach the line $y = 5$?

Solution: After experimenting, it is pretty easy to get a checker to $y = 2$, and a lot more work can get a checker to $y = 3$. But these examples shed no light on what is and what isn't possible.

The key idea: define a monovariant that assigns a number to each configuration of checkers, one that moves in one direction as we approach our goal.

Without loss of generality, let the destination point be $C = (0, 5)$. For each point (x, y) in the plane, compute its "taxi-cab distance" (length of the shortest path that stays on the lattice grid lines) to C. For example, the distance from $(3, 4)$ to $(0, 5)$ is $3 + 1 = 4$. Assign to each point at a distance d from C the value ζ^r, where

$$\zeta = \frac{-1 + \sqrt{5}}{2}.$$

There are three crucial things about ζ. First of all, it satisfies

$$\zeta^2 + \zeta - 1 = 0. \tag{2}$$

Also, it is positive. And finally, $\zeta < 1$.

For any (infinite) configuration of checkers, add up all the values assigned to each checker on the board. Let us call this the "Conway sum." This, we claim, is our monovariant. We need to check a number of things.

First of all, does this sum exist? Yes; we have infinitely many infinite geometric series to consider,[6] but luckily, they will converge. Let us compute the Conway sum for the starting configuration. The checkers on the y-axis (directly below C) have the values $\zeta^5, \zeta^6, \ldots$. In general, the checkers on the line $x = \pm r$ have the values $\zeta^{5+r}, \zeta^{6+r}, \ldots$. Hence the Conway sum for the whole "half-plane" is

$$(\zeta^5 + \zeta^6 + \cdots) + 2 \sum_{r=1}^{\infty} (\zeta^{5+r} + \zeta^{6+r} + \cdots) = \frac{\zeta^5}{1 - \zeta} + 2 \sum_{r=1}^{\infty} \frac{\zeta^{5+r}}{1 - \zeta}$$

$$= \frac{1}{1 - \zeta} \left(\zeta^5 + \frac{2\zeta^6}{1 - \zeta} \right)$$

$$= \frac{1}{1 - \zeta} \left(\frac{\zeta^5(1 - \zeta) + 2\zeta^6}{1 - \zeta} \right)$$

$$= \frac{1}{1 - \zeta} \left(\frac{\zeta^5 + \zeta^6}{1 - \zeta} \right).$$

Using (2), we observe that $1 - \zeta = \zeta^2$. Therefore the expression above simplifies to

$$\frac{\zeta^5 + \zeta^6}{\zeta^4} = \zeta + \zeta^2 = 1.$$

Thus, the starting configuration has Conway sum 1, and all other configurations will have computable Conway sums.

Next, we must show that the Conway sum is a monovariant. Consider a horizontal move that moves a checker *away* from the destination point C. For example, suppose we could jump a checker from $(9, 3)$, which has value ζ^{11}, to $(11, 3)$. What happens to the Conway sum? We remove the checkers at $(9, 3)$ and $(10, 3)$ and create a checker at $(11, 3)$. The net change in the Conway sum is

$$-\zeta^{11} - \zeta^{12} + \zeta^{13} = \zeta^{11}(-1 - \zeta + \zeta^2).$$

---

[6]See page 176 for information about infinite geometric series.

But $-1 - \zeta + \zeta^2 = -2\zeta$, by (2). Consequently, the net change is negative; the sum decreases. Clearly this is a general situation (check the other cases if you are not sure): whenever a checker moves away from C, the sum decreases.

On the other hand, if the checker moves toward C, the situation is different. For example, suppose we went from $(9, 3)$ to $(7, 3)$. Then we remove the checkers at $(9, 3)$ and $(8, 3)$ and create a checker at $(7, 3)$, changing the sum by

$$-\zeta^{11} - \zeta^{10} + \zeta^9 = \zeta^9(1 - \zeta - \zeta^2) = 0.$$

That, in fact, is where ζ came from in the first place! It was cleverly designed to make the sum decrease when you move away from the goal, and not change when you move toward it.

There is one more case: moves that leave the distance to C unchanged, such as, for example, a jump from $(1, 4)$ to $(-1, 4)$. It is easy to check (do it!) that moves of this kind decrease the Conway sum.

Thus the Conway sum is a monovariant which starts at a value of 1, and never increases. However, if a checker occupied C, its value would be $\zeta^0 = 1$, so the Conway sum would be strictly greater than 1 (since other checkers must be on the board, and $\zeta > 0$). We conclude that it is not possible to reach C.   ■

Problems and Exercises

3.4.17 Prove that if you add up the reciprocals of a sequence of consecutive positive integers, the numerator of the sum (in lowest terms) will always be odd. For example, $\frac{1}{7} + \frac{1}{8} + \frac{1}{9} = \frac{191}{504}$.

3.4.18 Prove the Power of a Point Theorem (Example 3.4.2 on page 102) by looking at similar triangles.

3.4.19 Inspired by the Divisibility by Nine example on page 103, discover and prove a similar statement for divisibility by 11.

3.4.20 Start with a set of lattice points. Each second, we can perform one of the following operations:

1. The point (x, y) "gives birth" to the point $(x + 1, y + 1)$.

2. If x and y are both even, the point (x, y) "gives birth" to the point $(x/2, y/2)$.

3. The pair of points (x, y) and (y, z) "gives birth" to (x, z).

For example, if we started with the single point $(9, 1)$, operation #1 yields the new point $(10, 2)$, and then operation #2 yields $(5, 1)$, and then 9 applications of operation #1 gives us $(14, 10)$, and then operation #3 applied to $(14, 10)$ and $(10, 2)$ gives us $(14, 2)$, etc.

If we start with the single point $(7, 29)$, is it possible to eventually get the point $(3, 1999)$?

3.4.21 Initially, we are given the sequence $1, 2, \ldots, 100$. Every minute, we erase any two numbers u and v and replace them with the value $uv + u + v$. Clearly, we will be left with just one number after 99 minutes. Does this number depend on the choices that we made?

3.4.22 Prove that it is impossible to choose three distinct integers a, b and c such that

$$a - b \,|\, b - c, \quad b - c \,|\, c - a, \quad c - a \,|\, a - b.$$

3.4.23 (Tom Rike) Start with the set $\{3, 4, 12\}$. You are then allowed to replace any two numbers a and b with the new pair $0.6a - 0.8b$ and $0.8a + 0.6b$. Can you transform the set into $\{4, 6, 12\}$?

3.4.24 Two people take turns cutting up a rectangular chocolate bar which is 6×8 squares in size. You are allowed to cut the bar only along a division between the squares and your cut can be only a straight line. For example, you can turn the original bar into a 6×2 piece and a 6×6 piece, and this latter piece can be turned into a 1×6 piece and a 5×6 piece. The last player who can break the chocolate wins (and gets to eat the chocolate bar). Is there a winning strategy for the first or second player? What about the general case (the starting bar is $m \times n$)?

3.4.25 Consider a row of $2n$ squares colored alternately black and white. A legal move consists of choosing any contiguous set of squares (1 or more squares, but if you pick more than 1 square your squares must be next to one another; i.e., no "gaps" allowed), and inverting their colors. What is the minimum number of moves required to make the entire row be one color? Clearly, n moves will work (for example, invert the 1st square, then the 3rd square, etc.), but can you do better than that?

3.4.26 Answer the same question as above, except now we start with a $2n \times 2n$ "checkerboard" and a legal move consists of choosing any subrectangle and inverting its colors.

3.4.27 Show that if every room in a house has an even number of doors, then the number of outside entrance doors must be even as well.

3.4.28 Twenty-three people, each with integral weight, decide to play football, separating into two teams of 11 people, plus a referee. To keep things fair, the teams chosen must have equal *total* weight. It turns out that no matter who is chosen to be the referee, this can always be done. Prove that the 23 people must all have the same weight.

3.4.29 (Russia, 1995) There are n seats at a merry-go-round. A boy takes n rides. Between each ride, he moves clockwise a certain number of places to a new horse. Each time he moves a different number of places. Find all n for which the boy ends up riding each horse.

3.4.30 Consider 9 lattice points in three-dimensional space. Show that there must be a lattice point on the interior of one of the line segments joining two of these points.

3.4.31 The first six terms of a sequence are $0, 1, 2, 3, 4, 5$. Each subsequent term is the last digit of the sum of the six previous terms. In other words, the seventh term is 5 (since $0 + 1 + 2 + 3 + 4 + 5 = 15$), the eighth term is 0 (since $1 + 2 + 3 + 4 + 5 + 5 =$

20), etc. Can the subsequence 13579 occur anywhere in this sequence?

3.4.32 The solution to the checker problem (Example 3.4.16) was rather slick, and almost seemed like cheating. Why couldn't you assign any value that you wanted to the point C, for example, 10^{100}. Then this would assure that you could never get there. What is wrong with this idea?

3.4.33 Make sure that you *really* understand how example 3.4.10 works. For example, exactly how many times is the pigeonhole principle used in this problem? After reading the solution, see how well you can explain this problem to another person (someone who asks intelligent questions!). If you cannot explain it to his satisfaction, go back and reread the solution, making note of the crux moves, auxiliary problems posed and solved, etc. A solid understanding of this example will help you with the next problem.

3.4.34 (IMO 1978) An international society has its members from six different countries. The list of members contains 1978 names, numbered $1, 2, \ldots, 1978$. Prove that there is at least one member whose number is the sum of the numbers of two members from his own country, or twice as large as the number of one member from his own country.

3.4.35 (IMO 1997) An $n \times n$ matrix (square array) whose entries come from the set $S = \{1, 2, \ldots, 2n-1\}$ is called a *silver* matrix if, for each $i = 1, \ldots, n$, the ith row and the ith column together contain all elements of S. Show that there is no silver matrix for $n = 1997$.

3.4.36 (Taiwan, 1995) Consider the operation which transforms the 8-term sequence $x_1, x_2, \ldots, x_8$ into the new 8-term sequence

$$|x_2 - x_1|, |x_3 - x_2|, \ldots, |x_8 - x_7|, |x_1 - x_8|.$$

Find all 8-term sequences of integers which have the property that after finitely many applications of this operation, one is left with a sequence, all of whose terms are equal.

3.4.37 *Euler's Formula.* Consider a polyhedron P. We wish to show that $v - e + f = 2$, where v, e, f are respectively the number of vertices, edges and faces of P. Imagine that P is made of white rubber, with the edges painted black and the vertices painted red. Carefully cut out a face (but don't remove any edges or vertices), and stretch the resulting object so that it lies on a plane. For example, here is what a cube looks like after such "surgery":

Now we want to prove that $v - e + f = 1$ for the new object. To do so, start erasing edges and/or vertices, one at a time. What does this do to the value of $v - e + f$? What must you end up with?

Chapter 4

Three Important Crossover Tactics

A **crossover** (first mentioned on page 60) is an idea that connects two or more different branches of math, usually in a surprising way. In this chapter, we will introduce perhaps the three most productive crossover topics: graph theory, complex numbers, and generating functions.

We will just scratch the surface of these three very rich topics. Our presentation will be a mix of exposition and problems for you to ponder on the spot. You may find it worthwhile to read the chapters of Part II first or at least concurrently, as some of the examples below involve relatively sophisticated mathematics.

4.1   Graph Theory

The concept of a **graph** is very simple: merely a finite collection of **vertices** and **edges**. The vertices are usually visualized as dots, and the edges as lines which connect some or all of the pairs of vertices. If two vertices are connected by an edge, they are called **neighbors**. By convention, graphs do not contain multiple edges (two or more edges connecting the same pair of vertices) or loops (an edge connecting a vertex to itself). If there are multiple edges or loops, we use the term **multigraph**. In the figure below, the object on the left is a graph, while its neighbor to the right is a multigraph.

You have already seen many examples of graphs. The Affirmative Action Problem (Example 2.1.9 on page 23) can be restated as a problem about graphs as follows.

> *Given an arbitrary graph, show that it is possible to color the vertices black and white in such a way that each white vertex has at least as many black neighbors as white neighbors, and vice versa.*

Likewise, the handshake problem (1.1.4), and the solution to the gallery problem (2.4.3) all can be reformulated as questions about graphs, rather than people and handshakes or networks of pipes.

This is what makes graph theory surprisingly useful. Just about any situation involving "relationships" between "objects" can be recast as a graph, where the vertices are the "objects" and we join vertices with edges if the corresponding objects are "related."

If you're not yet convinced, look at the following problem. Don't read the analysis immediately.

Example 4.1.1 (USAMO 1986) During a certain lecture, each of five mathematicians fell asleep exactly twice. For each pair of these mathematicians, there was some moment when both were sleeping simultaneously. Prove that, at some moment, some three of them were sleeping simultaneously.

Partial Solution: Let us call the mathematicians A, B, C, D, E, and denote the time intervals that each was asleep by A_1, A_2, B_1, B_2, etc. Now define a graph whose 10 vertices are these time intervals, with vertices connected by an edge if the time intervals overlapped. There are $\binom{5}{2} = 10$ pairs of mathematicians,[1] so this graph must have at least 10 edges. Here is one example, depicting the situation where mathematician A's first nap overlaps with C's first nap and E's second nap, etc. Notice that A_1 and A_2 cannot be joined with a vertex, nor can B_1 and B_2, etc., since each mathematician took two distinct naps.

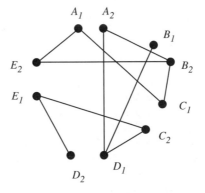

This particular graph contains the **cycle** $C_1 A_1 E_2 B_2 C_1$. By this we mean a closed path that "travels" along edges. A cycle is helpful, because it constrains the overlap times. It is easiest to see this by "stacking" the intervals. Here is one possibility. Time is measured horizontally. Notice that each nap interval must overlap with its vertical neighbor(s).

$$C_1$$
$$B_2$$
$$E_2$$
$$A_1$$
$$C_1$$

---

[1] Read Section 6.1 if you don't understand the notation $\binom{n}{r}$.

We will show that three different intervals must overlap. Since C_1 and B_2 overlap, and C_1 and A_1 overlap, we are done if B_2 and A_1 overlap *within* C_1. If they don't (as in the example above), there will be a gap between the end of B_2 and the start of A_1. However, E_2 must straddle this gap, since it overlaps both B_2 and A_1. Thus C_1, B_2, and E_2 overlap (likewise, C_1, E_2, and A_1 overlap).

This argument involved a 4-cycle (cycle with 4 vertices), but a little thought (get out some paper and do some experiments!) shows that it will work with a cycle of any finite length. We have therefore reduced the problem to a "pure" graph theory question:

If a graph has 10 vertices and 10 edges, must it contain a cycle?

Connectivity and Cycles

Now that we see how graph theory can completely recast a problem, we shall investigate some simple properties of graphs, in particular, the relationship between the number of vertices and edges and the existence of cycles. The number of edges emanating from a particular vertex is called the **degree** of the vertex. If the vertex is x, the degree is often denoted by $d(x)$. You should easily be able to verify the following important fact, often called the **Handshake Lemma** (if you want a hint, reread Example 3.4.7 on page 104)

> *In any graph, the sum of the degrees of all the vertices is equal to twice the number of edges.*

A graph is **connected** if every pair of vertices has a path between them. If a graph is not connected, one can always decompose the graph into **connected components**. For example, the graph below has 10 vertices, 11 edges, and two connected components. Observe that the handshake lemma does not require connectivity; in this graph the degrees (scanning from left to right, from top to bottom) are $1, 2, 1, 1, 3, 2, 3, 4, 3, 2$. The sum is $22 = 2 \cdot 11$.

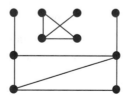

A connected graph that contains no cycles is called a **tree**.[2] For example, the following 8-vertex graph is a tree.

---

[2] A non-connected graph containing no cycles is called a **forest**; each of its connected components is a tree.

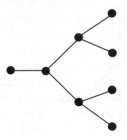

In a tree, we call the vertices with degree 1 **leaves**.[3]

It certainly seems plausible that trees must have leaves. In fact,

Every tree with two or more vertices has at least two leaves.

Informally, this is pretty obvious. Since a tree has no cycles, it has to have **paths** where the starting and ending vertex are different. These two vertices will have degree 1. But this is a little vague. Here is a rigorous proof that uses the extreme principle and argument by contradiction:

Given a tree, pick any vertex. Now consider all paths that include this vertex, and pick the *longest* one, i.e., the path which contains the most vertices. Since the graph is a tree, there are no cycles, so there is no ambiguity here—no path can cross back on itself. And also, since trees are connected, we are guaranteed that there are paths to begin with.

Let $P := x_1 x_2 \ldots x_n$ denote this longest path, where the x_i are vertices. We claim that x_1 and x_n must have degree 1. Suppose that x_1 had degree more than 1. Then x_1 has among its neighbors the vertices x_2 and y. Observe that y cannot be any of the vertices $x_3, x_4, \ldots, x_n$, because that would create a cycle! But if y is not one of these vertices, then we could create a *longer* path $y x_1 x_2 \ldots x_n$ which contradicts the maximality of P. Thus $d(x_1) = 1$, and a similar argument shows that $d(x_n) = 1$. ∎

When is a connected graph a tree? Intuitively, it seems clear that trees are rather poor in edges relative to vertices, and indeed, experimentation (do it!) suggests very strongly that

For trees, the number of edges is exactly one less than the number of vertices.

This conjecture is a natural candidate for mathematical induction. We shall induct on the number of vertices v, starting with $v = 2$. The only tree with two vertices consists of a single edge joining the two vertices, so the base case is true. Now, assume that we know that all trees with v vertices contain $v - 1$ edges. Consider a tree T with $v + 1$ vertices. We will show that T has v edges. Pick a leaf (we know that T has leaves). Remove this vertex, along with the edge emanating from it. What is left? A graph with v vertices that is still connected (since T was connected, and plucking a

---

[3]This terminology is not quite standard. It is customary to designate one of the degree-1 vertices as a "root," and indeed there is a whole theory about so-called "rooted trees" which is quite important in computer science, but we will not discuss this here. See [30] for details.

leaf cannot disconnect it), and has no cycles (since T had no cycles, and plucking a leaf cannot create a cycle). Hence the new graph is a tree. By the inductive hypothesis, it must have $v - 1$ edges. Thus T has $v - 1 + 1 = v$ edges.   ∎

4.1.2 Generalize the above by showing that if a forest has k connected components, e edges and v vertices, then $e = v - k$.

4.1.3 Conclude by showing that

> *If a graph has e edges and v vertices and $e \geq v$, then the graph must contain a cycle.*

Note that it does not matter whether the graph is connected or not.

Now we can finish up Example 4.1.1, the problem about the napping mathematicians. The graph in question has 10 edges and 10 vertices. By 4.1.3, it must contain a cycle.

Eulerian and Hamiltonian Paths

Problem 2.1.26 on page 28 has a simple graph theory formulation:

> *Find the conditions on a connected graph (or multigraph) for which it is possible to travel a path that traverses every edge exactly once.*[4]

Such paths are called **Eulerian**, in honor of the 18th-century Swiss mathematician who first studied graphs in a formal way. Here are two examples which appeared in Problem 2.1.26.

A   B

Graph A (the vertices are not marked with dots, but are simply the places where the line segments intersect) has an Eulerian path, while graph B does not. If you draw enough graphs (and multigraphs), you are inescapably led to focus on the vertices with odd degree. Let v be a vertex with odd degree in a graph that possesses an Eulerian path. There are three cases:

1. The path can start at v.
2. The path can end at v.

---

[4]More precise language distinguishes between "walks" that avoid repeated edges, called "trails," with walks that avoid repeated vertices, called "paths." These distinctions are not needed here, but for a very clear discussion, see Chapter 9 of [8].

3. The path starts and ends elsewhere, or is a closed path (no start or end). This is not possible, because whenever the path enters v along an edge, it will need to exit v along a different edge. This means that $d(v)$ is even.

Therefore if a graph has an Eulerian path, it must have either zero or exactly two odd-degree vertices. In fact, this is a necessary and sufficient condition. More precisely,

> *A connected graph (or multigraph[5]) possesses an Eulerian path if and only if it has zero or exactly two odd-degree vertices. In the former case, the path is a closed path. In the latter, the path must begin and end at the odd-degree vertices.*

It is possible to prove this by induction, and indeed, the argument below can easily be rewritten as an induction proof. But it is much more instructive to present a new type of argument, an **algorithmic** proof where we give a general recipe for the construction of an Eulerian path.

Consider first a graph with exactly two odd-degree vertices, which we will call s and f. Let us try to naively draw an Eulerian path, starting from s. We will travel randomly, figuring that we can't lose: if we enter an even-degree vertex, we can then leave it, and either this vertex will have no untraveled edges left, or an even number of them, in which case we can travel through later.

But this doesn't quite work. Consider the following graph (actually, it is a multigraph, but we will just use the term "graph" instead for now; there shouldn't be any confusion). The edges are labeled with upper-case letters, and vertices are labeled with lower-case letters.

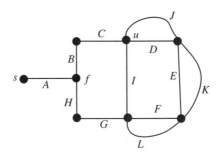

Starting at vertex s, what if we traversed, in order, edges A, B, C, D, E, F, G, H? We would be stuck at vertex f, with no way to "backtrack" and traverse the other edges. In this case, let us temporarily remove the edges that we have traveled. We are left with the **subgraph** containing the edges I, J, K, L. This subgraph has four vertices, each with even degree. Since the original graph was connected, the subgraph "intersected" some of the edges that we removed, for example, at the vertex labeled u. Now let us apply the "naive" algorithm to the subgraph, starting at u. We traverse, in order, J, K, L, I. We ended up back at u, and that is no coincidence. Since all the vertex

---

[5]Unless otherwise stated, we assume that graphs are not multigraphs. When investigating Eulerian paths, multigraphs may be relevant, but this is virtually the only place in the text where they are.

degrees of our subgraph are even, we cannot get "stuck" unless we returned to our starting point.

So now we can perform "reconstructive surgery" on our original path and get an Eulerian path for the entire graph.

1. Start at s, as before, and travel along edges A, B, C until we reach vertex u.
2. Now travel along the subgraph (edges J, K, L, I), returning to u.
3. Finish the trip along the edges D, E, F, G, H, reaching vertex f.

This method will work in general. We may have to repeat the "remove the traveled edges and travel the subgraph" step several times (since we could have gotten stuck back at the starting point without traversing all of the edges of the subgraph), but since the graph is finite, eventually we will finish. ∎

4.1.4 If you aren't convinced about the argument above, try the algorithm on the following multigraph. Then you will understand the algorithm.

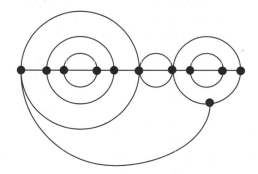

4.1.5 A **directed graph** (also called **digraph**) is a graph (or multigraph) where each edge is given a direction (usually indicated by an arrow). In other words, a directed graph is like a network of one-way streets. Find necessary and sufficient conditions for a directed graph or multigraph to have an Eulerian path.

The "dual" of an Eulerian path is a **Hamiltonian** path (named after a 19th- century Irish mathematician), a path that visits each *vertex* exactly once. If the path is closed, it is called a Hamiltonian cycle. While Eulerian paths possess a "complete theory," very little is known about Hamiltonian paths. At present, the necessary and sufficient conditions for Hamiltonian paths are unknown. This is unfortunate, because many practical problems involve Hamiltonian paths. For example, suppose we want to seat people around a table in such a way that no one sits next to someone that she dislikes. Then we can make a graph where the people are vertices and we connect edges between friends. A Hamilton path, if one exists, gives us a seating plan. Many problems involving scheduling and optimization of network paths can be recast as searches for Hamiltonian paths.

The following is a rather weak statement that gives a *sufficient* condition for Hamiltonian paths.

4.1.6 *Let G be a graph (not a multigraph) with v vertices. If each vertex has degree at least v/2, then G has a Hamiltonian cycle.*

This statement is weak, because the hypothesis is so strong. For example, suppose that G has 50 vertices. Then we need each vertex to have degree at least 25 in order to conclude that a Hamiltonian path exists.

We urge you to prove 4.1.6; note that one of the first things you need to do is show that the hypothesis forces G to be connected.

The Two Men of Tibet

Our goal has not been a comprehensive study of graph theory, but merely an introduction to the subject, to give you a new problem-solving tactic and to sensitize you to think about recasting problems in terms of graphs whenever possible. If you wish to learn more about this subject, there is a huge literature, but [11], [22], and [30] are all great places to start (especially [11]).

We will conclude this section with a classic problem which at first does not seem to have anything to do with graphs.

Example 4.1.7 *The Two Men of Tibet.* Two men are located at opposite ends of a mountain range, at the same elevation. If the mountain range never drops below this starting elevation, is it possible for the two men to walk along the mountain range and reach each other's starting place, while always staying at the same elevation?

Here is an example of a "mountain range." Without loss of generality, it is "piecewise linear," i.e., composed of straight line pieces. The starting positions of the two men is indicated by two dots.

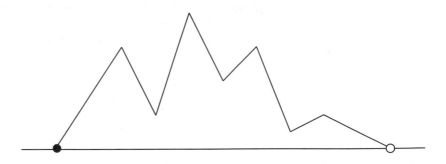

At first, it doesn't seem too hard. As long as it is legal to walk backward, it is pretty easy for the two men to stay at the same elevation. Let us label the "interesting" locations on the range (those with elevations equal to the peaks and troughs) with letters.

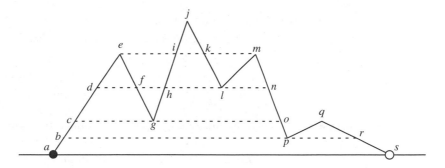

Then the black dot walks from a to c, while the white one goes from s to q. Next, the black dot walks backward from c to b while the white one goes from q to p, etc. It is pretty easy to write out the exact sequence of forward and backward steps to take.

But *why* does it work? And how can we guarantee that it will always work, even for really complicated mountain ranges (as long as the range does not have any valleys which drop below the starting elevation)? Before reading further, take some time to try to develop a *convincing* argument. It's not easy! Then you will enjoy our graph theory solution all the more.

Solution: As in the diagram above, label all the "interesting" locations. Let us call this set I, so in our example, $I = \{a, b, c, \ldots, s\}$. As the dots travel, we can keep track of their joint locations with an ordered pair of the form (x, y), where x indicates the black dot's location and y indicates the white dot's location. Using this notation, the progress of the two dots could be abbreviated as

$$(a, s) \to (c, q) \to (b, p) \to (e, m) \to (f, l) \to \cdots \to (s, a),$$

where the final configuration of (s, a) indicates that the two dots have switched places.

Now define a graph Γ whose vertices are all ordered pairs (x, y), where $x, y \in I$ and x and y are at the same elevation. In other words, the vertices of Γ consist of all possible legal configurations of where the two dots *could* be, although it may be that some of these configurations are impossible to reach from the starting locations. We shall join two vertices by an edge if it is possible to travel between the two configurations in one "step." In other words, the vertex (a, s) is not joined to (c, q), but we do join (a, s) to (b, r) and (b, r) to (c, q). Here is an incomplete picture of Γ, using a cartesian coordinate system ([so the starting configuration (a, s) is at the upper-left corner]. This picture is missing many vertices [for example, $(a, a), (b, b), (c, c), \ldots$] and not all of the edges are drawn from the vertices that are pictured.

If we can show that there is a path from (a, s) to (s, a), we'd be done. [Actually, the path from (a, s) to (j, j) does the trick, since the graph is symmetrical.] Verify the following facts.

1. The only vertices of Γ with degree 1 are (a, s) and (s, a).
2. If a vertex is of the form (peak, peak), it has degree 4. For example, (e, m) has degree 4.
3. If a vertex is of the form (peak, slope), it has degree 2. An example is (e, i).

4. If a vertex is of the form (slope, slope), it has degree 2. An example is (d, n).

5. If a vertex is of the form (peak, trough), it is isolated (has degree zero). The vertex (g, q) is an example of this.

Now consider the connected component of Γ which contains the vertex (a, s). This is a subgraph of Γ [it is not all of Γ, since (g, q) and (q, g) are isolated]. By the handshake lemma (page 121), the sum of the degrees of the vertices of this subgraph must be even. Since the only two vertices with odd degree are (a, s) and (s, a), this subgraph *must* contain (s, a) as well. Thus there is a path from (a, s) to (s, a). This argument will certainly generalize to any mountain range, so we are done. ∎

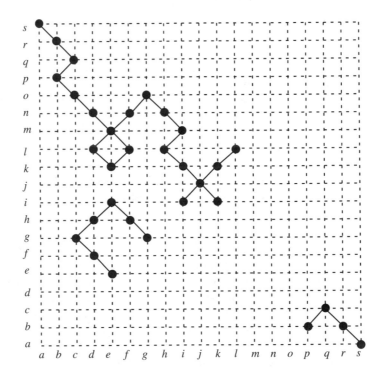

In the end, we solved this hard problem with a very simple parity analysis. Of course, we first needed the insight of constructing a graph, and the crux move of defining the vertices and edges in a very clever way. The moral of the story: just about anything can be analyzed with graphs!

Problems and Exercises

In these problems, a graph is not a multigraph (no loops or multiple edges) unless specifically so stated.

4.1.8 Show that every graph contains two vertices of equal degree.

4.1.9 Given six people, show that either three are mutual friends, or three are complete strangers to one another. (Assume that "friendship" is mutual; i.e., if you are my friend then I must be your friend.)

4.1.10 Seventeen people are at a party. It turns out that for each pair of people present, exactly one of the following statements is always true: "They haven't met," "They are good friends," or "They hate each other." Prove that there must be a trio (3) of people, all of whom are either mutual strangers, mutual good friends, or mutual enemies.

4.1.11 Show that if a graph has v vertices, each of degree at least $(v-1)/2$, then this graph is connected.

4.1.12 How many edges must a graph with n vertices have in order to guarantee that it is connected?

4.1.13 A large house contains a television set in each room that has an odd number of doors. There is only one entrance to this house. Show that it is always possible to enter this house and get to a room with a television set.

4.1.14 A **complete** graph is one where there is an edge between every two vertices. The complete graph with r vertices is denoted by K_r. Show that if the edges of K_{10} are colored blue or red, then there must be 4 vertices, such that all the edges connecting them are the same color. In other words, "a bi-colored K_{10} must possess a monochromatic K_4 subgraph."

4.1.15 A **bipartite** graph is one where the vertices can be partitioned into two sets U, V such that each edge has one end in U and one end in V. The figure below shows two bipartite graphs. The one on the right is a **complete** bipartite graph and is denoted $K_{4,3}$.

Show that a graph is bipartite if and only if it has no odd cycles.

4.1.16 A group of people play a round-robin chess tournament, which means that everyone plays a game with everyone else exactly once (chess is a one-on-one game, not a team sport). There are no draws.

(a) Prove that it is always possible to line up the players in such a way that the first player beat the second, who beat the third, etc. down to the last player. Hence it is always possible to declare not only a winner, but a meaningful ranking of all the players.

(b) Give a graph theoretic statement of the above.

(c) Must this ranking be unique?

4.1.17 (USAMO 1981) Every pair of communities in a county are linked directly by exactly one mode of transportation: bus, train or automobile. All three modes of transportation are used in the county with no community being serviced by all three modes and no three communities being linked pairwise by the same mode. Determine the maximum number of communities in the county.

4.1.18 A domino consists of two squares, each of which is marked with 0, 1, 2, 3, 4, 5, or 6 dots. Here is one example.

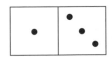

Verify that there are 28 different dominos. Is it possible to arrange them all in a circle so that the adjacent halves of neighboring dominos show the same number?

4.1.19 Is it possible for a knight to travel around a standard 8×8 chessboard, starting and ending at the same square, while making every single possible move that a knight can make on the chessboard, *exactly once*? We consider a move to be completed if it occurs in either direction.

4.1.20 Cities $C_1, C_2, \ldots, C_N$ are served by airlines $A_1, A_2, \ldots, A_n$. There is direct non-stop service between any two cities (by at least one airline), and all airlines provide service in both directions. If $N \geq 2^n + 1$, prove that at least one of the airlines can offer a round trip with an odd number of landings.

4.1.21 (IMO 1991) Let G be a connected graph with k edges. Prove that the edges can be labeled $1, \ldots, k$ in some fashion, such that for every vertex of degree greater than 1, the labels of those edges incident to that vertex have greatest common divisor 1.

4.1.22 (USAMO 1989) The 20 members of a local tennis club have scheduled exactly 14 two-person games among themselves, with each member playing in at least one game. Prove that within this schedule there must be a set of 6 games with 12 distinct players.

4.1.23 An *n-cube* is defined intuitively to be the graph you get if you try to build an n-dimensional cube out of wire. More rigorously, it is a graph with 2^n vertices labeled by the n-digit binary numbers, with two vertices joined by an edge if the binary digits differ by exactly one digit. Show that for every $n \geq 1$, the n-cube has a Hamiltonian cycle.

4.1.24 Consider a cube made out of 27 subcubes (so its dimensions are $3 \times 3 \times 3$). The subcubes are connected by doors on their faces (so every subcube has 6 doors, although of course some cubes have doors which open to the "outside"). Is is possible to start at the center cube and visit every other cube exactly once?

4.1.25 If you place the digits 0,1,1,0 clockwise on a circle, it is possible to read any two-digit binary number from 00 to 11 by starting at a certain digit and then reading clockwise. Is it possible to do this in general?

4.1.26 In a group of nine people, one person knows two of the others, two people each know four others, four each know five others, and the remaining two each know six others. Show that there are three people who all know each other.

4.1.27 Devise a graph-theoretic recasting of De Bruijn's rectangle problem (Example 3.4.11 on page 108).

4.2 Complex Numbers

Long ago you learned how to manipulate the complex numbers **C**, the set of numbers of the form $a + bi$, where a, b are real and $i = \sqrt{-1}$. What you might not have learned is that complex numbers are the crossover artist's dream: like light, which exists simultaneously as wave and particle, complex numbers are both algebraic and geometric. You will not realize their full power until you become comfortable with their geometric, physical nature. This in turn will help you to become fluent at translating between the algebraic and the geometric in a wide variety of problems.

We will develop the elementary properties of complex numbers below mostly as a sequence of exercises and problems. This section is brief, meant only to open your eyes to some interesting possibilities.[6]

Basic Operations

4.2.1 *Basic notation and representation of complex numbers.* A useful way to depict complex numbers is via the **Gaussian** or **Argand** plane. Take the usual cartesian plane, but replace the x- and y-axes with real and imaginary axes, respectively. We can view each complex number $z = a + bi$ as a *point* on this plane with coordinates (a, b). We call a the **real part** of z and write $a = \mathrm{Re}\, z$. Likewise, the **imaginary part** $\mathrm{Im}\, z$ is equal to b. We can also think of $\mathrm{Re}\, z$ and $\mathrm{Im}\, z$ as the real and imaginary **components** of the *vector* which starts at the origin and ends at (a, b). Hence the complex number $z = a + bi$ has a double meaning: it is both a *point* with coordinates (a, b) and simultaneously the *vector* which starts at the origin and ends at (a, b). Do keep in mind, though, that a vector can start anywhere, not just at the origin, and what defines a vector uniquely is its **magnitude** and **direction**. The magnitude of the complex number $z = a + bi$ is

$$|z| := \sqrt{a^2 + b^2},$$

which is, of course, the length of the vector from the origin to (a, b). Other terms for magnitude are **modulus** and **absolute value**, as well as the more informal term

---

[6]For much more information, we strongly urge you to read at least the first few chapters of our chief inspiration for this section, Tristan Needham's *Visual Complex Analysis* [17]. This trail-blazing book is fun to read, beautifully illustrated, and contains dozens of geometric insights that you will find nowhere else.

"length," which we will often use. The direction of this vector is conventionally defined to be the angle that it makes with the horizontal (real) axis, measured counterclockwise. This is called the **argument** of z, denoted $\arg z$. Informally, we also call this the "angle" of z. If $\theta = \arg z$, and $r = |z|$, we have

$$z = r(\cos\theta + i\sin\theta).$$

This is called the **polar form** of z. A handy abbreviation for $\cos\theta + i\sin\theta$ is $\operatorname{Cis}\theta$; thus we write

$$z = r\operatorname{Cis}\theta.$$

For example (all angles are in radians),

$$57 = 57\operatorname{Cis}0, \quad -12i = 12\operatorname{Cis}\frac{3\pi}{2}, \quad 1+i = \sqrt{2}\operatorname{Cis}\frac{\pi}{4}.$$

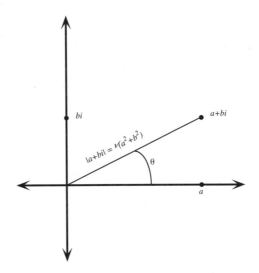

4.2.2 *Conjugation.* If $z = a + bi$, we define the **conjugate** of z to be

$$\bar{z} = a - bi.$$

Geometrically, $\bar{z}$ is just the reflection of z about the real axis.

4.2.3 *Addition and Subtraction.* Complex numbers add "componentwise" i.e.

$$(a + bi) + (c + di) = (a + c) + (b + d)i.$$

Geometrically, complex number addition obeys the "parallelogram rule" of vector addition: If z and w are complex numbers viewed as vectors, their sum $z + w$ is the diagonal of the parallelogram with sides z and w with one endpoint at the origin. Likewise, the difference $z - w$ is a vector which has the same magnitude and direction as the vector with starting point at w and ending point at z. Consequently, if $z_1, z_2, \ldots z_n$ are complex numbers which sum to zero, when drawn as vectors and placed *end-to-end* they form a closed polygon.

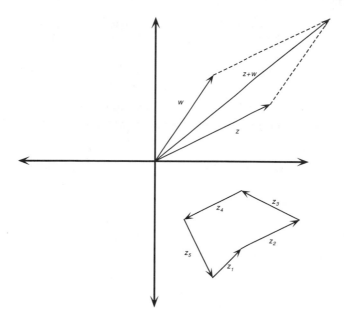

4.2.4 *Multiplication.* All algebraic manipulations of complex numbers follow the usual rules, with the additional proviso that $i^2 = -1$. Hence, for example,

$$(2 + 3i)(4 + 5i) = 8 + 12i + 10i + 15i^2 = -7 + 22i.$$

A straightforward use of trigonometric identities verifies that if $z = r\operatorname{Cis}\alpha$ and $w = s\operatorname{Cis}\beta$, then

$$zw = (r\operatorname{Cis}\alpha)(s\operatorname{Cis}\beta) = rs\operatorname{Cis}(\alpha + \beta);$$

i.e.,

> *The length of zw is the product of the lengths of z and w, and the angle of zw is the sum of the angles of z and w.*

This trigonometric derivation is a good exercise, but is not really illuminating. It doesn't really tell us *why* the multiplication of complex numbers has this satisfying geometric property. Here is a different way to see it. We will do a specific case: the geometric action of multiplying any complex number z by $3 + 4i$. In polar form, $3 + 4i = 5\operatorname{Cis}\theta$, where $\theta = \arctan(4/3)$, which is approximately 0.93 radians.

1. Since $i(a + bi) = -b + ai$, it follows (draw a picture!) that

   > *"Multiplication by i" means "rotate by $\pi/2$ counterclockwise."*

2. Likewise, if a is real,

   > *"Multiplication by a" means "expand by a factor of a."*

For example, multiplication of a complex vector z by 3 produces a new vector which has the same direction, but is three times as long. Multiplication of z by $1/5$ produces a vector with the same direction, but only $1/5$ as long.

3. Multiplying z by $3 + 4i = 5\operatorname{Cis}\theta$ means that z is turned into $(3+4i)z = 3z+4iz$. This is the sum of two vectors, $3z$ and $4iz$. The first vector is just z expanded by a factor of 3. The second vector is z rotated by 90 degrees counterclockwise, then expanded by a factor of 4. So the net result (draw a picture!!!) will be a vector with length $5|z|$ and angle $\theta + \arg z$.

Clearly this argument generalizes to multiplication by any complex number.

Multiplication by the complex number $r\operatorname{Cis}\theta$ is a counterclockwise rotation by θ followed by stretching by the factor r.

So we have a third way to think about complex numbers. Every complex number is *simultaneously* a point, a vector, and *a geometric transformation*, namely the rotation and stretching above!

4.2.5 *Division.* It is easy now to determine the geometric meaning of z/w, where $z = r\operatorname{Cis}\alpha$ and $w = s\operatorname{Cis}\beta$. Let $v = z/w = t\operatorname{Cis}\gamma$. Then $vw = z$. Using the rules for multiplication, we have

$$ts = r, \quad \gamma + \beta = \alpha,$$

and consequently

$$t = r/s, \quad \gamma = \alpha - \beta.$$

Thus

The geometric meaning of division by $r\operatorname{Cis}\theta$ is clockwise rotation by θ (counterclockwise rotation by $-\theta$) followed by stretching by the factor $1/r$.

4.2.6 *De Moivre's Theorem.* An easy consequence of the rules for multiplication and division is this lovely trigonometric identity, true for any integer n, positive or negative, and any real θ:

$$(\cos\theta + i\sin\theta)^n = \cos n\theta + i\sin n\theta.$$

4.2.7 *Exponential Form.* The algebraic behavior of argument under multiplication and division should remind you of exponentiation. Indeed, **Euler's formula** states that

$$\operatorname{Cis}\theta = e^{i\theta},$$

where $e = 2.71828\ldots$ is the familiar natural logarithm base that you have encountered in calculus. This is a useful notation, somewhat less cumbersome than $\operatorname{Cis}\theta$, and quite profound, besides. Most calculus and complex analysis textbooks prove Euler's formula using the power series for e^x, $\sin x$, and $\cos x$, but this doesn't really give much insight as to *why* it is true. This is a deep and interesting issue, which is beyond the scope of this book. Consult [17] for a thorough treatment, and try problem 4.2.28 below.

4.2.8 *Easy Practice Exercises.* Use the above to verify the following.

(a) $|zw| = |z||w|$ and $|z/w| = |z|/|w|$.

(b) $\operatorname{Re} z = \frac{1}{2}(z + \bar{z})$ and $\operatorname{Im} z = \frac{1}{2i}(z - \bar{z})$.

(c) $z\bar{z} = |z|^2$.

(d) The midpoint of the line segment joining the complex numbers z, w is $(z+w)/2$. Make sure you can visualize this!

(e) $\overline{z+w} = \overline{z} + \overline{w}$ and $\overline{zw} = \overline{z}\,\overline{w}$ and $\overline{z/w} = \overline{z}/\overline{w}$.

(f) $(1+i)^{10} = 32i$ and $(1 - i\sqrt{3})^5 = 32(1 + i\sqrt{3})$.

(g) Show by drawing a picture, that $z = \frac{\sqrt{2}}{2}(1+i)$ satisfies $z^2 = i$.

(h) Observe that if $a \in \mathbf{C}$ and $r \in \mathbf{R}$, the set of points $a + re^{it}$, where $0 \le t \le 2\pi$, describes a circle with center at a and radius r.

4.2.9 *Less Easy Practice Problems.* The following are somewhat more challenging. Draw careful pictures, and do not be tempted to resort to algebra (except for checking your work).

(a) It is easy to "simplify"

$$\frac{1}{a+bi} = \frac{a - bi}{a^2 + b^2}$$

by multiplying numerator and denominator by $a - bi$. But one can also verify this without any calculation. How?

(b) $|z + w| \le |z| + |w|$, with equality if and only if z and w have the same direction or point in opposite directions, i.e. if the angle between them is 0 or π.

(c) Let z lie on the **unit circle**; i.e., $|z| = 1$. Show that

$$|1 - z| = 2\sin\left(\frac{\arg z}{2}\right)$$

without computation.

(d) Let $P(x)$ be a polynomial with real coefficients. Show that if z is a zero of $P(x)$, then $\overline{z}$ is also a zero; i.e., the zeros of a polynomial with real coefficients come in complex-conjugate pairs.

(e) Without much calculation, determine the locus for z for each of

$$\operatorname{Re}\left(\frac{z-1-i}{z+1+i}\right) = 0, \qquad \text{and} \qquad \operatorname{Im}\left(\frac{z-1-i}{z+1+i}\right) = 0.$$

(f) Without solving the equation, show that all 9 roots of $(z - 1)^{10} = z^{10}$ lie on the line $\operatorname{Re}(z) = \frac{1}{2}$.

(g) Show that if $|z| = 1$ then

$$\operatorname{Im}\left(\frac{z}{(z+1)^2}\right) = 0.$$

(h) Let k be a real constant, and let a, b be fixed complex numbers. Describe the locus of points z which satisfy

$$\arg\left(\frac{z-a}{z-b}\right) = k.$$

4.2.10 *Grids to circles and vice versa.* The problems below will familiarize you with the lovely interplay between geometry, algebra, and analytic geometry when you ponder complex transformations. The transformation that we analyze below is an example of a **Möbius transformation** . See [17] for more details.

(a) Prove the following simple geometry proposition (use similar triangles).

> *Let AB be a diameter of a circle with diameter k. Consider a right triangle ABC with right angle at B. Let D be the point of intersection of line AC with the circle. Then*

$$AD \cdot AC = k^2.$$

(b) Consider the transformation $f(z) = \dfrac{z}{z-1}$. This is a function with complex domain and range. Here is a computer output (using *Mathematica*) of what $f(z)$ does to the domain $(0 \le \mathrm{Re}\,(z) \le 2, -2 \le \mathrm{Im}\,(z) \le 2)$. The plot shows how $f(z)$ transforms the rectangular grid lines. Notice that f appears to transform the rectangular cartesian grid of the Gaussian plane into circles, all tangent at 1 [although the point 1 is not in the range, nor is a neighborhood about 1, since z has to be very large in order for $f(z)$ to be close to 1].

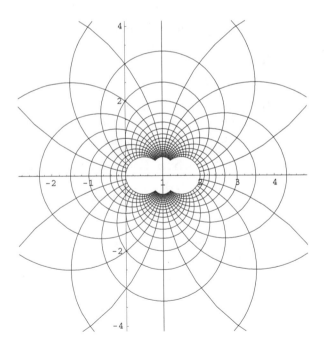

Verify this phenomenon explicitly, without calculation, for the imaginary axis. Prove that the function $f(z) = \dfrac{z}{z-1}$ transforms the imaginary axis into a circle with center at $(\frac{1}{2}, 0)$ and radius $\frac{1}{2}$. Do this two ways.

1. *Algebraically.* Let a point on the imaginary axis be it, where t is any real number. Find the real and imaginary parts of $f(it)$; i.e, put $f(it)$ into the form $x + yi$. Then show that $(x - \frac{1}{2})^2 + y^2 = \frac{1}{4}$.

2. *Geometrically.* First, show that $f(z)$ is the composition of four mappings, in the following order:

$$z \mapsto \frac{1}{z}, \quad z \mapsto -z, \quad z \mapsto z + 1, \quad z \mapsto \frac{1}{z}.$$

In other words, if you start with z and then reciprocate, negate, translate by 1 and then reciprocate again, you will get $f(z)$.

Next, use this "decomposition" of $f(z)$ plus the geometry lemma that you proved above to show that every point z on the imaginary axis is mapped to a circle with diameter 1 with center at $\frac{1}{2}$. Draw a good diagram!

(c) The "converse" of (b) is true; not only does $f(z)$ transform the cartesian grid to circles, it transforms certain circles to cartesian grid lines! Verify this explicitly for the unit circle (circle with radius 1 centered at the origin). Show that $f(z)$ turns the unit circle into the vertical line consisting of all points with real part equal to $\frac{1}{2}$. In fact, show that

$$f(e^{i\theta}) = \frac{1}{2} - \frac{1}{2}i \cot \frac{\theta}{2}.$$

As in (b), do this in two different ways—algebraically and geometrically. Make sure that you clearly show exactly how the unit circle gets mapped into this line. For example, it is obvious that -1 is mapped to $\frac{1}{2}$. What happens as you move counterclockwise, starting from -1 along the unit circle?

Roots of Unity

The zeros of the equation $x^n = 1$ are called the nth **roots of unity**. These numbers have many beautiful properties that interconnect algebra, geometry and number theory. One reason for the ubiquity of roots of unity in mathematics is symmetry: roots of unity, in some sense, epitomize symmetry, as you will see below. (We will be assuming some knowledge about polynomials and summation. If you are unsure about this material, consult Chapter 5.)

4.2.11 For each positive integer n, there are n different nth roots of unity, namely

$$1, \zeta, \zeta^2, \zeta^3, \ldots, \zeta^{n-1},$$

where

$$\zeta = \text{Cis} \frac{2\pi}{n}.$$

Geometrically, the nth roots of unity are the vertices of a regular n-gon which is inscribed in the unit circle (the set $\{z \in \mathbf{C} : |z| = 1\}$) with one vertex at 1.

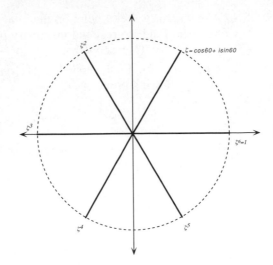

4.2.12 Let $\zeta = \mathrm{Cis}\,\frac{2\pi}{n}$ as above. Then for each positive integer n,

1. $x^n - 1 = (x - 1)(x - \zeta)(x - \zeta^2)\cdots(x - \zeta^{n-1})$,
2. $x^{n-1} + x^{n-2} + \cdots + x + 1 = (x - \zeta)(x - \zeta^2)\cdots(x - \zeta^{n-1})$,
3. $1 + \zeta + \zeta^2 + \cdots + \zeta^{n-1} = 0$. Can you see why this is true without using the formula for the sum of geometric series?

Some Applications

We will conclude this section with a few examples of interesting use of complex numbers in several branches of mathematics, including trigonometry, geometry, and number theory.

Example 4.2.13 Find a formula for $\tan(2a)$.

Solution: This can of course be done in many ways, but the complex numbers method is quite slick, and easily works with many other trig identities. The key idea is that if $z = x + iy$, then $\tan(\arg z) = y/x$. Let $t := \tan a$ and

$$z = 1 + it.$$

Now square z, and we get

$$z^2 = (1 + it)^2 = 1 - t^2 + 2it.$$

But

$$\tan(\arg z^2) = \frac{2t}{1 - t^2},$$

and of course $\arg z^2 = 2a$, so we conclude that

$$\tan 2a = \frac{2\tan a}{1 - \tan^2 a}.\qquad\blacksquare$$

Example 4.2.14 (Putnam 1996) Let C_1 and C_2 be circles whose centers are 10 units apart, and whose radii are 1 and 3. Find, with proof, the locus of all points M for which there exist points X on C_1 and Y on C_2 such that M is the midpoint of the line segment XY.

Solution: Our solution illustrates a useful application of viewing complex numbers as vectors for parametrizing curves in the plane. Consider the general case, illustrated in the figure below, of two circles situated in the complex plane, with centers at a, b and radii u, v, respectively. Notice that a and b are complex numbers, while u and v are real. We are assuming, as in the original problem, that v is quite a bit bigger than u.

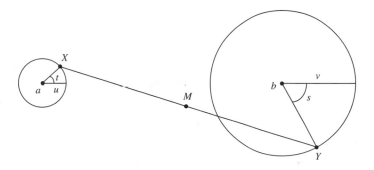

The locus we seek is the set of midpoints M of the line segments XY, where X can be any point on the left circle and Y can be any point on the right circle.

Thus, $X = a + ue^{it}$ and $Y = b + ve^{is}$, where t and s can be any values between 0 and 2π. We have

$$M = \frac{X+Y}{2} = \frac{a+b}{2} + \frac{ue^{it} + ve^{is}}{2}.$$

Let us interpret this geometrically, by first trying to understand what

$$ue^{it} + ve^{is}$$

looks like as $0 \leq s, t \leq 2\pi$. If we fix s, then $P = ve^{is}$ is a point on the circle with radius v centered at the origin. Now, when we add ue^{it} to P, and let t vary between 0 and 2π, we will get a circle with radius u, centered at P (shown as a dotted line below).

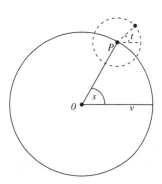

Now let s vary as well. The small dotted circle will travel all along the circumference of the large circle, creating an **annulus** or filled-in "ring." In other words, the locus of points

$$ue^{it} + ve^{is}, \quad 0 \le s, t \le 2\pi$$

is the annulus

$$v - u \le |z| \le v + u$$

centered at the origin, i.e. the set of points whose distance to the origin is between $v - u$ and $v + u$.

Now it is a simple matter to wrap up the problem. Since

$$M = \frac{a + b}{2} + \frac{ue^{it} + ve^{is}}{2},$$

our locus is an annulus centered at the midpoint of the line segment joining the two centers, i.e., the point $(a + b)/2$. Call this point c. Then the locus is the set of points whose distance from c is between $(v - u)/2$ and $(v + u)/2$. ∎

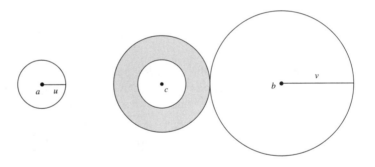

Example 4.2.15 Let m and n be integers such that each can be expressed as the sum of two perfect squares. Show that mn has this property as well. For example, $17 = 4^4 + 1$ and $13 = 2^2 + 3^2$ and sure enough,

$$17 \cdot 13 = 221 = 14^2 + 5^2. \tag{1}$$

Solution: Let $m = a^2 + b^2, n = c^2 + d^2$, where a, b, c, d are integers. Now consider the product

$$z := (a + bi)(c + di).$$

Note that

$$|z| = |(a + bi)||c + di| = \sqrt{(a^2 + b^2)(c^2 + d^2)}.$$

Therefore $mn = |z|^2$. But $|z|^2$ will be a sum of squares of integers, since $\operatorname{Re} z$ and $\operatorname{Im} z$ are integers. Not only does this prove what we were looking for; it gives us an algorithm for computing the values in the right-hand side of equations such as (1). ∎

Our final example is a surprising problem which shows how roots of unity can be used to create an invariant.

Example 4.2.16 (Jim Propp) Given a circle of n lights, exactly one of which is initially on, it is permitted to change the state of a bulb provided one also changes the state of every dth bulb after it (where d is a a divisor of n strictly less than n), provided that all n/d bulbs were originally in the same state as one another. For what values of n is it possible to turn all the bulbs on by making a sequence of moves of this kind?

Solution: The insight here is to realize that this is not a problem about lights, but about roots of unity

$$1, \zeta, \zeta^2, \ldots, \zeta^{n-1},$$

where $\zeta = \cos\frac{2\pi}{n} + i\sin\frac{2\pi}{n}$. Place each light on the unit circle located at a root of unity, and without loss of generality, let the light at 1 be on initially. Now, if $d < n$ is a divisor of n and the lights at

$$\zeta^a, \zeta^{a+d}, \zeta^{a+2d}, \ldots, \zeta^{a+(\frac{n}{d}-1)d}$$

have the same state, then we can change the state of these n/d lights. The sum of these is

$$\zeta^a + \zeta^{a+d} + \zeta^{a+2d} + \cdots + \zeta^{a+(\frac{n}{d}-1)d} = \zeta^a(1 + \zeta^d + \zeta^{2d} + \cdots + \zeta^{(\frac{n}{d}-1)d}).$$

The terms in parentheses in the right-hand side form a geometric series which sums easily and the right-hand side simplifies to

$$\zeta^a\left(\frac{1 - \zeta^{\frac{n}{d}d}}{1 - \zeta^d}\right) = \zeta^a\left(\frac{1 - \zeta^n}{1 - \zeta^d}\right) = \zeta^a\left(\frac{1 - 1}{1 - \zeta^d}\right) = 0.$$

This surprising fact tells us that if we add up all the roots of unity which are "on," the sum will never change, since whenever we change the state of a bunch of lights, they add up to zero! The original sum was equal to 1, and the goal is to get all the lights turned on. That sum will be

$$1 + \zeta + \zeta^2 + \cdots + \zeta^{n-1} = \frac{1 - \zeta^n}{1 - \zeta} = \frac{1 - 1}{1 - \zeta} = 0 \neq 1.$$

Hence we can never turn on all the lights! ∎

Problems and Exercises

4.2.17 Use complex numbers to derive identities for $\cos na$ and $\sin na$, for $n = 3, 4, 5$.

4.2.18 Test your understanding of Example 4.2.15. Given that $17 = 4^2 + 1$ and $101 = 10^2 + 1$, *mentally calculate* integers u, v such that $17 \cdot 101 = u^2 + v^2$.

4.2.19 Prove (without calculation, if you can!) that

$$e^{it} + e^{is} = 2\cos\left(\frac{t-s}{2}\right) e^{i\left(\frac{s+t}{2}\right)}.$$

4.2.20 Show that if $x + \dfrac{1}{x} = 2\cos a$, then for any integer n,

$$x^n + \frac{1}{x^n} = 2\cos na.$$

4.2.21 Find (fairly) simple formulas for $\sin a + \sin 2a + \sin 3a + \cdots + \sin na$ and $\cos a + \cos 2a + \cos 3a + \cdots + \cos na$.

4.2.22 Factor $z^5 + z + 1$.

4.2.23 Solve $z^6 + z^4 + z^3 + z^2 + 1 = 0$.

4.2.24 Let n be a positive integer. Find a closed-form expression for

$$\sin\frac{\pi}{n} \sin\frac{2\pi}{n} \sin\frac{3\pi}{n} \cdots \sin\frac{(n-1)\pi}{n}.$$

4.2.25 Consider a regular n-gon which is inscribed in a circle with radius 1. What is the product of the lengths of all $n(n-1)/2$ diagonals of the polygon (this includes the sides of the n-gon)?

4.2.26 (USAMO 1976) If $P(x)$, $Q(x)$, $R(x)$, $S(x)$ are all polynomials such that

$$P(x^5) + xQ(x^5) + x^2 R(x^5) = (x^4 + x^3 + x^2 + x + 1)S(x),$$

prove that $x - 1$ is a factor of $P(x)$.

4.2.27 The set of points (x, y) which satisfies $x^3 - 3xy^2 \geq 3x^2 y - y^3$ and $x + y = -1$ is a line segment. Find its length.

4.2.28 (T. Needham) Try to derive Euler's formula $e^{it} = \cos t + i\sin t$ in the following way:

(a) Assume that the function $f(t) = e^{it}$ can be differentiated with respect to t "in the way that you would expect"; in other words, that $f'(t) = ie^{it}$. (Note that this is not automatic, since the range of the function is complex; you need to define and check lots of things, but we will avoid that for now—this is an intuitive argument!)

(b) If you view the variable t as time, and the function $f(t)$ as tracing a curve in the complex plane, the equation $f'(t) = ie^{it}$ has a rate-of-change interpretation. Recall that multiplication by i means "rotate by 90 degrees counterclockwise." Show that this implies that $f(t)$ is a circular path.

(c) Think about the speed that this circular curve $t \mapsto f(t)$ is being traced out, and conclude that $e^{it} = \cos t + i\sin t$.

4.2.29 Let $R_a(\theta)$ denote the transformation of the plane that rotates everything about the center point a by θ radians counterclockwise. Prove the interesting fact that the composition of $R_a(\theta)$ and $R_b(\phi)$ is another rotation $R_c(\alpha)$. Find c, α in terms of a, b, θ, ϕ. Does this agree with your intuition?

4.2.30 Show that there do not exist any equilateral triangles in the plane whose vertices are lattice points (integer coordinates).

4.2.31 Show that the triangle with vertices a, b, c in the complex plane is equilateral if and only if

$$a^2 + b^2 + c^2 = ab + bc + ca.$$

4.2.32 Find necessary and sufficient conditions for the two roots of $z^2 + az + b = 0$, plus 0, to form the vertices of an equilateral triangle.

4.2.33 (T. Needham) Draw any quadrilateral, and on each side draw a square lying outside the given quadrilateral. Draw line segments joining the centers of opposite squares. Show that these two line segments are perpendicular and equal in length.

4.2.34 (T. Needham) Let ABC be a triangle, with points P, Q, R situated outside ABC so that triangles PAC, RCB, QBA are similar to each other. Then the centroids (intersection of medians) of ABC and PQR are the same point.

4.2.35 (T. Needham) Draw any triangle, and on each side draw an equilateral triangle lying outside the given triangle. Show that the centroids of these three equilateral triangles are the vertices of an equilateral triangle. (The **centroid** of a triangle is the intersection of its medians; it is also the center of gravity.) If the vertices of a triangle are the complex numbers a, b, c, the centroid is located at $(a + b + c)/3$.

4.2.36 (IMO proposal) Let n be a positive integer having at least two distinct prime factors. Show that there is a permutation $(a_1, a_2, \ldots, a_n)$ of $(1, 2, \ldots, n)$ such that

$$\sum_{k=1}^{n} k \cos \frac{2\pi a_k}{n} = 0.$$

4.2.37 For each positive integer n, define the polynomial

$$P_n(z) := 1^3 z + 2^3 z^2 + \cdots + n^3 z^n.$$

Do the zeros of P_n lie inside, outside, or on the unit circle $|z| = 1$?

4.3 Generating Functions

The crossover tactic of generating functions owes its power to two simple facts.

- When you multiply x^m by x^n, you get x^{m+n}.
- "Local" knowledge about the coefficients of a polynomial or power series $f(x)$ often provides "global" knowledge about the behavior of $f(x)$, and vice versa.

The first fact is trivial, but it is the technical "motor" that makes things happen, for it relates the addition of numbers and the multiplication of polynomials. The second fact is deeper, and provides the motivation for doing what we are about to do.

Introductory Examples

Before we do anything, though, we need to define our subject. Given a (possibly infinite) sequence $a_0, a_1, a_2, \ldots$, its **generating function** is

$$a_0 + a_1 x + a_2 x^2 + \cdots.$$

Here are a few simple examples. We are assuming that you have a basic understanding of sequences, polynomials, and simple summation formulas (Chapter 5), combinatorics and the binomial theorem (Chapter 6), and infinite series (Chapters 5 and 8). If you need to brush up in these areas, you may want to just skim this section now and then reread it again later. We do not recommend that you avoid this section altogether, because the idea of generating functions is so powerful. The sooner you are exposed to it, the better.

Example 4.3.1 Let $1 = a_0 = a_1 = a_2 = \cdots$. Then the corresponding generating function is just

$$1 + x + x^2 + x^3 + \cdots.$$

This is an infinite geometric series which converges to $\dfrac{1}{1-x}$, provided that $|x| < 1$ (see page 176). In general, we don't worry too much about convergence issues with generating functions. As long as the series converges for some values, we can usually get by, as you will see below.

   The infinite geometric series used above is ubiquitous in the world of generating functions. Make a note of it; we shall call it the **geometric series tool**. Remember that it works both ways: you probably have practice summing infinite geometric series, but here's an example of the reverse direction. Study it carefully.

$$\frac{x}{2+x} = \frac{x}{2\left(1 - (-\frac{1}{2}x)\right)} = \left(\frac{1}{2}\right)\left(x - \frac{1}{2}x + \frac{1}{2^2}x^2 - \frac{1}{2^3}x^3 + \cdots\right).$$

Example 4.3.2 For a fixed positive integer n, define the sequence $a_k = \dbinom{n}{k}$, $k = 0, 1, 2, \ldots, n$. The corresponding generating function is

$$\binom{n}{0} + \binom{n}{1}x + \binom{n}{2}x^2 + \cdots + \binom{n}{n}x^n = (x+1)^n \qquad (2)$$

by the binomial theorem. If we plug $x = 1$ into (2), we get the nice identity

$$\binom{n}{0} + \binom{n}{1} + \binom{n}{2} + \cdots + \binom{n}{n} = 2^n.$$

This identity can of course be proven in other ways (and you should know some of them; if not, consult Section 6.2), but notice that our method is both easy and easy to generalize. If we let $x = -1$, we get a completely new identity,

$$\binom{n}{0} - \binom{n}{1} + \binom{n}{2} - \cdots + (-1)^n\binom{n}{n} = 0.$$

These two examples of plugging in a value to get an identity are typical of the "local $\leftrightarrow$ global" point of view. Globally, the generating function is just the simple function $(1+x)^n$. We can plug in any values that we want. But each time we plug in a value, we get a new statement involving the coefficients $\binom{n}{k}$ (the local information). The key is to deftly move the focus back and forth between the function and its coefficients, to get useful information.

Example 4.3.3 Plugging in values is just one of the "global" things we can do. Let's try differentiation! If we differentiate both sides of (2), we get

$$\binom{n}{1} + 2\binom{n}{2}x + 3\binom{n}{3}x^2 + \cdots + n\binom{n}{n}x^{n-1} = n(x+1)^{n-1}.$$

Now, if we plug in $x = 1$, we get the interesting identity

$$\binom{n}{1} + 2\binom{n}{2} + 3\binom{n}{3} + \cdots + n\binom{n}{n} = n2^{n-1}.$$

Recurrence Relations

So far, we have not used the simple fact mentioned on page 143, that $x^m x^n = x^{m+m}$. Let us do so now, by using generating functions to analyze recurrence relations (see Section 6.4 for examples).

Example 4.3.4 Define the sequence (a_n) by $a_0 = 1$ and $a_n = 3a_{n-1} + 2$ for $n = 1, 2, \ldots$. Find a simple formula for a_n.

Solution: There are several ways to tackle this problem; indeed the simplest approach—one that any problem solver should try first—is to work out the first few terms and guess. The first few terms are (verify!)

$$1, 5, 17, 53, 161, 485, \ldots,$$

which may lead an inspired guesser (who knows her powers of 3) to conjecture that $a_n = 2 \cdot 3^n - 1$, and this is easy to prove by induction.

Let's look at an alternate solution, using generating functions. It is much less efficient for this particular problem than the "guess and check" method above, but it can be applied reliably in many situations where inspired guessing won't help. Let

$$f(x) := a_0 + a_1 x + a_2 x^2 + \cdots = 1 + 5x + 17x^2 + \cdots$$

be the generating function corresponding to the sequence (a_n). Now look at

$$xf(x) = a_0 x + a_1 x^2 + a_2 x^3 + \cdots.$$

This is the generating function of the original sequence, but *shifted*. In other words, the coefficient of x^n in $f(x)$ is a_n, while the coefficient of x^n in $xf(x)$ is a_{n-1}. Now we make use of the relationship between a_n and a_{n-1}. Since $a_n - 3a_{n-1} = 2$ for all $n \geq 1$, we have

$$f(x) - 3xf(x) = a_0 + 2(x + x^2 + x^3 + \cdots).$$

That looks ugly, but the expression in parentheses on the right-hand side is just an infinite geometric series. Thus we have (remember, $a_0 = 1$)

$$f(x) - 3xf(x) = 1 + \frac{2x}{1-x}.$$

The left-hand side is $f(x)(1 - 3x)$, so we can easily solve for $f(x)$, getting

$$f(x) = \frac{1}{1-3x} + \frac{2x}{(1-x)(1-3x)} = \frac{x+1}{(1-x)(1-3x)}.$$

Our goal is to somehow recover the coefficients when $f(x)$ is expanded in a power series. If the denominators were just $(1 - x)$ or $(1 - 3x)$, we could use the geometric series tool. Partial fractions[7] comes to the rescue, yielding

$$f(x) = \frac{x+1}{(1-x)(1-3x)} = \frac{2}{1-3x} - \frac{1}{1-x}.$$

Using the geometric series tool on the first term, we get

$$\frac{2}{1-3x} = 2(1 + 3x + 3^2 x^2 + 3^3 x^3 + \cdots).$$

And since

$$\frac{1}{1-x} = 1 + x + x^2 + x^3 + \cdots,$$

we get

$$f(x) = 2(1 + 3x + 3^2 x^2 + 3^3 x^3 + \cdots) - (1 + x + x^2 + x^3 + \cdots)$$
$$= 1 + (2 \cdot 3 - 1)x + (2 \cdot 3^2 - 1)x^2 + (2 \cdot 3^3 - 1)x^3 + \cdots,$$

from which it follows immediately that $a_n = 2 \cdot 3^n - 1$. ∎

This method was technically messy, since it involved using the geometric-series tool repeatedly as well as partial fractions. But don't get overwhelmed by the technical details—it worked because multiplying a generating function by x produced the generating function for the "shifted" sequence. Likewise, dividing by x will shift the sequence in the other direction. These techniques can certainly be used for many kinds of recurrences.

Partitions

Consider the following polynomial product.

$$P(x) := (x^2 + 3x + 1)(x^2 + 2x + 1) = x^4 + 5x^3 + 8x^2 + 5x + 1.$$

How did we compute the coefficient of x^k in $P(x)$? For example, the x^2 term is the sum

$$x^2 \cdot 1 + 3x \cdot 2x + 1 \cdot x^2 = 8x^2,$$

---

[7]If you don't remember this technique, consult any calculus text. The basic idea is to write $\frac{x+1}{(1-x)(1-3x)} = \frac{A}{1-3x} + \frac{B}{1-x}$ and solve for the unknown constants A and B.

so the coefficient is 8. In general, to find the x^k term of $P(x)$, we look at all pairings of terms, where one term comes from the first factor and the other comes from the second factor and the exponents add up to k. We multiply the pairs, and then add them up, and that is our answer.

Now let us rewrite $P(x)$ as

$$(x^2 + x + x + x + 1)(x^2 + x + x + 1). \tag{3}$$

The product will be unchanged, of course, so for example, the x^2 term is still $8x^2$. This corresponds to the sum

$$x^2 \cdot 1 + x \cdot x + x \cdot x + x \cdot x + x \cdot x + x \cdot x + x \cdot x + 1 \cdot x^2.$$

All that *really* matters here are the exponents; we can list the pairings of exponents by the sums

$$2 + 0, \quad 1 + 1, \quad 1 + 1, \quad 1 + 1, \quad 1 + 1, \quad 1 + 1, \quad 1 + 1, \quad 0 + 2,$$

and all we need to do is count the number of sums (there are 8). Here's another way to think of it. Imagine that the first factor of (3) is colored red, and second factor is colored blue. We can reformulate our calculation of the x^2 coefficient in the following way:

> *Assume that you have one red 2, three red 1's and one red zero; and one blue 2, two blue 1's and one blue zero. Then there are 8 different ways that you can add a red number and a blue number to get a sum of 2.*

At this point, you are probably saying, "So what?," so it is time for an example.

Example 4.3.5 How many distinct ordered pairs (a, b) of nonnegative integers satisfy

$$2a + 5b = 100?$$

Solution: Look at the general case: Let u_n be the number of nonnegative ordered pairs that solve $2a + 5b = n$. Thus $u_0 = 1, u_1 = 0, u_2 = 1$, etc. We need to find u_{100}. Now define

$$A(x) = 1 + x^2 + x^4 + x^6 + x^8 + \cdots,$$
$$B(x) = 1 + x^5 + x^{10} + x^{15} + x^{20} + \cdots,$$

and consider the product

$$A(x)B(x) = (1 + x^2 + x^4 + \cdots)(1 + x^5 + x^{10} + \cdots)$$
$$= 1 + x^2 + x^4 + x^5 + x^6 + x^7 + x^8 + x^9 + 2x^{10} + \cdots.$$

We claim that $A(x)B(x)$ is the generating function for the sequence $u_0, u_1, u_2, \ldots$. This isn't hard to see if you pondered the "red and blue" discussion above: After all, each term in $A(x)$ has the form x^{2a} where a is a nonnegative integer, and likewise, each term in $B(x)$ has the form x^{5b}. Hence all the terms in the product $A(x)B(x)$ will be terms with exponents of the form $2a + 5b$. Each different pair (a, b) that satisfies $2a + 5b = n$ will produce the monomial x^n in the product $A(x)B(x)$. Hence the coefficient of x^n will be the number of different solutions to $2a + 5b = n$.

Now we use the geometric-series tool to simplify

$$A(x) = \frac{1}{1-x^2} \quad \text{and} \quad B(x) = \frac{1}{1-x^5}.$$

Thus

$$\frac{1}{(1-x^2)(1-x^5)} = u_0 + u_1 x + u_2 x^2 + u_3 x^3 + \cdots. \tag{4}$$

In an abstract sense, we are "done," for we have a nice form for the generating function. But we haven't the slightest idea what u_{100} equals! This isn't too hard to find. By inspection, we can compute

$$u_0 = u_2 = u_4 = u_5 = u_6 = u_7 = 1 \quad \text{and} \quad u_1 = u_3 = 0. \tag{5}$$

Then we transform (4) into

$$1 = (1-x^2)(1-x^5)\left(u_0 + u_1 x + u_2 x^2 + u_3 x^3 + \cdots\right)$$
$$= (1-x^2-x^5+x^7)\left(u_0 + u_1 x + u_2 x^2 + u_3 x^3 + \cdots\right).$$

The x^k coefficient of the product of the terms on the right-hand side must be zero (if $k > 0$). Multiplying out, this coefficient is

$$u_k - u_{k-2} - u_{k-5} + u_{k-7}.$$

So for all $k > 7$, we have the recurrence relation

$$u_k = u_{k-2} + u_{k-5} - u_{k-7}. \tag{6}$$

It is a fairly simple, albeit tedious, exercise to compute u_{100} by using (5) and (6). For example,

$$u_8 = u_6 + u_3 - u_1 = 1, u_9 = u_7 + u_4 - u_2 = 1, u_{10} = u_8 + u_5 - u_3 = 2,$$

etc. If you play around with this, you will find some shortcuts (try working backward, and/or make a table to help eliminate some steps), and eventually you will get $u_{100} = 11$. ∎

The next example does not concern itself with computing the coefficient of a generating function, but rather solves a problem by equating two generating functions—one ugly, one pretty.

Example 4.3.6 Let n be any positive integer. Show that the set of weights

$$1, 3, 3^2, 3^3, 3^4, \ldots$$

grams can be used to weigh an n-gram weight (using both pans of a scale), and that this can be done in exactly *one* way.

Solution: For example, if $n = 10$, the n-gram weight is balanced by a 1-gram weight and a 9-gram weight. The corresponding arithmetic fact is

$$10 = 1 + 3^2.$$

If $n = 73$, the n-gram weight is joined by a 9-gram weight on one pan, which is balanced by an 81-gram weight and a 1-gram weight. This corresponds to the statement

$$73 + 3^2 = 3^4 + 1,$$

which is equivalent to

$$73 = 3^4 - 3^2 + 1.$$

We will be done if we can show that for any positive integer n it is possible to write n as a sum and/or difference of distinct powers of 3, and that this can be done in exactly one way. It is sort of like base-3, but not allowing the digit 2, and instead admitting the "digit" -1. Indeed, playing around with this idea can lead to an algorithm for writing n as a sum/and or difference of powers of 3, but it is hard to see why the representation will be unique.

Here's a generatingfunctionological[8] approach: Consider the function

$$f_1(x) := (1 + x + x^{-1})(1 + x^3 + x^{-3}).$$

The two factors of f_1 each contain the exponents $0, 1, -1$, and $0, 3, -3$, respectively. When f_1 is multiplied out, we get

$$f_1(x) = 1 + x + x^{-1} + x^3 + x^4 + x^2 + x^{-3} + x^{-2} + x^{-4}. \tag{7}$$

Every integer exponent from -4 to 4, inclusive, is contained in this product, and each term has a coefficient equal to 1. Each of these 9 terms is the result of choosing exactly one term from the first factor of f_1 and one term from the second factor, and then multiplying them (*adding* their exponents). In other words, the exponents in (7) are precisely all possible ways of combining the two numbers 1 and 3 with additions and/or subtractions and/or omissions. (By "omissions," we mean that we don't have to include either 1 or 3 in our combination. For example, the combination "$+3$" omits 1.)

Let's move on. Consider

$$f_2(x) := (1 + x + x^{-1})(1 + x^3 + x^{-3})(1 + x^{3^2} + x^{-3^2}).$$

When multiplied out, each of the resulting $3 \cdot 3 \cdot 3 = 27$ terms will be of the form x^a, where a is a sum and/or difference of powers of 3. For example, if we multiply the third term of the first factor by the first term of the second factor by the second term of the third factor, the corresponding term in the product is

$$(x^{-1})(1)(x^9) = x^{9-1}.$$

---

[8]The term "generatingfunctionology" was coined by Herbert Wilf, in his book of the same name [33]. We urge the reader to at least browse through this beautifully written textbook, which among its many other charms, has the most poetic opening sentence we've ever seen (in a math book). Incidentally, [30] and [26] also contain excellent and very clear material on generating functions.

What we would like to show, of course, is that the terms in the expansion of f_2 all have coefficient 1 (meaning no duplicates) and range from $-(1+3+9)$ to $+(1+3+9)$ inclusive (meaning that every positive integer between 1 and 13 can be represented as a sum/difference of powers of 3). We can certainly verify this by multiplying out, but we seek a more general argument. Recall the factorization (see page 163)

$$u^3 - v^3 = (u-v)(u^2 + uv + v^2).$$

Applying this to the factors of f_2, we have

$$f_2(x) = \left(\frac{x^2 + x + 1}{x}\right)\left(\frac{x^6 + x^3 + 1}{x^3}\right)\left(\frac{x^{18} + x^9 + 1}{x^9}\right)$$

$$= \frac{1}{x \cdot x^3 \cdot x^9}\left(\frac{x^3 - 1}{x - 1}\right)\left(\frac{x^9 - 1}{x^3 - 1}\right)\left(\frac{x^{27} - 1}{x^9 - 1}\right).$$

After canceling, we are left with

$$f_2(x) = \frac{1}{x^{13}}\left(\frac{x^{27} - 1}{x - 1}\right) = \frac{x^{26} + x^{25} + \cdots + x + 1}{x^{13}},$$

and thus

$$f_2(x) = x^{13} + x^{12} + \cdots + x + 1 + x^{-1} + x^{-2} + \cdots + x^{-13},$$

which is just what we wanted. We have shown that the weights $1, 3, 9$ allow us to weigh any positive (or negative!) integer n less than than or equal to 13, and in exactly one way (since the coefficients all equal 1).

The argument generalizes, of course. For example, if

$$f_3(x) := (1 + x + x^{-1})(1 + x^3 + x^{-3})(1 + x^{3^2} + x^{-3^2})(1 + x^{3^3} + x^{-3^3}),$$

then

$$f_3(x) = f_2(x)(1 + x^{3^3} + x^{-3^3})$$

$$= \frac{1}{x^{13}}\left(\frac{x^{27} - 1}{x - 1}\right)\left(\frac{x^{2\cdot 27} + x^{27} + 1}{x^{27}}\right)$$

$$= \frac{1}{x^{13} \cdot x^{27}}\left(\frac{x^{27} - 1}{x - 1}\right)\left(\frac{x^{81} - 1}{x^{27} - 1}\right)$$

$$= \frac{1}{x^{40}}\left(\frac{x^{81} - 1}{x - 1}\right).$$

Now it is clear that the weights $1, 3, 3^2, \ldots, 3^r$ can be used to weigh any integral value from 1 to $(3^r - 1)/2$, in exactly one way. As $r \to \infty$, we get the limiting case, a beautiful identity of generating functions:

$$\prod_{n=0}^{\infty}(1 + x^{3^n} + x^{-3^n}) = \sum_{n=-\infty}^{\infty} x^n. \qquad \blacksquare$$

Our final example is from the theory of partitions of integers, a subject first investigated by Euler. Given a positive integer n, a **partition** of n is a representation of n as a sum of positive integers. The order of the summands does not matter, so they are conventionally placed in increasing order. For example, $1 + 1 + 3$ and $1 + 1 + 1 + 1 + 4$ are two different partitions of 5.

Example 4.3.7 Show that for each positive integer n, the number of partitions of n into unequal parts is equal to the number of partitions of n into odd parts. For example, if $n = 6$, there are 4 partitions into unequal parts, namely

$$1 + 5, \quad 1 + 2 + 3, \quad 2 + 4, \quad 6.$$

And there are also 4 partitions into odd parts,

$$1 + 1 + 1 + 1 + 1 + 1, \quad 1 + 1 + 1 + 3, \quad 1 + 5, \quad 3 + 3.$$

Solution: Let u_n and v_n denote the number of partitions of n into unequal parts and odd parts, respectively. It takes practice thinking in a generatingfunctionological way, but by now you should have no trouble verifying (even if you had trouble coming up with it) that

$$U(x) := (1 + x)(1 + x^2)(1 + x^3)(1 + x^4)(1 + x^5) \cdots$$

is the generating function for (u_n). For example, the x^6 term in $U(x)$ is equal to

$$x \cdot x^5 + x \cdot x^2 \cdot x^3 + x^2 \cdot x^4 + x^6 = 4x^6.$$

Notice that it is impossible to get a term like $x^5 \cdot x^5$; i.e., the generating function prevents repeated parts in the partition.

If you are comfortable with $U(x)$ (please do ponder it until it becomes "obvious" to you that it is the correct generating function), you should try to construct $V(x)$, the generating function for (v_n). The parts can be duplicated, but they all must be odd. For example, if we wanted to include the possibility of zero, one or two 3's in a partition, the factor $(1 + x^3 + x^6)$ would do the trick, since $x^6 = x^{2 \cdot 3}$ plays the role of "two 3's." Following this reasoning, we define the generating function

$$V(x) := (1 + x + x^2 + x^3 + \cdots)(1 + x^3 + x^6 + x^9 + \cdots)(1 + x^5 + x^{10} + x^{15} + \cdots) \cdots.$$

Now, all that "remains" to be done is show that $U(x) = V(x)$. The geometric series tool provides an immediate simplification, yielding

$$V(x) = \left(\frac{1}{1 - x} \right) \left(\frac{1}{1 - x^3} \right) \left(\frac{1}{1 - x^5} \right) \left(\frac{1}{1 - x^7} \right) \cdots.$$

On the other hand, we can write

$$U(x) = (1 + x)(1 + x^2)(1 + x^3)(1 + x^4)(1 + x^5) \cdots$$
$$= \left(\frac{1 - x^2}{1 - x} \right) \left(\frac{1 - x^4}{1 - x^2} \right) \left(\frac{1 - x^6}{1 - x^3} \right) \left(\frac{1 - x^8}{1 - x^4} \right) \left(\frac{1 - x^{10}}{1 - x^5} \right) \cdots.$$

Notice that in this last expression, we can cancel out all the terms of the form $(1 - x^{2k})$, leaving only

$$\frac{1}{(1-x)(1-x^3)(1-x^5)(1-x^7)\cdots} = V(x).$$

∎

Problems and Exercises

4.3.8 Prove that for any positive integer n,

$$\binom{n}{0}^2 + \binom{n}{1}^2 + \binom{n}{2}^2 + \cdots + \binom{n}{n}^2 = \binom{2n}{n}.$$

4.3.9 Prove that for any positive integers $k < m, n$,

$$\sum_{j=0}^{k} \binom{n}{j}\binom{m}{k-j} = \binom{n+m}{k}.$$

4.3.10 Use generating functions to prove the formula for the Fibonacci series given in Problem 1.3.18 on page 12.

4.3.11 Reread Example 4.3.6. Prove formally by induction that if

$$f_r(x) := (1 + x + x^{-1})(1 + x^3 + x^{-3})(1 + x^{3^2} + x^{-3^2}) \cdots (1 + x^{3^r} + x^{-3^r}),$$

then

$$f_r(x) = x^a + x^{a-1} + \cdots + x^2 + x + 1 + x^{-1} + x^{-2} + \cdots + x^{-a},$$

where $a = (3^{r+1} - 1)/2$.

4.3.12 Here is an alternate way to end Example 4.3.7: Show that

$$(1 - x)(1 + x)(1 - x^3)(1 + x^2)(1 - x^5)(1 + x^3)(1 - x^7)(1 + x^4) \cdots = 1. \tag{8}$$

Here are suggestions for two different arguments:

1. Try induction by looking at the expansion of the first $2n$ terms. It won't equal 1, but it should equal something that does not have any x^k terms for "small" k. The idea is to show that as n grows, you can guarantee that there are no x^k terms for $k = 1, 2, \ldots, L$, where L is something that grows as n grows. That does it!

2. Show that the left-hand side of (8) is invariant under the substitution $x \mapsto x^2$. Then keep iterating this substitution (you may also want to note that the expressions only converge when $|x| < 1$).

4.3.13 (Putnam 1992) For nonnegative integers n and k, define $Q(n, k)$ to be the coefficient of x^k in the expansion of $(1 + x + x^2 + x^3)^n$. Prove that

$$Q(n, k) = \sum_{j=0}^{k} \binom{n}{j}\binom{n}{k-2j}.$$

4.3.14 Show that every positive integer has a *unique* binary (base-2) representation. For example, 6 is represented by 110 in binary, since $1 \cdot 2^2 + 1 \cdot 2^1 + 0 \cdot 2^0 = 6$. (This uniqueness can be proven in several ways; you are urged to try generating functions here, of course.)

4.3.15 *The Root of Unity Filter.*

Let $\zeta = \text{Cis } \frac{2\pi}{n}$ be an nth root of unity (see page 137).

(a) Show that the sum

$$1 + \zeta^k + \zeta^{2k} + \zeta^{3k} + \cdots + \zeta^{(n-1)k}$$

equals k or 0, according to whether k is a multiple of n or not.

(b) Find a simple formula for the sum $\displaystyle\sum_{j=0}^{\lfloor n/2 \rfloor} \binom{n}{2j}$.

(c) Find a simple formula for the sum $\displaystyle\sum_{j=0}^{\lfloor n/3 \rfloor} \binom{n}{3j}$.

(d) Generalize!

4.3.16 (Leningrad Mathematical Olympiad, 1991) A finite sequence $a_1, a_2, \ldots, a_n$ is called p-balanced if any sum of the form $a_k + a_{k+p} + a_{k+2p} + \cdots$ is the same for any $k = 1, 2, \ldots, p$. Prove that if a sequence with 50 members is p-balanced for $p = 3, 5, 7, 11, 13, 17$, then all its members are equal to zero.

4.3.17 Let $p(n)$ denote the number of unrestricted partitions of n. Here is a table of the first few values of $p(n)$.

n	$p(n)$	The different sums
1	1	1
2	2	$1+1, 2$
3	3	$1+1+1, 1+2, 3$
4	5	$1+1+1+1, 1+1+2, 1+3, 2+2, 4$
5	7	$1+1+1+1+1, 1+1+1+2,$
		$1+1+3, 1+2+2, 2+3, 1+4, 5$

Let $f(x)$ be the generating function for $p(n)$ [in other words, the coefficient of x^k in $f(x)$ is $p(k)$]. Explain why

$$f(x) = \prod_{n=1}^{\infty} \frac{1}{1 - x^n}.$$

4.3.18 Show that the number of partitions of a positive integer n into parts that are not multiples of three is equal to the number of partitions of n in which there are at most two repeats. For example, if $n = 6$, then there are 7 partitions of the first kind, namely

$$1+1+1+1+1+1, \quad 1+1+1+1+2, \quad 1+1+2+2,$$

$$1+1+4, \quad 1+5, \quad 2+2+2, \quad 2+4;$$

and there are also 6 partitions of the second kind,

$$1+1+4, \quad 1+1+2+2, \quad 1+2+3, \quad 1+5, \quad 2+4, \quad 3+3, \quad 6.$$

Can you generalize this problem?

4.3.19 In how many ways can you make change for a dollar, using pennies, nickels, dimes, quarters, and half-dollars? For example, 100 pennies is one way; 20 pennies, 2 nickels, and 7 dimes is another. Order doesn't matter.

4.3.20 The function $(1 - x - x^2 - x^3 - x^4 - x^5 - x^6)^{-1}$ is the generating function for what easily stated sequence?

4.3.21 A standard die is labeled $1, 2, 3, 4, 5, 6$ (one integer per face). When you roll two standard dice, it is easy to compute the probability of the various sums. For example, the probability of rolling two dice and getting a sum of 2 is just $1/36$, while the probability of getting a 7 is $1/6$.

Is it possible to construct a pair of "nonstandard" dice (possibly different from one another) with positive integer labels that nevertheless are indistinguishable from a pair of standard dice, if the sum of the dice is all that matters? For example, one of these nonstandard dice may have the label 8 on one of its faces, and two 3's. But the probability of rolling the two and getting a sum of 2 is still $1/36$, and the probability of getting a sum of 7 is still $1/6$.

Part II

Specifics

Chapter 5

Algebra

You probably consider yourself an old hand at algebra. Nevertheless, you may have picked up some bad habits or missed a few tricks in your mathematical education. The purpose of this chapter is reeducation: We shall relearn algebra, from the problem solver's perspective.

> *Algebra, combinatorics, and number theory are intimately connected.*
> *Please read the first few sections of Chapters 6 and 7 concurrently with*
> *this chapter.*

5.1 Sets, Numbers, and Functions

This first section contains a review of basic set and function notation, and can probably be skimmed (but make sure that you understand the function examples which begin on page 159).

Sets

Sets are collections of **elements**. If an element x belongs to (is an element of) a set A we write $x \in A$. Sets can be collections of anything (including other sets). One way to define a set is by listing the elements inside brackets, for example,

$$A = \{2, 3, 8, \sqrt{2}\}.$$

A set can contain no elements at all; this is the **empty set** $\emptyset = \{\}$.

Recall the set operations $\cup$ (**union**) and $\cap$ (**intersection**). We define $A \cup B$ to be the set each of whose elements is contained either in A or in B (or in both). For example,

$$\{1, 2, 5\} \cup \{1, 3, 8\} = \{1, 2, 3, 5, 8\}.$$

Similarly, we define $A \cap B$ to be the set whose elements are contained in both A and B, so for example,

$$\{1, 2, 5\} \cap \{1, 3, 8\} = \{1\}.$$

If all elements of a set A are contained in a set B, we say that A is a **subset** of B and write $A \subset B$. Note that $A \subset A$ and $\emptyset \subset A$ for all sets A.

We can define "subtraction" for sets in the following natural way:

$$A - B := \{a \in A : a \notin B\};$$

in other words, $A - B$ is the set of all elements of A which are not elements of B.

Two fundamental sets are the **natural numbers** $\mathbf{N} := \{1, 2, 3, 4, \ldots\}$ and the **integers** $\mathbf{Z} := \{0, \pm 1, \pm 2, \pm 3, \pm 4, \ldots\}$.[1]

Usually, there is a larger "universal" set U which contains all the sets under our consideration. This is usually understood by context. For example, if the sets that we are looking at contain numbers, then $U = \mathbf{Z}$ or $U = \mathbf{R}$. When the universal set U is known, we can define the **complement** $\overline{A}$ of the set A to be "everything" not in A; i.e.,

$$\overline{A} := U - A.$$

For example, if $U = \mathbf{Z}$ and A consisted of all even integers, then $\overline{A}$ would consist of all the odd integers. (Without knowledge of U, the idea of a set complement is meaningless; for example, if U was unknown and A was the set of even integers, then the elements that are "not in A" would include the odd integers, the imaginary numbers, the inhabitants of Paris, the rings of Saturn, etc.)

A common way to define a set is with "such that" notation. For example, the set of **rational numbers** $\mathbf{Q}$ is the set of all quotients of the form a/b such that $a, b \in \mathbf{Z}$ and $b \neq 0$. We abbreviate "such that" by "|" or ":"; hence

$$\mathbf{Q} := \left\{ \frac{a}{b} : a, b \in \mathbf{Z}, b \neq 0 \right\}.$$

Not all numbers are rational. For example, $\sqrt{2}$ is not rational, which we proved on page 47. This proof can be extended, with some work, to produce many (in fact, infinitely many) other irrational numbers. Hence there is a "larger" set of values on the "number line" which includes $\mathbf{Q}$. We call this set the **real numbers** $\mathbf{R}$. One can visualize $\mathbf{R}$ intuitively as the entire "continuous" set of points on the number line, while $\mathbf{Q}$ and $\mathbf{Z}$ are respectively "grainy" and "discrete" subsets of $\mathbf{R}$.[2]

Frequently we refer to intervals of real numbers. We use the notation $[a, b]$ to denote the **closed** interval $\{x \in \mathbf{R} : a \leq x \leq b\}$. Likewise, the **open** interval (a, b) is defined to be $\{x \in \mathbf{R} : a < x < b\}$. The hybrids $[a, b)$ and $(a, b]$ are defined analogously.

Finally, we extend the real numbers by adding the new element i, defined to be the square root of -1; i.e., $i^2 = -1$. Including i among the elements of $\mathbf{R}$ produces the set of **complex numbers** $\mathbf{C}$, defined formally by

$$\mathbf{C} := \{a + bi : a, b \in \mathbf{R}\}.$$

The complex numbers possess the important property of **algebraic closure**. This means that any finite combination of additions, subtractions, multiplications, divisions

---

[1] The letter "Z" comes from the German *zahlen*, which means "number."

[2] There are many rigorous ways of defining the real numbers carefully as an "extension" of the rational numbers. See, for example, Chapter 1 in [23].

(except by zero) and extractions of roots, when applied to a complex number, will result in another complex number. None of the smaller sets **N**, **Z**, **Q** or **R** has algebraic closure. The natural numbers **N** are not closed under subtraction, **Z** is not closed under division, and neither **Q** nor **R** is closed under square roots.

Given two sets A and B (which may or may not be equal), the **cartesian product** $A \times B$ is defined to be the set of all ordered pairs of the form (a, b) where $a \in A$ and $b \in B$. Formally, we define

$$A \times B := \{(a, b) : a \in A, b \in B\}.$$

For example, if $A = \{1, 2, 3\}$ and $B = \{\text{Paris}, \text{London}\}$, then

$$A \times B = \{(1, \text{Paris}), (2, \text{Paris}), (3, \text{Paris}), (1, \text{London}), (2, \text{London}), (3, \text{London})\}.$$

Functions

Given two sets A and B, we can assign a specific element of B to each element of A. For example, with the sets above, we can assign Paris to both 1 and 2, and London to 3. In other words, we are specifying the subset

$$\{(1, \text{Paris}), (2, \text{Paris}), (3, \text{London})\}$$

of $A \times B$.

Any subset of $A \times B$ with the property that each $a \in A$ is paired with *exactly one* $b \in B$ is called a **function** from A to B. Typically, we write $f : A \to B$ to indicate the function whose name is f with **domain** set A and **range** in the set B, and we write $f(a)$ to be the element in B which corresponds to $a \in A$, and we often call $f(a)$ the **image** of a. Informally, a function f is just a "rule" which assigns a B-value $f(a)$ to each A-value a. Here are several important examples, which also develop a few more concepts and notations.

Squaring Define $f : \mathbf{R} \to \mathbf{R}$ by $f(x) = x^2$ for each $x \in \mathbf{R}$. An alternate notation is to write $x \mapsto x^2$ or $x \overset{f}{\mapsto} x^2$. Notice that the range of f is not all of $\mathbf{R}$, but just the non-negative real numbers. Also notice that $f(x) = 9$ has two solutions $x = \pm 3$. The set $\{3, -3\}$ is called the **inverse image** of 9 and we write $f^{-1}(9) = \{3, -3\}$. Notice that an inverse image is not an element, but a set, because in general, as in this example, the inverse image may have more than one element.

Cubing Define $g : \mathbf{R} \to \mathbf{R}$ by $x \mapsto x^3$ for each $x \in \mathbf{R}$. In this case, the range is *all* of $\mathbf{R}$. We call such functions **onto**. Moreover, each inverse image contains just one element (since the cube root of a negative number is negative and the cube root of a positive number is positive). Functions with this property are called **1-to-1** (the function f above is 2-to-1, except at 0, where it is 1-to-1). A function like g, which is both one-to-one and onto, is also called a **1-1 correspondence** or a **bijection**.

Exponentiation and Logarithms Fix a positive real number b. Define $h : \mathbf{R} \to \mathbf{R}$ by $h(x) = b^x$ for each $x \in \mathbf{R}$. The range is all positive real numbers (but not zero), so h is not onto. On the other hand, h is 1-to-1, for if $y > 0$, then there is exactly one solution x to the equation $b^x = y$. We call this solution the **logarithm** $\log_b y$. For example, if $b = 3$, then $\log_3 81 = 4$ because $x = 4$ is the unique[3] solution to $3^x = 81$.

Now consider the function $x \mapsto \log_b x$. Verify that the domain is the positive reals, and the range is all reals.

Floors and Ceilings   For each $x \in \mathbf{R}$, define the **floor** function $\lfloor x \rfloor$ to be the greatest integer less than or equal to x (another notation for $\lfloor x \rfloor$ is $[x]$, but this is somewhat old-fashioned). For example, $\lfloor 3.7 \rfloor = 3$, $\lfloor 2 \rfloor = 2$, $\lfloor -2.4 \rfloor = -3$. Likewise, the **ceiling** function $\lceil x \rceil$ is defined to be the smallest integer greater than or equal to x. For example, $\lceil 3.1 \rceil = 4$, $\lceil -1.2 \rceil = -1$. For both functions, the domain is $\mathbf{R}$ and the range is $\mathbf{Z}$. Both functions are onto and neither is 1-to-1. In fact, these functions are ∞-to-1!

Sequences   If the domain of a function f is the natural numbers $\mathbf{N}$, then the range values will be $f(1), f(2), f(3), \ldots$. Sometimes it is more convenient to use the notation

$$f_1, f_2, f_3, \ldots$$

in which case the function is called a **sequence**. The domain need not be $\mathbf{N}$ precisely; it may start with zero, and it may be finite. Sometimes an infinite sequence is denoted by $(f_i)_1^\infty$, or sometimes just by (f_i). Since the subscript takes on integer values, the conventional letters used are i, j, k, l, m, n.[4]

Indicator Functions   Let U be a set with subset A. The **indicator function of** A is denoted by $\mathbf{1}_A$ and is a function with domain U and range $\{0, 1\}$ defined by

$$\mathbf{1}_A(x) = \begin{cases} 0 & \text{if } x \notin A, \\ 1 & \text{if } x \in A, \end{cases}$$

for each $x \in U$. For example, if $U = \mathbf{N}$ and A is the set of primes, then $\mathbf{1}_A(9) = 0$ and $\mathbf{1}_A(17) = 1$.

Problems and Exercises

5.1.1 Let A, B respectively denote the even and odd integers (remember, 0 is even).

(a) Is there a bijection from A to B?

(b) Is there a bijection from $\mathbf{Z}$ to A?

5.1.2 Prove that for any sets A, B,

(a) $\mathbf{1}_A(x)\mathbf{1}_B(x) = \mathbf{1}_{A \cap B}(x)$,

(b) $1 - \mathbf{1}_A(x) = \mathbf{1}_{\overline{A}}(x)$.

---

[3]If we include complex numbers, the solution is no longer unique. See [17] for more information.

[4]In the pioneering computer programming language FORTRAN, integer variables had to begin with the letters I, J, K, L, M, or N (the mnemonic aid was "INteger.")

In other words, the product of two indicator functions is the indicator function of the intersection of the two sets and the indicator function of a set's complement is just one subtracted from the indicator function of that set.

5.1.3 True or false and why: $\emptyset = \{\emptyset\}$.

5.1.4 Prove the following "dual" statements which show that in some sense, both the rational and irrational numbers are "grainy."

(a) Between any two rational numbers there is an irrational number.

(b) Between any two irrational numbers there is a rational number.

5.1.5 The number of elements of a set is called its **cardinality**. The cardinality of A is usually indicated by $|A|$ or $\#A$. If $|A| = m$ and $|B| = n$, then certainly $|A \times B| = mn$. How many different functions are there from A to B?

5.1.6 (AIME 1984) The function $f : \mathbf{Z} \to \mathbf{Z}$ satisfies $f(n) = n - 3$ if $n \geq 1000$ and $f(n) = f(f(n+5))$ if $n < 1000$. Find $f(84)$.

5.1.7 (AIME 1984) A function f is defined for all real numbers and satisfies

$$f(2 + x) = f(2 - x) \quad \text{and} \quad f(7 + x) = f(7 - x)$$

for all real x. If $x = 0$ is a root of $f(x) = 0$, what is the least number of roots $f(x) = 0$ must have in the interval $-1000 \leq x \leq 1000$?

5.1.8 (AIME 1985) How many of the first 1000 positive integers can be expressed in the form

$$\lfloor 2x \rfloor + \lfloor 4x \rfloor + \lfloor 6x \rfloor + \lfloor 8x \rfloor?$$

5.1.9 True or false and why: $\lfloor \sqrt{\lfloor x \rfloor} \rfloor = \lfloor \sqrt{x} \rfloor$ for all non-negative x.

5.1.10 Prove that for all $n \in \mathbf{N}$,

$$\left\lfloor \sqrt{n} + \sqrt{n+1} \right\rfloor = \left\lfloor \sqrt{4n+2} \right\rfloor.$$

5.1.11 Try Problem 2.4.14 on page 66.

5.1.12 Find a formula for the nth element of the sequence

$$1, 2, 2, 3, 3, 3, 4, 4, 4, 4, 5, 5, 5, 5, 5, \ldots$$

where the integer m occurs exactly m times.

5.1.13 Prove that

$$\left\lfloor \frac{n + 2^0}{2^1} \right\rfloor + \left\lfloor \frac{n + 2^1}{2^2} \right\rfloor + \left\lfloor \frac{n + 2^2}{2^3} \right\rfloor + \cdots + \left\lfloor \frac{n + 2^{n-1}}{2^n} \right\rfloor = n$$

for any positive integer n.

5.2 Algebraic Manipulation Revisited

Algebra is commonly taught as a series of computational techniques. We say "computational" because there really is no conceptual difference between these two exercises:

1. Compute 42×57.
2. Write $(4x + 2)(5x + 7)$ as a trinomial.

Both are exercises of routine, boring algorithms. The first manipulates pure numbers while the second manipulates both numbers and symbols. We call such mind-numbing (albeit useful) algorithms "computations." Algebra is full of these algorithms, and you have undoubtedly practiced many of them. What you may not have learned, however, is that algebra is also an aesthetic subject. Sometimes one has to slog through messy thickets of algebraic expressions to solve a problem. But these unfortunate occasions are pretty rare. A good problem solver takes a more confident approach to algebraic problems. The wishful thinking strategy teaches her to look for an *elegant* solution. Cultivate this mindset: employ a light, almost delicate touch, keeping watch for opportunities that avoid ugly manipulations in favor of elegant, often *symmetrical* patterns. Our first example illustrates this.

Example 5.2.1 If $x + y = xy = 3$, find $x^3 + y^3$.

Solution: One way to do this problem—the bad way—is to solve the system $xy = 3, x + y = 3$ for x and y (this would use the quadratic formula, and the solutions will be complex numbers) and then substitute these values into the expression $x^3 + y^3$. This will work, but it is ugly and tedious and messy and surely error-prone.

Instead, we keep a light touch. Our goal is $x^3 + y^3$, so let's try for $x^2 + y^2$ as a penultimate step. How to get $x^2 + y^2$? Try by squaring $x + y$.

$$3^2 = (x + y)^2 = x^2 + 2xy + y^2,$$

and since $xy = 3$, we have $x^2 + y^2 = 3$. Consequently,

$$3 \cdot 3 = (x + y)(x^2 + y^2) = x^3 + x^2y + xy^2 + y^3 = x^3 + y^3 + xy(x + y).$$

From this, we conclude that

$$x^3 + y^3 = 3 \cdot 3 - 3 \cdot 3 = 0,$$

which is rather surprising.

Incidentally, what if we really wanted to find out the values of x and y? Here is an elegant way to do it. The equation $x + y = 3$ implies $(x + y)^2 = 3^2$, or

$$x^2 + 2xy + y^2 = 9.$$

Since $xy = 3$, we subtract $4xy = 12$ from this last equation, getting

$$x^2 - 2xy + y^2 = -3.$$

This is a perfect square, and taking square roots gives us

$$x - y = \pm i\sqrt{3}.$$

This equation is particularly useful, since it is given that $x + y = 3$. Adding these two equations immediately gives us $x = (3 \pm i\sqrt{3})/2$, and subtraction yields $y = (3 \mp i\sqrt{3})/2$. Our two solutions for (x, y) are

$$\left(\frac{3 + i\sqrt{3}}{2}, \frac{3 - i\sqrt{3}}{2} \right), \quad \left(\frac{3 - i\sqrt{3}}{2}, \frac{3 + i\sqrt{3}}{2} \right).$$ ∎

The Factor Tactic

Multiplication rarely simplifies things. Instead, you should

Factor relentlessly.

The following are basic formulas that you learned in an algebra class. Make sure that you know them actively, rather than passively. Notice how formula 5.2.4 *instantly* solves example 5.2.1!

5.2.2 $(x + y)^2 = x^2 + 2xy + y^2$.

5.2.3 $(x - y)^2 = x^2 - 2xy + y^2$.

5.2.4 $(x + y)^3 = x^3 + 3x^2 y + 3xy^2 + y^3 = x^3 + y^3 + 3xy(x + y)$.

5.2.5 $(x - y)^3 = x^3 - 3x^2 y + 3xy^2 - y^3 = x^3 - y^3 - 3xy(x - y)$.

5.2.6 $x^2 - y^2 = (x - y)(x + y)$.

5.2.7 $x^n - y^n = (x - y)(x^{n-1} + x^{n-2}y + x^{n-3}y^2 + \cdots + y^{n-1})$ for all n.

5.2.8 $x^n + y^n = (x + y)(x^{n-1} - x^{n-2}y + x^{n-3}y^2 - \cdots + y^{n-1})$ for all odd n (the terms of the second factor alternate in sign).

Many problems involve combinations of these formulas, along with basic strategies (for example, wishful thinking), awareness of symmetry, and the valuable **add zero creatively** tool.[5] Here is an example.

Example 5.2.9 Factor $x^4 + 4$ into two polynomials with real coefficients.

Solution: If it weren't for the requirement that the factors have real coefficients, we could just treat $x^4 + 4$ as a difference of two squares (formula 5.2.6) and obtain

$$x^4 + 4 = x^2 - (-4) = x^4 - (2i)^2 = (x^2 + 2i)(x^2 - 2i).$$

While we cannot use the difference-of-two-squares method directly, we should not abandon it just yet, since the expression at hand contains two perfect squares. Unfortunately, it is not a difference of two perfect squares. But there are other possible perfect squares, and our expression nearly contains them. Use **wishful thinking** to make more perfect squares appear, by adding zero creatively.

$$x^4 + 4 = x^4 + 4x^2 + 4 - 4x^2.$$

---

[5]The sister to the add zero creatively tool is the **multiply cleverly by one** tool.

This was the crux move, for now we have

$$x^4 + 4x^2 + 4 - 4x^2 = (x^2 + 2)^2 - (2x)^2 = (x^2 + 2x + 2)(x^2 - 2x + 2).$$

This instructive example shows that you should always look for perfect squares, and try to create them if they are not already there.

Manipulating Squares

Also well worth remembering is how to square a trinomial, not to mention more complicated polynomials.

Know how to create and recognize perfect squares.

Toward this end, please learn the following formulas, actively, not passively!

5.2.10 $(x + y + z)^2 = x^2 + y^2 + z^2 + 2xy + 2xz + 2yz.$

5.2.11 $(x + y + z + w)^2 = x^2 + y^2 + z^2 + w^2 + 2xy + 2xz + 2xw + 2yz + 2yw + 2zw.$

5.2.12 *Completing the square.*

$$x^2 + ax = x^2 + ax + \frac{a^2}{4} - \frac{a^2}{4} = \left(x + \frac{a}{2}\right)^2 - \left(\frac{a}{2}\right)^2.$$

Ponder the completing-the-square formula above. One way to "discover" it is by recognizing the perfect square which begins with $x^2 + ax$, and then **adding zero creatively**. Another approach uses simple factoring, followed by an attempt to **symmetrize** the terms, plus adding zero creatively:

$$x^2 + ax = x(x + a) = \left(x + \frac{a}{2} - \frac{a}{2}\right)\left(x + \frac{a}{2} + \frac{a}{2}\right) = \left(x + \frac{a}{2}\right)^2 - \left(\frac{a}{2}\right)^2.$$

The tactic of **extracting squares** includes many tools in addition to completing the square. Here are a few important ideas.

5.2.13 $(x - y)^2 + 4xy = (x + y)^2.$

5.2.14 Replacing the variables in the equation above by squares yields

$$(x^2 - y^2)^2 + 4x^2y^2 = (x^2 + y^2)^2,$$

which produces infinitely many **Pythagorean triples**; i.e. integers (a, b, c) that satisfy $a^2 + b^2 = c^2$. (In a certain sense, this method generates *all* pythagorean triples. See Example 7.4.3 on page 265.)

5.2.15 The following equation shows that if each of two integers can be written as the sum of two perfect squares, then so can their product:

$$(x^2 + y^2)(a^2 + b^2) = (xa - by)^2 + (ya + bx)^2.$$

For example, $29 = 2^2 + 5^2$ and $13 = 2^2 + 3^2$ and indeed,

$$29 \cdot 13 = 11^2 + 16^2.$$

It is easy enough to see *how* this works, but *why* is another matter. For now, hindsight will work: remember that many useful squares lurk about and come to light when you manipulate "cross-terms" appropriately (making them cancel out or making them survive as you see fit). For a "natural" explanation of this example, see Example 4.2.15 on page 140.

Substitutions and Simplifications

The word "fractions" strikes fear into the hearts of many otherwise fine mathematics students. This is because most people, including those few who go on to enjoy and excel at math, are subjected to fraction torture in grade school, where they are required to complete long and tedious computations such as

$$\text{``Simplify''} \quad \frac{1}{x-1} + \frac{10}{17-x} - \frac{x^2}{1-5x} + \frac{11}{3}.$$

You have been taught that "simplification" is to combine things in "like terms." This sometimes simplifies an expression, but the good problem solver has a more focused, task-oriented approach, motivated by the wishful thinking strategy.

> *Avoid mindless combinations unless this makes your expressions simpler. Always move in the direction of greater simplicity and/or symmetry and/or beauty (the three are often synonymous).*

(There are, of course, exceptions. Sometimes you may want to make an expression uglier because it then yields more information. Example 5.5.10 on page 191 is a good illustration of this.)

An excellent example of a helpful substitution (inspired by symmetry) was Example 3.1.10 on page 76, in which the substitution $y = x + 1/x$ reduced the 4th-degree equation $x^4 + x^3 + x^2 + x + 1 = 0$ to two quadratic equations. Here are a few more examples.

Example 5.2.16 (AIME 1983) What is the product of the real roots of the equation

$$x^2 + 18x + 30 = 2\sqrt{x^2 + 18x + 45}?$$

Solution: This is not a very hard problem. The only real obstacle to immediate solution is the fact that there is a square root. The first thing to try, then, is to eliminate this obstacle by boldly substituting

$$y = \sqrt{x^2 + 18x + 45}.$$

Notice that if x is real, then y must be non-negative. The equation immediately simplifies to

$$y^2 - 15 = 2y,$$

which factors nicely into $(y-5)(y+3) = 0$. Reject the root $y = -3$ (since y must be non-negative); substituting the root $y = 5$ back into the original substitution yields

$$x^2 + 18x + 45 = 5^2$$

or

$$x^2 + 18x + 20 = 0.$$

Hence the product of the roots is 20, using the relationship between zeros and coefficients formula (see page 183). ∎

Example 5.2.17 (AIME 1986) Simplify

$$\left(\sqrt{5} + \sqrt{6} + \sqrt{7}\right)\left(\sqrt{5} + \sqrt{6} - \sqrt{7}\right)\left(\sqrt{5} - \sqrt{6} + \sqrt{7}\right)\left(-\sqrt{5} + \sqrt{6} + \sqrt{7}\right).$$

We could multiply out all the terms, but it would take a long time, and we'd probably make a mistake. We need a strategy. If this expression is to simplify, we will probably be able to eliminate radicals. If we multiply any two terms, we can use the difference of two squares formula (5.2.6) and get expressions which contain only one radical. For example, the product of the first and second terms is

$$\left(\sqrt{5} + \sqrt{6} + \sqrt{7}\right)\left(\sqrt{5} + \sqrt{6} - \sqrt{7}\right) = \left(\sqrt{5} + \sqrt{6}\right)^2 - \left(\sqrt{7}\right)^2$$
$$= 5 + 6 + 2\sqrt{30} - 7$$
$$= 4 + 2\sqrt{30}.$$

Likewise, the product of the last two terms is

$$\left(\sqrt{7} + \left(\sqrt{5} - \sqrt{6}\right)\right)\left(\sqrt{7} - \left(\sqrt{5} - \sqrt{6}\right)\right) = 7 - \left(5 - 2\sqrt{30} + 6\right) = -4 + 2\sqrt{30}.$$

The final product, then, is

$$\left(4 + 2\sqrt{30}\right)\left(-4 + 2\sqrt{30}\right) = 4 \cdot 30 - 16 = 104.$$ ∎

Example 5.2.18 (AIME 1986) Solve the system of equations

$$2x_1 + x_2 + x_3 + x_4 + x_5 = 6$$
$$x_1 + 2x_2 + x_3 + x_4 + x_5 = 12$$
$$x_1 + x_2 + 2x_3 + x_4 + x_5 = 24$$
$$x_1 + x_2 + x_3 + 2x_4 + x_5 = 48$$
$$x_1 + x_2 + x_3 + x_4 + 2x_5 = 96.$$

Solution: The standard procedure for solving systems of equations by hand is to substitute for and/or eliminate variables in a systematic (and tedious) way. But notice that each equation is almost symmetric, and that the system is symmetric as a whole. Just add together all five equations; this will serve to symmetrize all the coefficients:

$$6(x_1 + x_2 + x_3 + x_4 + x_5) = 6(1 + 2 + 4 + 8 + 16),$$

so

$$x_1 + x_2 + x_3 + x_4 + x_5 = 31.$$

Now we can subtract this quantity from each of the original equations to immediately get $x_1 = 6 - 31$, $x_2 = 12 - 31$, etc. ∎

You have seen the **define a function** tool in action in Example 3.3.8 on page 98 and in Example 5.4.2 on page 182. Here is another example, one that also employs a large dose of symmetry.

Example 5.2.19 Show, *without multiplying out*, that

$$\frac{b-c}{a} + \frac{c-a}{b} + \frac{a-b}{c} = \frac{(a-b)(b-c)(a-c)}{abc}.$$

Solution: Even though it is easy to multiply out, let us try to find a more elegant approach. Notice how the right-hand side factors. We can deduce this factorization by defining

$$f(x) := \frac{b-c}{x} + \frac{c-x}{b} + \frac{x-b}{c}.$$

Notice that $f(b) = f(c) = 0$. By the factor theorem, if we write $f(x)$ as a quotient of polynomials

$$f(x) = \frac{P(x)}{xbc},$$

then $P(x)$ must have $x - b$ and $x - c$ as factors. Also, it is clear that $P(x)$ has degree 3. Plugging in $x = a$ into $f(x)$, we conclude that

$$\frac{b-c}{a} + \frac{c-a}{b} + \frac{a-b}{c} = \frac{(a-b)(a-c)R(x)}{abc},$$

where $R(x)$ is a linear polynomial. By symmetry, we could also define the function

$$g(x) := \frac{x-c}{a} + \frac{c-a}{x} + \frac{a-x}{c},$$

and we have $g(a) = g(c) = 0$, yielding the factorization

$$\frac{b-c}{a} + \frac{c-a}{b} + \frac{a-b}{c} = \frac{(b-a)(b-c)Q(x)}{abc},$$

where $Q(x)$ is a different linear polynomial. We conclude that

$$\frac{b-c}{a} + \frac{c-a}{b} + \frac{a-b}{c} = K\left(\frac{(a-b)(b-c)(c-a)}{abc}\right),$$

for some constant K. Plugging in values (for example, $a = 1$, $b = 2$, $c = 3$) establishes that $K = -1$. ∎

Example 5.2.20 (Putnam 1939) Let $x^3 + bx^2 + cx + d = 0$ have integral coefficients and roots r, s, t. Find a polynomial equation with integer coefficients whose roots are r^3, s^3, t^3.

Solution: An incredibly ugly way to do this would be to solve for r, s, t, in terms of a, b, c and then construct the cubic polynomial $(x - r^3)(x - s^3)(x - t^3)$. Instead we define $p(x) := x^3 + bx^2 + cx + d$ and note that

$$p(\sqrt[3]{x}) = 0$$

is satisfied by $x = r^3, s^3, t^3$. We must thus convert

$$x + b\sqrt[3]{x^2} + c\sqrt[3]{x} + d = 0$$

into an equivalent polynomial equation. Cubing comes to mind, but what should we cube? Cubing anything but a binomial is too painful. If we put the radicals on one side and the non-radicals on the other, we have

$$-(x + d) = b\sqrt[3]{x^2} + c\sqrt[3]{x} \tag{1}$$

and now cubing both sides will remove *all* of the radicals. We shall employ the more useful form of 5.2.4, which states that

$$(x + y)^3 = x^3 + y^3 + 3xy(x + y),$$

and cubing both sides of (1) yields

$$-(x + d)^3 = (b\sqrt[3]{x^2})^3 + (c\sqrt[3]{x})^3 + 3b\sqrt[3]{x^2}c\sqrt[3]{x}(b\sqrt[3]{x^2} + c\sqrt[3]{x})$$
$$= b^3x^2 + c^3x + 3bcx(b\sqrt[3]{x^2} + c\sqrt[3]{x}).$$

On the surface, this does not look like much of an improvement, since the right-hand side still contains radicals. But (1) allows us to substitute $-(x + d)$ for those pesky radicals! Our equation becomes

$$-(x + d)^3 = b^3x^2 + c^3x - 3bcx(x + d),$$

a cubic with integral coefficients. ∎

Example 5.2.21 (AIME 1986) The polynomial $1 - x + x^2 - x^3 + \cdots + x^{16} - x^{17}$ may be written in the form $a_0 + a_1y + a_2y^2 + a_3y^3 + \cdots + a_{16}y^{16} + a_{17}y^{17}$, where $y = x + 1$ and the a_i's are constants. Find the value of a_2.

Solution: Using our *active* knowledge of the factorization formulas, we recognize immediately that

$$1 - x + x^2 - x^3 + \cdots + x^{16} - x^{17} = -\frac{x^{18} - 1}{x - (-1)}.$$

(Alternatively, we could have used the formula for the sum of a geometric series.) Substituting $y = x + 1$, we see that the polynomial becomes[6]

$$-\frac{(y - 1)^{18} - 1}{y} = -\frac{1}{y}\left(y^{18} - \binom{18}{1}y^{17} + \binom{18}{2}y^{16} - \cdots + 1 - 1\right)$$

$$= -y^{17} + \binom{18}{1}y^{16} - \cdots + \binom{18}{15}y^2 - \binom{18}{16}y + \binom{18}{17}.$$

---

[6]If you are not familiar with the binomial theorem, read Section 6.1.

Thus

$$a_2 = \binom{18}{15} = \binom{18}{3} = 816. \qquad \blacksquare$$

The following problem appeared in the 1972 IMO. Its solution depends on symmetry and the careful extraction of squares, but more than anything else, on *confidence* that a reasonably elegant solution exists. It is rather contrived, but quite instructive.

Example 5.2.22 Find all solutions $(x_1, x_2, x_3, x_4, x_5)$ of the system of inequalities

$$(x_1^2 - x_3x_5)(x_2^2 - x_3x_5) \leq 0$$
$$(x_2^2 - x_4x_1)(x_3^2 - x_4x_1) \leq 0$$
$$(x_3^2 - x_5x_2)(x_4^2 - x_5x_2) \leq 0$$
$$(x_4^2 - x_1x_3)(x_5^2 - x_1x_3) \leq 0$$
$$(x_5^2 - x_2x_4)(x_1^2 - x_2x_4) \leq 0$$

where x_1, x_2, x_3, x_4, x_5 are real numbers.

Solution: This problem is pretty intimidating, but notice that it is cyclically symmetric: each inequality has the form

$$(x_i^2 - x_{i+2}x_{i+4})(x_{i+1}^2 - x_{i+2}x_{i+4}),$$

where the indices are read modulo 5, for example, if $i = 3$; then the inequality becomes

$$(x_3^2 - x_5x_2)(x_4^2 - x_5x_2) \leq 0.$$

When the left-hand sides are multiplied out, we will get a total of 20 terms: all $\binom{5}{2} = 10$ "perfect square" terms of the form $x_i^2x_j^2$ ($i \neq j$) as well as 10 "cross-terms," five of the form $-x_i^2x_{i+1}x_{i+3}$ and five of the form $-x_i^2x_{i+2}x_{i+4}$.

These terms look suspiciously like they came from squares of binomials. For example,

$$(x_1x_2 - x_1x_4)^2 = x_1^2x_2^2 - 2x_1^2x_2x_4 + x_1^2x_4^2$$

produces two of the perfect square terms and one cross-term. Our strategy: Write the sum of left-hand sides in the form

$$\frac{1}{2}\left(y_1^2 + y_2^2 + \cdots + y_{10}^2\right),$$

where each y_k produces a different cross-term, and all the perfect square terms are exactly duplicated. And indeed, after some experimentation, we have

$$0 \geq \sum_{i=1}^{5}(x_i^2 - x_{i+2}x_{i+4})(x_{i+1}^2 - x_{i+2}x_{i+4})$$

$$= \frac{1}{2}\sum_{i=1}^{5}\left((x_ix_{i+1} - x_ix_{i+3})^2 + (x_{i-1}x_{i+1} - x_{i-1}x_{i+3})^2\right).$$

Since we have written 0 as greater than or equal to a sum of squares, the only solution is when all the squares are zero, and this implies that

$$x_1 = x_2 = x_3 = x_4 = x_5.$$

Consequently, the solution set to the system of inequalities is

$$\{(u, u, u, u, u) : u \in \mathbf{R}\}. \qquad \blacksquare$$

The final example looks at a tricky inequality. We don't solve it, but a nice substitution makes it at least a little bit nicer.

Example 5.2.23 (IMO 1995) Let a, b, c be positive real numbers such that $abc = 1$. Prove that

$$\frac{1}{a^3(b+c)} + \frac{1}{b^3(c+a)} + \frac{1}{c^3(a+b)} \geq \frac{3}{2}.$$

Partial Solution: We will not solve this problem just yet, but point out an algebraic simplification that must be done. What is the worst thing about this problem? It is an inequality involving fairly ugly fractions. Wishful thinking tells us that it would be nicer if the fractions either were less ugly or did not exist at all. How can this be achieved? There is a pretty obvious **substitution** — but only obvious if you have the idea of substitution in the forefront of your consciousness. The substitution is

$$x = 1/a, y = 1/b, z = 1/c,$$

which transforms the original inequality (use the fact that $xyz = 1$) into

$$\frac{x^2}{y+z} + \frac{y^2}{z+x} + \frac{z^2}{x+y} \geq \frac{3}{2}.$$

This inequality is still not that easy to deal with, but the denominators are much less complicated, and the problem has been reduced in complexity. See Example 5.5.23 for continuation.

Problems and Exercises

5.2.24 (AIME 1987) Find $3x^2y^2$ if x, y are integers such that $y^2 + 3x^2y^2 = 30x^2 + 517$.

5.2.25 Find all positive integer solutions (x, y) to $x^2 - y^2 = 20$.

5.2.26 Find all positive integer solutions (x, y) to $xy + 5x + 3y = 200$.

5.2.27 (AIME 1988) Find the smallest positive integer whose cube ends in 888 (of course, do this without a calculator or computer).

5.2.28 Find the minimum value of $xy + yz + xz$, given that x, y, z are real and $x^2 + y^2 + z^2 = 1$. No calculus, please!

5.2.29 (AIME 1991) Find $x^2 + y^2$ if $x, y \in \mathbf{N}$ and

$$xy + x + y = 71, \quad x^2y + xy^2 = 880.$$

5.2.30 Find all integer solutions (n, m) to

$$n^4 + 2n^3 + 2n^2 + 2n + 1 = m^2.$$

5.2.31 (AIME 1989) Assume that $x_1, x_2, \ldots, x_7$ are real numbers such that

$$x_1 + 4x_2 + 9x_3 + 16x_4 + 25x_5 + 36x_6 + 49x_7 = 1$$
$$4x_1 + 9x_2 + 16x_3 + 25x_4 + 36x_5 + 49x_6 + 64x_7 = 12$$
$$9x_1 + 16x_2 + 25x_3 + 36x_4 + 49x_5 + 64x_6 + 81x_7 = 123.$$

Find the value of

$$16x_1 + 25x_2 + 36x_3 + 49x_4 + 64x_5 + 81x_6 + 100x_7.$$

5.2.32 Show that each number in the sequence

$$49, 4489, 444889, 44448889, \ldots$$

is a perfect square.

5.2.33 (*Crux Mathematicorum*, June/July 1978) Show that $n^4 - 20n^2 + 4$ is composite when n is any integer.

5.2.34 If $x^2 + y^2 + z^2 = 49$ and $x + y + z = x^3 + y^3 + z^3 = 7$, find xyz.

5.2.35 Find all real values of x that satisfy $(16x^2 - 9)^3 + (9x^2 - 16)^3 = (25x^2 - 25)^3$.

5.2.36 Find all ordered pairs of positive integers (x, y) that satisfy $x^3 - y^3 = 721$.

5.2.37 (*Crux Mathematicorum*, April 1979) Determine the triples of integers (x, y, z) satisfying the equation

$$x^3 + y^3 + z^3 = (x + y + z)^3.$$

5.2.38 (AIME 87) Compute

$$\frac{(10^4 + 324)(22^4 + 324)(34^4 + 324)(46^4 + 324)(58^4 + 324)}{(4^4 + 324)(16^4 + 324)(28^4 + 324)(40^4 + 324)(52^4 + 324)}.$$

5.3 Sums and Products

Notation

The upper-case Greek letters Σ (sigma) and Π (pi) are used respectively for sums and products. We abbreviate the sum $x_1 + x_2 + \cdots + x_n$ by $\sum_{i=1}^{n} x_i$. Likewise, $\prod_{i=1}^{n} x_i$ indicates the product $x_1 x_2 \cdots x_n$. The variable i is called the **index**, and can of course be denoted by any symbol and assume any upper and lower limits, including infinity. If the indices are not consecutive integers, one can specify them in other ways. Here are a few examples.

- $\sum\limits_{d|10} d^2 = 1^2 + 2^2 + 5^2 + 10^2$, since $d|10$ means "d ranges through all divisors of 10."

- $\prod\limits_{p \text{ prime}} \dfrac{p^2}{p^2-1} = \dfrac{4}{3} \cdot \dfrac{9}{8} \cdot \dfrac{25}{24} \cdot \dfrac{49}{48} \cdots$, an infinite product.[7]

- $\sum\limits_{3 \le i < j \le 5} f(i,j) = f(3,4) + f(3,5) + f(4,5)$.

If the index specifications are understood in the context of a problem, they can certainly be omitted. In fact, often the indices get in the way of an informal, but clear, argument. For example,

$$\left(\sum x_i\right)^2 = \sum x_i^2 + 2\left(\sum x_i x_j\right)$$

is a reasonable, albeit technically incorrect way to write the square of a multinomial. The precise notation is

$$\left(\sum_{i=1}^{n} x_i\right)^2 = \sum_{i=1}^{n} x_i^2 + 2\left(\sum_{1 \le i < j \le n} x_i x_j\right).$$

Make sure that you understand the subscript $1 \le i < j \le n$. Carefully verify (look at examples where $n = 2, 3$, etc.) that

$$\sum_{\substack{i \ne j \\ 1 \le i,j \le n}} x_i x_j = 2\left(\sum_{1 \le i < j \le n} x_i x_j\right).$$

Also verify that a summation with subscript $1 \le i < j \le n$ has $\binom{n}{2}$ terms (you have been reading Chapter 6, right?).

Arithmetic Series

An **arithmetic sequence** is a sequence of consecutive terms with a constant difference; i.e. a sequence of the form

$$a, a+d, a+2d, \cdots.$$

An **arithmetic series** is a sum of an arithmetic sequence. The sum of an arithmetic sequence is a simple application of the Gaussian pairing tool (see page 75 in Section 3.1). Consider an arithmetic series of n terms with first term a and last term ℓ. We write the sum twice (d is the common difference):

$$S = a + (a+d) + \cdots + (\ell - d) + \ell,$$
$$S = \ell + (\ell - d) + \cdots + (a+d) + a.$$

---

[7]There are infinitely many primes. See Problem 2.3.21 on page 56 and Section 7.1. Incidentally, the value of this infinite product is $\pi^2/6$. See Example 8.4.8 on page 314.

Upon adding, we immediate deduce that

$$S = n\left(\frac{a + \ell}{2}\right),$$

the intuitively reasonable fact that the sum is equal to the average value of the terms multiplied by the number of terms. It is no coincidence that another term for "average" is **arithmetic mean**.

Geometric Series and the Telescope Tool

A **geometric sequence** is exactly like an arithmetic sequence except that now the consecutive terms have a common *ratio*; i.e., the sequence has the form

$$a, ar, ar^2, ar^3, \cdots.$$

The Gaussian pairing tool is no help for summing geometric series, because the terms are not additively symmetric. However, the wonderful **telescope** tool comes to the rescue. Consider a geometric series of n terms with first term a and common ratio r (so the last term is ar^{n-1}). Rather than write the sum S twice, we look at S and rS:

$$S = a + ar + ar^2 + \cdots + ar^{n-1},$$
$$rS = ar + ar^2 + ar^3 + \cdots + ar^n.$$

Observe that S and rS are nearly identical, and hence subtracting the two quantities will produce something very simple. Indeed,

$$S - rS = a - ar + ar - ar^2 + ar^2 - ar^3 + \cdots + ar^{n-1} - ar^n,$$

and all terms cancel except for the first and the last. (That's why it's called "telescoping," because the expression "contracts" the way some telescopes do.) We have

$$S - rS = a - ar^n,$$

and solving for S yields

$$S = \frac{a - ar^n}{1 - r}.$$

Geometric series crop up so frequently that it is probably worth memorizing this formula. In any event, the crux move—the telescope tool—must be mastered.

There are many ways to telescope a series. With the geometric series above, we created two series that were virtually the same. The next series, one that you first saw as Example 1.1.2 on page 3, requires different treatment.

Example 5.3.1 Write

$$\frac{1}{1 \cdot 2} + \frac{1}{2 \cdot 3} + \frac{1}{3 \cdot 4} + \cdots + \frac{1}{99 \cdot 100}$$

as a fraction in lowest terms.

Solution: Notice that each term can be written as

$$\frac{1}{k(k+1)} = \frac{1}{k} - \frac{1}{k+1}.$$

The entire sum is

$$\left(1 - \frac{1}{2}\right) + \left(\frac{1}{2} - \frac{1}{3}\right) + \left(\frac{1}{3} - \frac{1}{4}\right) + \cdots + \left(\frac{1}{99} - \frac{1}{100}\right),$$

and all terms cancel except for the first and last. Our sum reduces to $1 - \dfrac{1}{100}$. ∎

The hard part above was discovering that each term could be written in a way that telescopes. Will this always work? Sadly, no. The important thing is to be aware of the possibility for telescoping, which is really just an application of the adding zero creatively tool. And quite often, a telescoping attempt won't work perfectly, but will reduce the complexity of a problem.

Example 5.3.2 Find a formula for the sum of the first n squares.

Solution: In other words, we seek a formula for

$$1^2 + 2^2 + 3^2 + \cdots + n^2 = \sum_{j=1}^{n} j^2.$$

If we were to get lucky as we did with the previous example, we'd discover a magical sequence $u_1, u_2, \ldots$ with the property that

$$u_{k+1} - u_k = k^2.$$

Then we'd be done; telescoping yields

$$\sum_{j=1}^{n} j^2 = \sum_{j=1}^{n} (u_{j+1} - u_j) = (u_2 - u_1) + (u_3 - u_2) + \cdots + (u_{n+1} - u_n) = u_{n+1} - u_1.$$

But we need not get perfect telescoping. We just need to find a sequence u_k so that consecutive differences look more or less like what we want. This is really nothing more than persuing a wishful thinking strategy. In this vein, let's experiment with some simple sequences. The first thing to try is the naive guess $u_k := k^2$. We get

$$u_{k+1} - u_k = k^2 + 2k + 1 - k^2 = 2k + 1.$$

The quadratic terms canceled, leaving only a linear expression. The next guess, of course, is to try $u_k := k^3$, which yields

$$u_{k+1} - u_k = (k+1)^3 - k^3 = k^3 + 3k^2 + 3k + 1 - k^3 = 3k^2 + 3k + 1.$$

This is not quite k^2, but it will do quite nicely, for now the telescope methods yields

$$\sum_{j=1}^{n} (3j^2 + 3j + 1) = \sum_{j=1}^{n} (u_{j+1} - u_j) = u_{n+1} - u_1 = (n+1)^3 - 1^3 = n^3 + 3n^2 + 3n.$$

In other words,

$$3\left(\sum_{j=1}^{n} j^2\right) + \sum_{j=1}^{n} (3j + 1) = n^3 + 3n^2 + 3n,$$

and we can solve for $\sum\limits_{j=1}^{n} j^2$. We still need to sum the arithmetic series $\sum\limits_{j=1}^{n}(3j+1)$, but we already have a formula for this! Verify that

$$\sum_{j=1}^{n} j^2 = \frac{n(n+1)(2n+1)}{6}.$$

Sometimes telescoping won't work with what you start with, but the introduction of a single new term will instantly transform the problem. We call this the **catalyst** tool. Once you see it, you will never forget it and will easily apply it to other problems.

Example 5.3.3 Simplify the product

$$\left(1+\frac{1}{a}\right)\left(1+\frac{1}{a^2}\right)\left(1+\frac{1}{a^4}\right)\cdots\left(1+\frac{1}{a^{2^{100}}}\right).$$

Solution: Call the product P and consider what happens when we multiply P by $1-\frac{1}{a}$. The "catalyst" is the simple difference of two squares formula $(x-y)(x+y) = x^2 - y^2$.

$$
\begin{aligned}
\left(1-\frac{1}{a}\right)P &= \left(1-\frac{1}{a}\right)\left(1+\frac{1}{a}\right)\left(1+\frac{1}{a^2}\right)\cdots\left(1+\frac{1}{a^{2^{100}}}\right) \\
&= \left(1-\frac{1}{a^2}\right)\left(1+\frac{1}{a^2}\right)\left(1+\frac{1}{a^4}\right)\cdots\left(1+\frac{1}{a^{2^{100}}}\right) \\
&= \left(1-\frac{1}{a^4}\right)\left(1+\frac{1}{a^4}\right)\left(1+\frac{1}{a^8}\right)\cdots\left(1+\frac{1}{a^{2^{100}}}\right) \\
&\ \ \vdots \\
&= \left(1-\frac{1}{a^{2^{100}}}\right)\left(1+\frac{1}{a^{2^{100}}}\right) \\
&= \left(1-\frac{1}{a^{2^{101}}}\right).
\end{aligned}
$$

Hence

$$P = \frac{1 - 1/a^{2^{101}}}{1 - 1/a}.$$ ∎

Infinite Series

Series with infinitely many terms is more properly a calculus topic, and you will find more information in Chapter 8. For now, let us note a few elementary ideas. An infinite series **converges** or **diverges** if its sum is finite or infinite, respectively. You may recall the formula for the convergent infinite geometric series:

$$a + ar + ar^2 + \cdots = \frac{a}{1 - r},$$

valid if and only if $|r| < 1$. This is a simple consequence of the formula for a finite geometric series.

There are many ways to determine whether a given series converges or diverges. The simplest principles, however, are

- If $\sum a_k < \infty$ (i.e., the series converges) and the a_k **dominate** all but a finite number of the b_k (i.e., $a_k \geq b_k$ for all but a finite number of values of k), then $\sum b_k < \infty$.
- Likewise, if $\sum a_k = \infty$ (i.e., the series diverges) and the b_k dominate all but a finite number of the a_k, then $\sum b_k = \infty$.

In other words, the simplest strategy when dealing with an unknown infinite series is to find a known series to compare it to. One fundamental series that you should know well is the **harmonic** series

$$1 + \frac{1}{2} + \frac{1}{3} + \frac{1}{4} + \cdots.$$

Example 5.3.4 Show that the harmonic series diverges.

Solution: We will find some crude approximations for partial sums of this series. Notice that

$$\frac{1}{3} + \frac{1}{4} \geq \frac{1}{4} + \frac{1}{4} = \frac{2}{4} = \frac{1}{2},$$

since $\frac{1}{3}$ and $\frac{1}{4}$ both dominate $\frac{1}{4}$. Likewise,

$$\frac{1}{5} + \frac{1}{6} + \frac{1}{7} + \frac{1}{8} \geq \frac{4}{8} = \frac{1}{2}$$

and

$$\frac{1}{9} + \frac{1}{10} + \cdots + \frac{1}{16} \geq \frac{8}{16} = \frac{1}{2}.$$

In general, for each $n > 1$,

$$\frac{1}{2^n + 1} + \frac{1}{2^n + 2} + \cdots + \frac{1}{2^n + 2^n} \geq \frac{1}{2},$$

since each of the 2^n terms are greater than or equal to $\frac{1}{2^{n+1}}$. Therefore, the entire harmonic series is greater than or equal to

$$1 + \frac{1}{2} + \frac{1}{2} + \frac{1}{2} + \frac{1}{2} + \cdots,$$

which diverges.   ∎

The key idea used above combines the obvious fact that

$$a \geq b \quad \Longrightarrow \quad \frac{1}{a} \leq \frac{1}{b}$$

with the nice trick of replacing a "complicated" denominator with a "simpler" one. This is an example of the many-faceted **massage tool**—the technique of fiddling with an expression, with whatever method works (adding zero, multiplying by one, adding or subtracting a bit, etc.) in order to make it more manageable.[8] Here is another example.

Example 5.3.5 The **zeta function** $\zeta(s)$ is defined by the infinite series

$$\zeta(s) := \frac{1}{1^s} + \frac{1}{2^s} + \frac{1}{3^s} + \cdots.$$

When $s = 1$, this becomes the harmonic series and diverges.[9]

Show that $\zeta(s)$ converges for all $s \geq 2$.

Solution: This is a routine exercise using the integral test from calculus, but it is much more instructive to use first principles. First of all, we note that even though the problem asks us to prove a statement about infinitely many values of s, we need only show that $\zeta(2) < \infty$, since if $s > 2$ we have

$$\frac{1}{k^2} \leq \frac{1}{k^s}$$

for all positive integers k and consequently the convergence of $\zeta(2)$ will imply the convergence of $\zeta(s)$ for all larger values of s.

But how to show that $\zeta(2)$ converges? The general term is $1/k^2$; we must search for a similar series that we know something about. We already "own" a nicely telescoping series whose terms are the reciprocals of quadratics, namely Example 5.3.1 on page 173, the series

$$\sum_{k=1}^{n} \frac{1}{k(k+1)} = 1 - \frac{1}{n+1}.$$

Certainly the infinite series converges (to 1), and now we are done, because

$$\frac{1}{k^2} < \frac{1}{k(k+1)}$$

for all positive integers k. ∎

---

[8] Yet another idea inspired by the Wishful Thinking strategy.

[9] It turns out that $\zeta(s)$ has many rich properties, with wonderful connections to combinatorics and number theory. Consult Chapter 2 of [33] as a starting point.

Problems and Exercises

5.3.6 Find a formula for the product of the terms of a geometric sequence.

5.3.7 Find a formula for the sum $r + 2r^2 + 3r^3 + \cdots + nr^n$, and generalize.

5.3.8 Find a formula for

$$\frac{1}{1 \cdot 2 \cdot 3} + \frac{1}{2 \cdot 3 \cdot 4} + \cdots + \frac{1}{n(n+1)(n+2)}.$$

Can you generalize this?

5.3.9 Find a formula for

$$1 \cdot 2 \cdot 3 + 2 \cdot 3 \cdot 4 + \cdots + n(n+1)(n+2).$$

Can you generalize this?

5.3.10 (AIME 1983) For $\{1, 2, 3, \ldots, n\}$ and each of its nonempty subsets a unique **alternating sum** is defined as follows: Arrange the numbers in the subset in decreasing order and then, beginning with the largest, alternately subtract and add successive numbers. (For example, the alternating sum for $\{1, 2, 4, 6, 9\}$ is $9 - 6 + 4 - 2 + 1 = 6$ and for $\{5\}$ it is simply 5.) For each n, find a formula for the sum of all of the alternating sums of all the subsets.

5.3.11 Prove that

$$\sum_{x \in U} \mathbf{1}_A(x) = |A|.$$

This is just a fancy way of saying that if you consider each x in U and write down a "1" whenever x lies in A, then the sum of these "1"s will of course be the number of elements in A.

5.3.12 Find the sum $1 \cdot 1! + 2 \cdot 2! + \cdots + n \cdot n!$.

5.3.13 Find a formula for the sum

$$\sum_{k=1}^{n} \frac{k}{(k+1)!}.$$

5.3.14 Evaluate the product $\displaystyle\prod_{k=0}^{n} \cos\left(2^k \theta\right)$.

5.3.15 Find the sum $\displaystyle\sum_{k=2}^{n} \frac{1}{\log_k u}$.

5.3.16 (AIME 1996) For each permutation $a_1, a_2, a_3, \ldots, a_{10}$ of the integers

$$1, 2, \ldots, 10,$$

form the sum

$$|a_1 - a_2| + |a_3 - a_4| + |a_5 - a_6| + |a_7 - a_8| + |a_9 - a_{10}|.$$

Find the average value of all of these sums.

5.3.17 Try Problem 1.3.8 on page 11, if you haven't done it already.

5.3.18 (Canada, 1989) Given the numbers $1, 2, 2^2, \ldots, 2^{n-1}$, for a specific permutation $\sigma = x_1, x_2, \ldots, x_n$ of these numbers we define $S_1(\sigma) = x_1$, $S_2(\sigma) = x_1 + x_2, \ldots$ and $Q(\sigma) = S_1(\sigma)S_2(\sigma) \cdots S_n(\sigma)$. Evaluate $\sum 1/Q(\sigma)$, where the sum is taken over all possible permutations.

5.3.19 A 2-inch elastic band is fastened to the wall at one end, and there's a bug at the other end. Every minute (beginning at time 0), the band is instantaneously and uniformly stretched by 1 inch, and then the bug walks 1 inch toward the fastened end. Will the bug ever reach the wall?

5.3.20 (E. Johnston) Let S be the set of positive integers which do not have a zero in their base-10 representation; i.e.,

$$S = \{1, 2, \ldots, 9, 11, 12, \ldots, 19, 21, \ldots\}.$$

Does the sum of the reciprocals of the elements of S converge or diverge?

5.3.21 Example 5.3.5 on page 177 showed that $\zeta(2) < \infty$. Use "massage" to show that, in fact, $\zeta(2) < 2$. Then improve your estimate further to show that $\zeta(2) < 7/4$. (The exact value for $\zeta(2)$ is $\pi^2/6$. See Example 8.4.8 on page 314 for a sketch of a proof.)

5.3.22 (Putnam 1977) Evaluate the infinite product

$$\prod_{n=2}^{\infty} \frac{n^3 - 1}{n^3 + 1}.$$

5.3.23 Can you generalize the idea used in Example 5.3.2 on page 174?

5.3.24 (AIME 1995) Let $f(n)$ be the integer closest to $\sqrt[4]{n}$. Find $\displaystyle\sum_{k=1}^{1995} \frac{1}{f(k)}$.

5.4 Polynomials

There is much more to polynomials than the mundane operations of adding, subtracting, multiplying and dividing. This section contains a few important properties of polynomials to review and/or learn.

First, some notation and definitions. Let A be a set of numbers which is closed under addition and multiplication. Define

$$A[x] = \{a_0 + a_1x + a_2x^2 + \cdots + a_nx^n : a_i \in A, n = 0, 1, 2, 3, \ldots\}$$

to be the set of **polynomials** with **coefficients** in A. The most common coefficient sets that we use are $\mathbf{Z}, \mathbf{Q}, \mathbf{R}$ and $\mathbf{C}$. Occasionally we may use $\mathbf{Z}_n$, the integers modulo n (see page 252). We call each expression of the form a_ix^i a **term** or **monomial**.

When writing an arbitrary polynomial, follow the convention of labeling a_i as the coefficient of x^i. Consistent notation is clear and helps to avoid errors and confusion with complicated manipulations. We define the **degree** of a polynomial to be the highest exponent with a non-zero exponent. This coefficient is also called the *leading coefficient*. If this coefficient is 1, the polynomial is called **monic.** The coefficient a_0 is called the **constant** term.

Polynomial Operations

Much of your early algebra education was devoted to adding, subtracting, multiplying, and dividing polynomials. We won't insult your intelligence by reviewing the first two operations, but it is worthwhile to think about multiplication and division. Multiplication is pretty easy, but it is important to use good notation. Make sure that you understand the following notation, by multiplying out a few examples by hand.

If $A(x) = \sum a_i x^i$, $B(x) = \sum b_i x^i$ and $C(x) = \sum c_i x^i = A(x)B(x)$, then

$$c_j = a_0 b_j + a_1 b_{j-1} + \cdots + a_j b_0 = \sum_{\substack{s+t=j; \\ s,t \geq 0}} a_s b_t.$$

Polynomials can be divided just like integers, and the result will be a **quotient** and **remainder**. More formally, polynomials with coefficients in $\mathbf{Z}, \mathbf{Q}, \mathbf{R}, \mathbf{C}$ and $\mathbf{Z}_n$ all have a **Division Algorithm** that is analogous to the integer version (3.2.15):

Let $f(x)$ and $g(x)$ be polynomials in $K[x]$, where K is one of $\mathbf{Z}, \mathbf{Q}, \mathbf{R}, \mathbf{C}$ or $\mathbf{Z}_n$. Then

$$f(x) = Q(x)g(x) + R(x),$$

where $Q(x), R(x) \in K[x]$ and the degree of $R(x)$ is less than the degree of $g(x)$. We call $Q(x)$ the **quotient** *and $R(x)$ the* **remainder**.

For example, let $f(x) = x^3 + x^2 + 7$ and $g(x) = x^2 + 3$. Both polynomials are in $\mathbf{Z}[x]$. By doing "long division", we get

$$x^3 + x^2 + 7 = (x^2 + 3)(x + 1) + (-3x + 4),$$

so $Q(x) = x + 1$ and $R(x) = -3x + 4$. The important thing is that the quotient $Q(x)$ is also in $\mathbf{Z}[x]$, i.e., also has integer coefficients. We may take the division algorithm for granted, but it is a very important property of polynomials, as well as integers.

Example 5.4.1 (AIME 1986) What is the largest integer n for which $n^3 + 100$ is divisible by $n + 10$?

Solution: By the division algorithm, $n^3 + 100 = (n+10)(n^2 - 10n + 100) - 900$, so

$$\frac{n^3 + 100}{n + 10} = n^2 - 10n + 100 - \frac{900}{n + 10}.$$

If $n^3 + 100$ is to be divisible by $n + 10$, then $\frac{900}{n+10}$ must be an integer. The largest positive n for which this is true is $n = 890$, of course.   ∎

The Zeros of a Polynomial

It is always nice to solve a polynomial equation; undoubtedly you know the quadratic formula, which states that if

$$ax^2 + bx + c = 0,$$

then

$$x = \frac{-b \pm \sqrt{b^2 - 4ac}}{2a}.$$

While this formula is useful, it is far more important to remember how it was derived, by using the **complete the square** tool. We will review this with a simple example.

$$x^2 + 6x - 5 = 0 \quad \Longleftrightarrow \quad x^2 + 6x = 5 \quad \Longleftrightarrow \quad x^2 + 6x + 9 = 14.$$

Thus $(x + 3)^2 = 14$, so $x + 3 = \pm\sqrt{14}$, etc.

But often the exact zeros of a polynomial are difficult or impossible to determine,[10] and in fact, sometimes the exact zeros are not all that important, but rather indirect information is what is needed. Thus it is important to understand as much as possible about the relationship between the zeros of a polynomial and other properties. Here are a few useful principles.

The Remainder Theorem

If the polynomial $P(x)$ is divided by $x - a$ the remainder will be $P(a)$.

For example, divide $x^3 - 2x + 3$ by $x + 2$ and get (after some work)

$$\frac{x^3 - 2x + 3}{x + 2} = x^2 - 2x + 2 - \frac{1}{x + 2};$$

i.e., the quotient is $x^2 - 2x + 2$ and the remainder is -1. And indeed,

$$(-2)^3 - 2(-2) + 3 = -1.$$

To see why the Remainder Theorem is true in general, divide the polynomial $P(x)$ by $x - a$, getting quotient $Q(x)$ with remainder r. Using the division algorithm, we write

$$P(x) = Q(x)(x - a) + r.$$

The above equation is an **identity**; i.e., it is true for all values of x. Therefore we are free to substitute in the most convenient value of x, namely $x = a$. This yields $P(a) = r$, as desired. Please make a note of this **substitute convenient values** tool. It has many applications!

---

[10]Formulas for the zeros of any cubic or quartic polynomial were discovered in the 16th century, and in the 19th century, it was proven by Abel that it is *impossible*, in general, to find an "elementary" formula for the zeros of all 5th- or higher-degree polynomials. See [28] for a very readable account of this.

The Factor Theorem

If a is a zero of a polynomial $P(x)$, then $x - a$ must be a factor; i.e., $P(x)$ is a product of $x - a$ and another polynomial.

This follows immediately from the Remainder Theorem.

The Fundamental Theorem of Algebra

The Factor Theorem above tells us that $x - a$ is a factor of the polynomial $P(x)$ if a is a zero. But how do we know if a polynomial even has a zero? The Fundamental Theorem of Algebra guarantees this:

Every polynomial in $\mathbf{C}[x]$ has at least one complex zero.

This theorem is quite deep and surprisingly hard to prove. Its proof is beyond the scope of this book.[11]

A corollary of the Fundamental Theorem (use the result of problem 5.4.6) is that any nth degree polynomial has exactly n complex zeros, although some of the zeros may not be distinct. Thus we have the following factored form for any polynomial:

$$a_n x^n + a_{n-1} x^{n-1} + \cdots + a_0 = a_n(x - r_1)(x - r_2) \cdots (x - r_n), \tag{2}$$

where the r_i are the zeros, possibly not all distinct.

If zeros are not distinct, we say that they have **multiplicity** greater than 1. For example, the 8th-degree polynomial

$$(x - 1)(x - 2i)(x + 2i)(x - 7)^3(x + 6)^2$$

has 8 zeros, but only 5 distinct zeros. The zero 7 appears with multiplicity 3 and the zero -6 has multiplicity 2.

Here is an example that uses an analysis of zeros combined with clever substitution and the **define a function** tool which you encountered in Example 3.3.8 on page 98.

Example 5.4.2 (USAMO 1975) If $P(x)$ denotes a polynomial of degree n such that $P(k) = k/(k + 1)$ for $k = 0, 1, 2, \ldots, n$, determine $P(n + 1)$.

Solution: Go back to the Factor Theorem. A reinterpretation of this theorem from a problem solver's perspective is

To know the zeros of a polynomial is to know the polynomial.

In other words, if you don't know the zeros of the polynomial under consideration, either expend some effort to find them, *or shift your focus to a new polynomial* whose zeros are apparent. In our case, knowing that $P(k) = k/(k + 1)$ does not tell us anything about the zeros of $P(x)$. There are two difficulties: the right-hand side of $P(k) = k/(k + 1)$ is neither zero nor a polynomial. We eliminate both difficulties simultaneously by multiplying by $(k + 1)$ and subtracting:

$$(k + 1)P(k) - k = 0.$$

---

[11]For an elementary but difficult proof, see [5]. For a much simpler but less elementary argument, see [17].

So now we have information about the zeros of another polynomial, namely the $(n+1)$-degree polynomial

$$Q(x) := (x+1)P(x) - x.$$

Clearly the zeros of $Q(x)$ are just $0, 1, 2, \ldots, n$, so we can write

$$(x+1)P(x) - x = Cx(x-1)(x-2)\cdots(x-n),$$

where C is a constant that must be determined. Since the above equation is an identity, true for all x values, we can plug in any convenient value. The values $x = 0, 1, \ldots, n$ don't work, since they make the right-hand side equal to zero. The left-hand side contains the troublesome term $(x+1)P(x)$, so clearly our choice should be $x = -1$. Plugging this in yields

$$1 = C(-1)(-2)(-3)\cdots(-(n+1)),$$

so

$$C = \frac{(-1)^{n+1}}{(n+1)!}.$$

Finally, we can plug in $x = n+1$, and we have

$$(n+2)P(n+1) - n - 1 = \frac{(-1)^{n+1}}{(n+1)!}(n+1)n\cdots 1 = \frac{(-1)^{n+1}}{(n+1)!}(n+1)! = (-1)^{n+1},$$

so

$$P(n+1) = \frac{n+1+(-1)^{n+1}}{n+2}. \qquad \blacksquare$$

Relationship between Zeros and Coefficients

If we multiply out the right-hand side of equation (2) on page 182, we can get a series of expressions for the coefficients of the polynomial in terms of its zeros. This seems like a pretty complicated and tedious job, so let us approach it gingerly. To see what is going on, let us look at a very simple example, a quadratic monic (without loss of generality, all of the polynomials we will consider will be monic) with zeros r and s. Then following equation (2), we can write our polynomial as

$$x^2 + a_1 x + a_0 = (x-r)(x-s).$$

The right-hand side is equal to $x^2 - rx - sx + rs$, and if we equate terms with those on the left-hand side, we get

$$a_1 = -(r+s), \quad a_0 = rs.$$

Since we need to multiply out more complicated expressions, let us think about how we just did this easy one. We used "FOIL," which really just means "multiply every monomial in $(x-r)$ with every monomial in $(x-s)$." In other words, we computed

$$(x-r)(x-s) = (x+(-r))(x+(-s)) = x\cdot x + (-r)\cdot x + x\cdot(-s) + (-r)\cdot(-s).$$

The same procedure works when we multiply out more complicated expressions. For example, consider

$$x^3 + a_2 x^2 + a_1 x + a_0 = (x - q)(x - r)(x - s).$$

After multiplying out the right-hand side, but *before* collecting like terms, we will have $2 \cdot 2 \cdot 2 = 8$ terms, since we multiply each monomial in $(x - q)$ with each monomial in $(x - r)$ with each monomial in $(x - s)$, and each term has just 2 monomials in it. In other words, each of the 8 terms in $(x - q)(x - r)(x - s)$ represents a three-element choice, one element chosen from x or $-q$, one chosen from x or $-r$ and one chosen from x or $-s$.

So what kind of terms can we get? If our three choices are all x, we end up with the term x^3. There are three ways that we can choose two x's and one constant, producing the terms $-qx^2$, $-rx^2$, $-sx^2$. Likewise, there are three ways in which we can choose just one x and two constants, producing qrx, qsx, rsx. Finally, there is just one way to chose no x's, the term $-qrs$. This is 8 terms in all, and collecting like terms we have

$$
\begin{aligned}
x^3 + a_2 x^2 + a_1 x + a_0 &= (x - q)(x - r)(x - s) \\
&= x^3 - (q + r + s)x^2 + (qr + qs + rs)x - qrs.
\end{aligned}
$$

Equating like terms, we see that

$$a_2 = -(q + r + s), \quad a_1 = qr + qs + rs, \quad a_0 = -qrs.$$

Let us do one more example, this time a monic quartic polynomial with zeros p, q, r, s. We write our polynomial as

$$x^4 + a_3 x^3 + a_2 x^2 + a_1 x + a_0 = (x - p)(x - q)(x - r)(x - s).$$

Using the same reasoning, the right-hand side will have 16 monomial terms (before collecting like terms), each formed by one choice of x or $-p$, x or $-q$, etc. For example, the terms using exactly two x's will also have exactly two constants. How many such terms will there be?[12] The number of ways that you can pick two different elements from the set $\{p, q, r, s\}$; i.e. $\binom{4}{2} = 6$ terms. Working out all the terms, we have

$$
\begin{aligned}
(x - p)(x - q)(x - r)(x - s) = x^4 &- (p + q + r + s)x^3 \\
&+ (pq + pr + ps + qr + qs + rs)x^2 \\
&- (pqr + pqs + prs + qrs)x + pqrs.
\end{aligned}
$$

Equating like terms, we have

$$
\begin{aligned}
a_3 &= -(\text{sum of the zeros}) \\
a_2 &= +(\text{sum of all products of two different zeros}) \\
a_1 &= -(\text{sum of all products of three different zeros}) \\
a_0 &= +(\text{product of the zeros}),
\end{aligned}
$$

---

[12] You've looked at Section 6.1, right?

where it is understood that "different" here has a purely symbolic meaning; i.e. we multiply only zeros with different labels, such as p and q, even if their numerical values are the same.

Finally, we see the pattern, and can write the formulas in general:

Let $r_1, r_2, \ldots, r_n$ be the zeros of the monic polynomial

$$x^n + a_{n-1}x^{n-1} + \cdots + a_0 = 0.$$

Then for $k = 1, 2, \ldots, n$,

$$\frac{a_k}{a_n} = (-1)^{n-k}(\text{sum of all products of } n - k \text{ different zeros})$$

$$= (-1)^{n-k} \sum_{1 \le i_1 < i_2 < \cdots < i_{n-k} \le n} r_{i_1} r_{i_2} \cdots r_{n-k}.$$

These formulas are very important, and should be committed to memory. The imprecise language "sum of all products ..." is easier to remember, but do take the time to understand how the careful use of subscripts rigorously formulates the sums. Also note the role that the power of -1 plays. We use the convenient fact that $(-1)^k$ is equal to $+1$ if k is even, and -1 if k is odd.

Let's come down to earth from this abstract discussion by looking at a concrete example.

Example 5.4.3 (USAMO 1984) The product of two of the four zeros of the quartic equation

$$x^4 - 18x^3 + kx^2 + 200x - 1984 = 0$$

is -32. Find k.

Solution: Let the zeros be a, b, c, d. Then the relationship between zeros and coefficients yields

$$a + b + c + d = 18,$$
$$ab + ac + ad + bc + bd + cd = k,$$
$$abc + abd + acd + bcd = -200,$$
$$abcd = -1984.$$

Without loss of generality, let $ab = -32$. Substituting this into $abcd = -1984$ yields $cd = 62$, and substituting this in turn yields the system

$$a + b + c + d = 18 \tag{3}$$
$$30 + ac + ad + bc + bd = k \tag{4}$$
$$-32c - 32d + 62a + 62b = -200 \tag{5}$$

Let us think strategically. We need to compute k, *not* the values a, b, c, d. A penultimate step is evaluating $ac + ad + bc + bd$. Notice that this factors:

$$ac + ad + bc + bd = a(c + d) + b(c + d) = (a + b)(c + d).$$

While we're at it, let's factor (5) as well:

$$-32(c+d) + 62(a+b) = -200.$$

Now it should be clear how to proceed. We need only find the two values $u := a+b$ and $v := c+d$. Equations (3) and (5) become the system

$$u + v = 18,$$
$$62u - 32v = -200,$$

which can be easily solved ($u = 4$, $v = 14$). Finally, we have

$$k = 30 + 4 \cdot 14 = 86. \qquad \blacksquare$$

Rational Roots Theorem

Suppose that $P(x) \in \mathbf{Z}[x]$ has the zero $x = 2/3$. Does this give you any information about $P(x)$? By the Factor Theorem,

$$P(x) = \left(x - \frac{2}{3}\right) Q(x),$$

where $Q(x)$ is a polynomial. But what kind of coefficients does $Q(x)$ have? All that we know for sure are that the coefficients must be rational. However, if $x - \frac{2}{3}$ is a factor, then $3(x - \frac{2}{3}) = 3x - 2$ will also be a factor, so we can write

$$P(x) = (3x - 2)S(x),$$

where $S(x) = Q(x)/3$. We know that $P(x)$ has *integer* coefficients; can we say the same thing about $S(x)$? Indeed we can; this is **Gauss's Lemma**:

> *If a polynomial with integer coefficients can be factored into polynomials with rational coefficients, it can also be factored into primitive polynomials with integer coefficients.*

(A polynomial with integer coefficients is called **primitive** if its coefficients share no factors. For example, $3x^2 + 9x + 7$ is primitive while $10x^2 - 5x + 15$ is not.) See problem 7.1.27 for some hints on proving Gauss's Lemma.

Since $P(x)$ factors into the product of $(3x - 2)$ and another polynomial with integer coefficients, the coefficient of the leading term of $P(x)$ must be a multiple of 3 and the coefficient of the final term must be a multiple of 2.

In general, assume that a polynomial $P(x)$ with integral coefficients has a rational zero $x = a/b$, where a and b are in lowest terms. By the Factor Theorem and Gauss's Lemma,

$$P(x) = (bx - a)Q(x),$$

where $Q(x)$ is a polynomial with *integer* coefficients. This immediately gives us the **Rational Root Theorem**:

> *If a polynomial $P(x)$ with integral coefficients has a rational zero $x =$ a/b, where a and b are in lowest terms, then the leading coefficient of $P(x)$ is a multiple of b, and the constant term of $P(x)$ is a multiple of a.*

In practice, the Rational Root Theorem is not just used to find zeros but also to prove that zeros are irrational.

Example 5.4.4 If $x^2 - 2$ has any rational zeros a/b (in lowest terms), we must have $b|1$ and $a|2$. Therefore the only possible rational zeros are ± 2. Since neither 2 nor -2 are zeros, we can conclude that $x^2 - 2$ has no rational zeros. This is another way to prove that $\sqrt{2}$ is irrational!

We can generalize the above reasoning when applied to *monic* polynomials. It is an interesting criterion for irrationality, and should be noted as a tool:

> *Any rational zero of a monic polynomial must be an integer. Conversely, if a number is not an integer but is a zero of a monic polynomial, it must be irrational.*[13]

We shall conclude the section with a rather hard problem which uses the monic polynomial tool above plus several other ideas.

Example 5.4.5 Prove that the sum

$$\sqrt{1001^2 + 1} + \sqrt{1002^2 + 1} + \cdots + \sqrt{2000^2 + 1}$$

is irrational.

Solution: Our strategy is two-pronged: first, to show that the sum in question is not an integer, and second, to show that it is a zero of a monic polynomial.

For the first step, observe that if $n > 1$, then $n < \sqrt{n^2 + 1} < n + 1/n$. The first inequality is obvious, and the second follows from $n^2 + 1 < n^2 + 2 < (n + 1/n)^2$. Let us call the sum in question S. Then

$$S = 1001 + \theta_1 + 1002 + \theta_2 + \cdots + 2000 + \theta_{1000},$$

where each θ_i lies between 0 and $1/1001$. Consequently,

$$0 < \theta_1 + \theta_2 + \cdots + \theta_{1000} < 1,$$

so S is not an integer.

Next, we will show that S is a zero of a monic polynomial.

More generally, we shall prove that for all positive integers n, the quantity

$$\sqrt{a_1} + \sqrt{a_2} + \cdots + \sqrt{a_n}$$

is a zero of a monic polynomial if each a_i is an integer that is not a perfect square. We proceed by induction. If $n = 1$, the assertion is true because $\sqrt{a_1}$ is a zero of the monic

---

[13]This statement can also be proven directly, without using the rational roots theorem (Problem 5.4.11). If you are stumped, look at Example 7.1.7 on page 247.

polynomial $x^2 - a_1$. Now assume that

$$y = \sqrt{a_1} + \sqrt{a_2} + \cdots + \sqrt{a_n}$$

is a zero of the monic polynomial $P(x) = x^r + c_{r-1}x^{r-1} + \cdots + c_0$. We will produce a monic polynomial that has $x = y + \sqrt{a_{n+1}}$ as a zero. We have

$$0 = P(y) = P(x - \sqrt{a_{n+1}}) = (x - \sqrt{a_{n+1}})^r + c_{r-1}(x - \sqrt{a_{n+1}})^{r-1} + \cdots + c_0.$$

Notice that the expansion of each $(x - \sqrt{a_{n+1}})^k$ term can be separated into terms that have integer coefficients and terms with coefficients equal to an integer times $\sqrt{a_{n+1}}$. Thus we have

$$0 = P(x - \sqrt{a_{n+1}}) = x^r + Q(x) + \sqrt{a_{n+1}} R(x),$$

where $Q(x)$ and $R(x)$ are polynomials with integer coefficients, each of degree at most $r - 1$. Putting the radicals on one side of the equation yields

$$x^r + Q(x) = -\sqrt{a_{n+1}} R(x),$$

and squaring both sides leads to

$$x^{2r} + 2x^r Q(x) + (Q(x))^2 - a_{n+1}(R(x))^2 = 0.$$

The term with highest degree is x^{2r}. Since all coefficients are now integers, we have produced a monic polynomial with $x = y + \sqrt{a_{n+1}}$ as a zero, as desired.   ∎

Problems and Exercises

5.4.6 Prove that a polynomial of degree n can have at most n distinct zeros.

5.4.7 Use problem 5.4.6 to prove a nice application called the **identity principle**, which states that if two degree-d polynomials $f(x)$, $g(x)$ are equal for $d + 1$ different x-values, then the two polynomials are equal.

5.4.8 Prove that if a polynomial has real coefficients, then its zeros come in complex conjugate pairs; i.e., if $a + bi$ is a zero, then $a - bi$ is also a zero.

5.4.9 Find the remainder when you divide $x^{81} + x^{49} + x^{25} + x^9 + x$ by $x^3 - x$.

5.4.10 Find a polynomial with integral coefficients whose zeros include $\sqrt{2} + \sqrt{5}$.

5.4.11 We call a polynomial with integer coefficients **monic** if the term with highest degree has coefficient equal to 1. For example, $x^6 - 1729x^3 + 314$ is monic. Prove that if a monic polynomial has a rational zero, then this zero must in fact be an integer.

5.4.12 Let $p(x)$ be a polynomial with integer coefficients satisfying $p(0) = p(1) = 1999$. Show that p has no integer zeros.

5.4.13 Let $p(x)$ be a 1999-degree polynomial with integer coefficients that is equal to ± 1 for 1999 different integer values of x. Show that $p(x)$ cannot be factored into the product of two polynomials with integer coefficients.

5.4.14 (Hungary, 1899) Let r and s be the roots of

$$x^2 - (a+d)x + (ad - bc) = 0.$$

Prove that r^3 and s^3 are the roots of

$$y^2 - (a^3 + d^3 + 3abc + 3bcd)y + (ad - bc)^3 = 0.$$

5.4.15 Let a, b, c be distinct integers. Can the polynomial $(x-a)(x-b)(x-c) - 1$ be factored into the product of two polynomials with integer coefficients?

5.4.16 Let $p(x)$ be a polynomial of degree n, not necessarily with integer coefficients. For how many consecutive integer values of x must a $p(x)$ be an integer in order to guarantee that $p(x)$ is an integer for *all* integers x?

5.4.17 (IMO 1993) Let $f(x) = x^n + 5x^{n-1} + 3$ where $n > 1$ is an integer. Prove that $f(x)$ cannot be expressed as the product of two polynomials, each of which has all its coefficients integers and degree at least 1.

5.4.18 (USAMO 1977) If a and b are two roots of $x^4 + x^3 - 1 = 0$, prove that ab is a root of $x^6 + x^4 + x^3 - x^2 - 1 = 0$.

5.4.19 (Canada, 1970) Let $P(x) = x^n + a_{n-1}x^{n-1} + \cdots + a_1 x + a_0$ be a polynomial with integral coefficients. Suppose that there exist four distinct integers a, b, c, d with $P(a) = P(b) = P(c) = P(d) = 5$. Prove that there is no integer k with $P(k) = 8$.

5.4.20 (USAMO 84) $P(x)$ is a polynomial of degree $3n$ such that

$$
\begin{aligned}
P(0) &= P(3) = \cdots = P(3n) &= 2, \\
P(1) &= P(4) = \cdots = P(3n-2) &= 1, \\
P(2) &= P(5) = \cdots = P(3n-1) &= 0, \quad \text{and} \\
& P(3n+1) &= 730.
\end{aligned}
$$

Determine n.

5.4.21 (*American Mathematical Monthly*, Oct. 1962) Let $P(x)$ be a polynomial with real coefficients. Show that there exists a nonzero polynomial $Q(x)$ with real coefficients such that $P(x)Q(x)$ has terms which are all of a degree divisible by 10^9.

5.5   Inequalities

Inequalities are important because many mathematical investigations involve estimations, optimizations, best-case and worst-case scenarios, limits, etc. Equalities are nice, but are really quite rare in the "real world" of mathematics. A typical example was the use of rather crude inequalities to establish the divergence of the harmonic series (Example 5.3.4 on page 176). Another example was Example 2.3.1 on page 46, where we proved that the equation $b^2 + b + 1 = a^2$ had no positive integer solutions by showing that the alleged equality was, in fact, an inequality.

   Here is another solution to that problem, one that uses the tactic of looking for perfect squares: the equation $b^2 + b + 1 = a^2$ asserts that $b^2 + b + 1$ is a perfect square. But

$$b^2 < b^2 + b + 1 < b^2 + 2b + 1 = (b+1)^2,$$

so $b^2 + b + 1$ lies between two *consecutive* perfect squares. An impossibility!

These examples used very simple inequalities. It is essential that you master them, as well as a few more sophisticated ideas.

Fundamental Ideas

Let us begin with a brief review of the basic ideas, many of which we will present as a series of statements which are problems (exercises?) for you to verify before moving on.

Basic Arithmetic

The following are pretty simple, but you should ponder them carefully to make sure that you really understand why they are true. Do pay attention to the signs of your variables!

5.5.1 *Addition.* If $x \geq y$ and $a \geq b$, then $x + a \geq y + b$.

5.5.2 *Multiplication.* If $x \geq y$ and $a \geq 0$, then $ax \geq ay$. Conversely, if $a < 0$, then $ax \leq ay$.

5.5.3 *Reciprocals.* If $x \geq y$, then $1/x \leq 1/y$, provided that both x and y have the same sign.

5.5.4 *Distance interpretation of absolute value.* The set

$$\{x : |x - a| = b\}$$

consists of all points x on the real number line[14] that lie within a distance b of the point a.

Growth Rates of Functions

It is important to understand the hierarchy of growth rates for the most common functions. The best way to learn about this is to draw lots of graphs.

5.5.5 Any quadratic function of x will dominate any linear function of x, provided that x is "large enough." For example,

$$0.001x^2 > 100000x + 20000000$$

is true for all $x > 10^9$.

5.5.6 By the same reasoning, x^a will "eventually dominate" x^b provided that $a > b > 0$.

5.5.7 Likewise, if a is any positive number, and $b > 1$, then b^x eventually dominates x^a. (In other words, exponential functions grow faster than polynomial functions.)

---

[14]You have probably noticed that we are restricting our attention to real numbers. That is because it is not possible to define $z > 0$ in a meaningful way when z is complex. See Problem 2.3.16 on page 56.

5.5.8 Conversely, if a is any positive number, and $b > 1$, then x^a eventually dominates $\log_b x$.

In summary, the hierarchy of growth rates, from slowest to highest, is

logarithms, powers, exponents.

Simple Proofs

Of the many ways of proving inequalities, the simplest is to perform operations that create logically equivalent but simpler inequalities. More sophisticated variants include a little massage, as well. Here are some examples.

Example 5.5.9 Which is larger, $\sqrt{19} + \sqrt{99}$, or $\sqrt{20} + \sqrt{98}$?

Solution: We shall use the convention of writing a question mark (?) for an alleged inequality. Then we can keep track. If the algebra preserves the direction of the alleged inequality, we keep using the question mark. If instead we do something that reverses the direction of the alleged inequality (for example, taking reciprocals of both sides), we change the question mark to an upside-down question mark (¿). So we start with

$$\sqrt{19} + \sqrt{99} \quad ? \quad \sqrt{20} + \sqrt{98}.$$

Squaring both sides yields

$$19 + 2\sqrt{19 \cdot 99} + 99 \quad ? \quad 20 + 2\sqrt{20 \cdot 98} + 98,$$

which reduces to

$$\sqrt{19 \cdot 99} \quad ? \quad \sqrt{20 \cdot 98}.$$

This of course is equivalent to

$$19 \cdot 99 \quad ? \quad 20 \cdot 98.$$

At this point we can just do the calculation, but let's use our factoring skills: Subtract $19 \cdot 98$ from both sides to get

$$19 \cdot 99 - 19 \cdot 98 \quad ? \quad 98,$$

which reduces to

$$19 \quad ? \quad 98.$$

Finally, we can replace the "?" with "¿" and we conclude that

$$\sqrt{19} + \sqrt{99} < \sqrt{20} + \sqrt{98}. \qquad \blacksquare$$

Example 5.5.10 Which is bigger, $\dfrac{1998}{1999}$ or $\dfrac{1999}{2000}$?

Solution: This can be done in many ways; here is an argument that uses the define a function tool: Let

$$f(x) := \frac{x}{x+1}.$$

Now our problem is equivalent to determining the relative order of $f(1998)$ and $f(1999)$. How does this function grow? We have

$$f(x) = \frac{x}{x+1} = \frac{1}{1+\frac{1}{x}}, \tag{6}$$

and now it is easy to check that as $x > 0$ increases, the $1/x$ term decreases, causing $f(x)$ to increase (all we are using is the fundamental principle 5.5.3 that says informally, "If the denominator increases, the fraction decreases, and vice versa"). In other words, $f(x)$ is **monotonically increasing** for positive x.

In any event, $f(1998) < f(1999)$.   ∎

Notice how we actually made an expression *uglier* in (6). The right-hand side of (6) is certainly unpleasant from a typographical standpoint, but it is much easier to analyze the behavior of the function.

Here are a few more for you to try on your own. Most are fairly easy exercises.

5.5.11 Let $a_1, a_2, \ldots, a_n$ be real numbers. Prove that if $\sum a_i^2 = 0$, then $a_1 = a_2 = \cdots = a_n = 0$.

5.5.12 *The average principle.* Let $a_1, a_2, \ldots, a_n$ be real numbers with sum S. Prove (as rigorously as you can!) that either the a_i are all equal, or else at least one of the a_i is strictly greater than the average value S/n and at least one of the a_i is strictly less than the average value.

5.5.13 The notation $n!^{(k)}$ means take the factorial of n k times. For example, $n!^{(3)}$ means $((n!)!)!$. Which is bigger, $1999!^{(2000)}$ or $2000!^{(1999)}$?

5.5.14 Which is bigger, $\dfrac{10^{1999}+1}{10^{2000}+1}$ or $\dfrac{10^{1998}+1}{10^{1999}+1}$?

5.5.15 Which is bigger, 2000! or 1000^{2000}?

5.5.16 Which is larger, 1999^{1999} or 2000^{1998}?

The AM-GM Inequality

Return to Example 5.5.10 on page 191. The alleged inequality

$$\frac{1998}{1999} \quad ? \quad \frac{1999}{2000}$$

is equivalent (after multiplying both sides by $(1999 \cdot 2000)$ to

$$1998 \cdot 2000 \quad ? \quad 1999^2.$$

Your intuition probably tells you that the question mark should be replaced with "$<$," for the left-hand side is the area of a rectangle that is not quite a square, while the right hand side is the area of a square with the same perimeter as the rectangle (namely $4 \cdot 1999$). It makes sense that given a fixed amount of fencing, the rectangle of maximum area is a square. Indeed, you have probably done this problem as an easy calculus exercise. The underlying principle is very simple mathematically—calculus is definitely not needed—yet amazingly fruitful.

Consider the following equivalent formulation. Let x, y be positive real numbers with sum $S = x + y$. Then the maximum value of the product xy is attained when x and y are equal, i.e., when $x = y = S/2$. In other words, we assert that

$$\left(\frac{S}{2}\right)^2 = \frac{(x+y)^2}{4} \geq xy.$$

It is a simple matter to prove this. We will place a question mark over the "$\leq$" to remind ourselves that it is an "alleged" inequality until we reduce it to an equivalent one that we know is true. The algebra is simple stuff that we have seen before:

$$\frac{(x+y)^2}{4} \overset{?}{\geq} xy$$

is equivalent to

$$(x+y)^2 \overset{?}{\geq} 4xy.$$

Thus

$$x^2 + 2xy + y^2 \overset{?}{\geq} 4xy,$$

and subtracting $4xy$ from both sides yields

$$x^2 - 2xy + y^2 \overset{?}{\geq} 0,$$

and now we can remove the question mark, for the left-hand side is the square $(x - y)^2$, hence always non-negative. We have proven the inequality, and moreover, this argument also shows that equality occurs only when $x = y$, for only then will the square equal zero.

This inequality is called the **Arithmetic-Geometric Mean Inequality**, often abbreviated as **AM-GM** or **AGM**, and is usually written in the form

$$\frac{x+y}{2} \geq \sqrt{xy}.$$

Recall that the left-hand side is the arithmetic mean of x and y, while the right-hand side is called the **geometric mean**. A succinct way to remember this inequality is that

> *The arithmetic mean of two positive real numbers is greater than or equal to the geometric mean, with equality only if the numbers are equal.*

We can extract more information from our algebraic proof of AM-GM. Since $(x + y)^2 \geq 4xy$ and $(x + y)^2 - 4xy = (x - y)^2$, we can write

$$S^2 - 4P = D^2,$$

where S, P, D are respectively the sum, product, and difference of x and y. If we let x and y vary so that their sum S is fixed, we see that the product of x and y is a strictly decreasing function of the distance between x and y. (By distance, we mean $|x - y|$.) This is so useful, we shall name it the **symmetry-product principle**:

> *As the distance between two positive numbers decreases, their product increases, provided that their sum stays constant.*

This agrees with our intuition: As a rectangle becomes more "squarish," i.e., more symmetrical, it encloses area more "efficiently."

Here is a nice geometric proof of AM-GM. Let AC be the diameter of a circle, and let B be any point on the circle. Recall that ABC will be a right triangle. Now locate point D so that BD is perpendicular to AC.

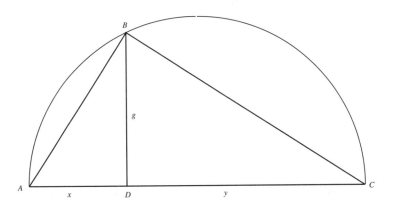

Then triangles ABD and BCD are similar; hence

$$\frac{x}{g} = \frac{g}{y}.$$

Thus $g = \sqrt{xy}$, the geometric mean of x and y. Indeed, that's why it is called a *geometric* mean!

Now, let the point B move along the circle, with D moving as well so that BD stays perpendicular to AC. It is clear that BD is largest when D is at the center of the circle, in which case x and y are equal (to the length of the radius). Moreover, as D moves towards the center, x and y become closer in distance, and BD increases.

The AM-GM inequality is true for any finite number of variables. Let

$$x_1, x_2, \ldots, x_n$$

be positive real numbers, and define the arithmetic mean A_n and geometric mean G_n respectively by

$$A_n := \frac{a_1 + a_2 + \cdots + a_n}{n}, \quad G_n := \sqrt[n]{a_1 a_2 \cdots a_n}.$$

The general version of AM-GM asserts that $A_n \geq G_n$, with equality if and only if $x_1 = x_2 = \cdots = x_n$.

There are many ways to prove this (see Problem 5.5.26 on page 202 for hints about an ingenious induction proof). We shall present a simple argument which uses two strategic ideas: an **algorithmic** proof style, plus a deliberate appeal to physical intuition. We begin by restating AM-GM with its equivalent "sum and product" formulation.[15]

5.5.17 *AM-GM Reformulated.* Let $x_1, x_2, \ldots, x_n$ be positive real numbers with product $P = x_1 x_2 \cdots x_n$ and sum $S = a_1 + a_2 + \cdots + a_n$. Prove that the largest value of P is attained when all the x_i are equal, i.e., when

$$x_1 = x_2 = \cdots = x_n = \frac{S}{n}.$$

Solution: Imagine the n positive numbers $x_1, x_2, \ldots, x_n$ as "physical" points on the number line, each with unit weight. The balancing point (center of mass) of these weights is located at the arithmetic mean value $A := S/n$. Notice that if we move the points around in such a way that they continue to balance at A, that is equivalent to saying that their sum stays constant.

Our strategy, inspired by the symmetry-product principle, is to consider situations where the x_i are not all equal and show that we can make them "more equal" and increase their product *without changing their sum*. If the points are not all clustering at A, then at least one will be to the left of A (call it L) and another (which we call R) will be to the right of A. [16] Of these two points, move the one which is closest to A right up to A, and move the other so that the balancing point of the two points hasn't changed. In the figure below, the arrowpoints indicate the new positions of the points.

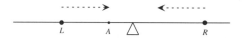

Notice that the distance between the two points has decreased, but their balancing point is unchanged. By the symmetry-product principle, the product of the two points increased. Since the sum of the two points was unchanged, the sum of all n points has not changed. We have managed to change two of the n numbers in such a way that

- one number that originally was not equal to A became equal to A;
- the sum of all n numbers did not change;
- the product of the n numbers increased.

Since there are finitely many numbers, this process will end when all of them are equal to A; then the product will be maximal.   ∎

---

[15]This simple idea of reformulating AM-GM is not very well known. Our treatment here is inspired by Kazarinoff's wonderful monograph [16]; this short book is highly recommended.

[16]Observe that this is a neat "physical" proof of the average principle (5.5.12).

The proof is called "algorithmic" because the argument used describes a concrete procedure which optimizes the product in a step-by-step way, ending after a finite number of steps. Another distinctive feature of this proof was that we altered our point of view and recast an inequality as an *optimization* problem. This is a powerful strategy, well worth remembering.

The AM-GM inequality is the starting point for many interesting inequalities. Here is one example (see the problems for several more).

Example 5.5.18 Let $a_1, a_2, \ldots, a_n$ be a sequence of positive numbers. Show that

$$(a_1 + a_2 + \cdots + a_n)\left(\frac{1}{a_1} + \frac{1}{a_2} + \cdots + \frac{1}{a_n}\right) \geq n^2,$$

with equality holding if and only if the a_i are equal.

Solution: First, make it easier by examining a simpler case. Let's try to prove

$$(a+b)\left(\frac{1}{a} + \frac{1}{b}\right) \overset{?}{\geq} 4.$$

Multiplying out, we get

$$1 + \frac{a}{b} + \frac{b}{a} + 1 \overset{?}{\geq} 4,$$

or

$$\frac{a}{b} + \frac{b}{a} \overset{?}{\geq} 2.$$

This inequality is true because of AM-GM:

$$\frac{a}{b} + \frac{b}{a} \geq 2\sqrt{\left(\frac{a}{b}\right)\left(\frac{b}{a}\right)} = 2.$$

It is worth remembering this result in the following form:

If $x > 0$, then $x + \dfrac{1}{x} \geq 2$, with equality if and only if $x = 1$.

Returning to the general case, we proceed in exactly the same way. When we multiply out the product

$$\sum_{j=1}^{n} a_j \sum_{k=1}^{n} \frac{1}{a_k},$$

we get n^2 terms, namely all the terms of the form

$$\frac{a_j}{a_k}, \quad 1 \leq j, k \leq n.$$

For n of these, $j = k$, and the term equals 1. The remaining $n^2 - n$ terms can be paired up in the form

$$\frac{a_j}{a_k} + \frac{a_k}{a_j}, \quad 1 \leq j < k \leq n.$$

(Note that the expression "$1 \le j < k \le n$" ensures that we get every pair with no duplications.) Applying AM-GM to each of these pairs yields

$$\frac{a_j}{a_k} + \frac{a_k}{a_j} \ge 2\sqrt{\left(\frac{a_j}{a_k}\right)\left(\frac{a_k}{a_j}\right)} = 2.$$

Thus the n^2 terms can be decomposed into n terms that equal 1 and $(n^2 - n)/2$ pairs of terms, with each pair greater than or equal to 2. Consequently, the entire sum is greater than or equal to

$$n \cdot 1 + \frac{n^2 - n}{2} \cdot 2 = n^2.$$ ∎

Massage, Cauchy-Schwarz, and Chebyshev

There are many, many kinds of inequalities, with literally hundreds of different theorems and specialized techniques. We will briefly look at three important "intermediate" ideas: more about massage, the Cauchy-Schwarz inequality, and Chebyshev's inequality.

Perhaps the most important inequality tactic is **massage**, which we encountered earlier (for example, the discussion of harmonic series in Example 5.3.4 on page 176). The philosophy of massage is to "loosen up" an expression in such a way that it *eventually* becomes easier to deal with. This is not restricted to inequalities, of course. Sometimes the first stage of massage seemingly worsens the difficulty. But that is temporary, much like physical massage, which can be rather painful until your muscles magically relax. Here is an instructive example which combines massage with its frequent partner, telescoping.

Example 5.5.19 Let $A = \sum_{n=1}^{10000} \frac{1}{\sqrt{n}}$. Find $\lfloor A \rfloor$ without a calculator.

Solution: In other words, we must estimate A to the nearest integer. We don't need an exact value, so we can massage the terms, perturbing them a bit, so that they telescope. We have

$$\frac{1}{\sqrt{n}} = \frac{2}{2\sqrt{n}} < \frac{2}{\sqrt{n} + \sqrt{n-1}} = 2(\sqrt{n} - \sqrt{n-1}),$$

where we rationalized the denominator in the last step. Likewise,

$$\frac{1}{\sqrt{n}} > \frac{2}{\sqrt{n+1} + \sqrt{n}} = 2(\sqrt{n+1} - \sqrt{n}).$$

In other words,

$$2(\sqrt{n+1} - \sqrt{n}) < \frac{1}{\sqrt{n}} < 2(\sqrt{n} - \sqrt{n-1}).$$

That was the crux move, for we have bounded the term $1/\sqrt{n}$ above and below by terms that will telescope when summed. Indeed, we have

$$\sum_{n=1}^{10000} 2(\sqrt{n+1} - \sqrt{n}) = 2\sqrt{10001} - 2$$

and

$$\sum_{n=1}^{10000} 2(\sqrt{n} - \sqrt{n-1}) = 2\sqrt{10000} - \sqrt{0}.$$

We conclude that

$$2\sqrt{10001} - 2 < \sum_{n=1}^{10000} \frac{1}{\sqrt{n}} < 2\sqrt{10000}.$$

This tells us that $\lfloor A \rfloor$ is either 198 or 199. We can easily refine this estimate, because the original trick of bracketing $1/\sqrt{n}$ between $2(\sqrt{n+1} - \sqrt{n})$ and $2(\sqrt{n} - \sqrt{n-1})$ was pretty crude for *small* values of n. For example, when $n = 1$, we ended up using the "estimate" that

$$2(\sqrt{2} - 1) < 1 < 2\sqrt{1}.$$

The lower limit is not too bad, but the upper limit is a silly overestimate by exactly 1. So let's not use it! Start the summation at $n = 2$ and write

$$A = 1 + \sum_{n=2}^{10000} \frac{1}{\sqrt{n}}.$$

Now we estimate

$$2\sqrt{10001} - 2\sqrt{2} < \sum_{n=1}^{10000} \frac{1}{\sqrt{n}} < 2\sqrt{10000} - 2\sqrt{1}.$$

The lower limit in the above expression is a little bit larger than 197, while the upper limit is 198. Thus we conclude that A is between 198 and 199, so $\lfloor A \rfloor = 198$. ∎

The Cauchy-Schwarz Inequality

Let $a_1, a_2, \ldots, a_n$ and $b_1, b_2, \ldots, b_n$ be sequences of real numbers. The **Cauchy-Schwarz inequality** states that

$$\left(\sum a_i b_i \right)^2 \leq \left(\sum a_i^2 \right) \left(\sum b_i^2 \right),$$

with equality holding if and only if $a_1/b_1 = a_2/b_2 = \cdots = a_n/b_n$. If $n = 1$, this inequality reduces to 2-variable AM-GM. If $n = 3$, the inequality (using friendlier variables) is

$$(ax + by + cz)^2 \leq (a^2 + b^2 + c^2)(x^2 + y^2 + z^2) \tag{7}$$

for all real a, b, c, x, y, z.

5.5.20 Prove (7), by multiplying out efficiently and looking at cross-terms and using 2-variable AM-GM whenever possible. This method generalizes to any value of n.

Another way to prove (7) uses the simple but important tool that

> *A sum of squares of real numbers is non-negative, and equal to zero if and only if all the numbers are zero.*

Thus, for any real a, b, c, x, y, z, we have

$$0 \leq (ay - bx)^2 + (az - cx)^2 + (bz - cy)^2.$$

This is equivalent to

$$2(abxy + acxz + bcyz) \leq a^2 y^2 + b^2 x^2 + a^2 z^2 + c^2 x^2 + b^2 z^2 + c^2 y^2. \qquad (8)$$

Adding

$$(ax)^2 + (by)^2 + (cz)^2$$

to both sides of (8) yields (7). This argument generalizes; it is not just true for $n = 3$.

5.5.21 Convince yourself that this method generalizes, by writing it out for the $n = 4$ case. Your starting point will be a sum of 6 squares.

Even though the Cauchy-Schwarz inequality is a fairly simple consequence of AM-GM, it is a powerful tool, because it has so many "degrees of freedom." For example, if we let $a = b = c = 1$ in (7), we get the appealing inequality

$$\frac{(x + y + z)^2}{3} \leq x^2 + y^2 + z^2.$$

We can derive another useful inequality from Cauchy-Schwarz if the variables are positive. For example, if $a, b, c, x, y, z > 0$, then

$$(\sqrt{ax} + \sqrt{by} + \sqrt{cz})^2 \leq (a + b + c)(x + y + z) \qquad (9)$$

is a simple consequence of (7)—just replace a by $\sqrt{a}$, etc. This inequality (which of course generalizes to any n) comes in quite handy sometimes, because it is a surprising way to find the lower bound of the product of two often unrelated sums.

Here is a more interesting example that uses (9).

Example 5.5.22 (Titu Andreescu) Let P be a polynomial with positive coefficients. Prove that if

$$P\left(\frac{1}{x}\right) \geq \frac{1}{P(x)}$$

holds for $x = 1$ then it holds for every $x > 0$.

Solution: Write $P(x) = u_0 + a_1 x + u_2 x^2 + \cdots + u_n x^n$. When $x = 1$, the inequality is just $P(1) \geq 1/P(1)$, or

$$(u_0 + u_1 + u_2 + \cdots + u_n)^2 \geq 1,$$

which reduces to

$$u_0 + u_1 + u_2 + \cdots + u_n \geq 1$$

since the coefficients are positive. We wish to show that

$$\left(u_0 + \frac{u_1}{x} + \cdots + \frac{u_n}{x^n}\right)\left(u_0 + a_1 x + \cdots + u_n x^n\right) \geq 1$$

for all positive x. Since x and the u_i are positive, we can define the real sequences

$$a_0 = \sqrt{u_0}, a_1 = \sqrt{u_1/x}, a_2 = \sqrt{u_2/x^2}, \ldots, a_n = \sqrt{u_n/x^n}$$

and

$$b_0 = \sqrt{u_0}, b_1 = \sqrt{u_1 x}, b_2 = \sqrt{u_2 x^2}, \ldots, b_n = \sqrt{u_n x^n}.$$

Notice that

$$a_i^2 = u_i/x^i, \quad b_i^2 = u_i x^i, \quad a_i b_i = u_i$$

for each i. Hence when we apply the Cauchy-Schwarz inequality to the a_i and b_i sequences, we get

$$u_0 + u_1 + u_2 + \cdots + u_n \leq P\left(\frac{1}{x}\right) P(x).$$

But $u_0 + u_1 + u_2 + \cdots + u_n \geq 1$, so we conclude that

$$P\left(\frac{1}{x}\right) P(x) \geq 1. \qquad \blacksquare$$

Here is another example, a solution to the IMO problem started with Example 5.2.23 on page 170.

Example 5.5.23 (IMO 1995) Let a, b, c be positive real numbers such that $abc = 1$. Prove that

$$\frac{1}{a^3(b+c)} + \frac{1}{b^3(c+a)} + \frac{1}{c^3(a+b)} \geq \frac{3}{2}.$$

Solution: Recall that the substitutions $x = 1/a, y = 1/b, z = 1/c$ transform the original problem into showing that

$$\frac{x^2}{y+z} + \frac{y^2}{z+x} + \frac{z^2}{x+y} \geq \frac{3}{2}, \tag{10}$$

where $xyz = 1$.

Denote the left-hand side of (10) by S. Notice that

$$S = \left(\frac{x}{\sqrt{y+z}}\right)^2 + \left(\frac{y}{\sqrt{z+x}}\right)^2 + \left(\frac{z}{\sqrt{x+y}}\right)^2,$$

and thus Cauchy-Schwarz implies that

$$S(u^2 + v^2 + w^2) \geq \left(\frac{xu}{\sqrt{y+z}} + \frac{yv}{\sqrt{z+x}} + \frac{zw}{\sqrt{x+y}} \right)^2, \qquad (11)$$

for *any* choice of u, v, w. Is there a helpful choice?

Certainly $u = \sqrt{y+z}$, $v = \sqrt{z+x}$, $w = \sqrt{x+y}$ is a natural choice to try, since this immediately simplifies the right-hand side of (11) to just $(x + y + z)^2$. But better yet, with this choice we also get

$$u^2 + v^2 + w^2 = 2(x + y + z).$$

Thus (11) reduces to

$$2S(x + y + z) \geq (x + y + z)^2,$$

which in turn is equivalent to

$$2S \geq (x + y + z).$$

But by AM-GM, we have

$$x + y + z \geq 3\sqrt[3]{xyz} = 3,$$

since $xyz = 1$. We conclude that $2S \geq 3$, and we can rest. ∎

Chebyshev's Inequality

Let $a_1, a_2, \ldots, a_n$ and $b_1, b_2, \ldots, b_n$ be sequences of real numbers which are monotonic in the same direction. In other words, we have $a_1 \leq a_2 \leq \cdots \leq a_n$ and $b_1 \leq b_2 \leq \cdots \leq b_n$ (or we could reverse all the inequalities). **Chebyshev's inequality** states that

$$\frac{1}{n} \sum a_i b_i \geq \left(\frac{1}{n} \sum a_i \right) \left(\frac{1}{n} \sum b_i \right).$$

In other words, if you order two sequences, then the average of the products of corresponding elements is at least as big as the product of the averages of the two sequences.

Let us try to prove Chebyshev's inequality, by looking at a few simple cases. If $n = 2$, we have (using nicer variables) the alleged inequality

$$2(ax + by) \geq (a + b)(x + y).$$

This is equivalent to

$$ax + by \geq ay + bx,$$

which is equivalent to

$$(a - b)(x - y) \geq 0,$$

which is true, since the sequences are ordered (thus $a - b$ and $x - y$ will be the same sign and their product is non-negative).

If $n = 3$, we are faced with verifying the truth of

$$3(ax + by + cz) \geq (a + b + c)(x + y + z),$$

where $a \leq b \leq c$ and $x \leq y \leq z$. Inspired by the previous case, we subtract the right-hand side from the left- hand side and do some factoring in order to show that the difference is positive. When we subtract, three terms cancel and we are left with the 12 terms

$$ax - ay + ax - az + by - bx + by - bz + cz - cx + cz - cy.$$

Rearrange these into the sum

$$(ax + by) - (ay + bx) + (ax + cz) - (ax + cx) + (by + cz) - (bz + cy),$$

which equals

$$(a - b)(x - y) + (a - c)(x - z) + (b - c)(y - z),$$

and this is positive. If you ponder this argument, you will realize that it generalizes to any value of n.

Chebyshev's inequality is much less popular than either AM-GM or Cauchy-Schwarz, but it deserves a place on your toolbelt. In particular, the argument that we used to prove it can be adapted to many other situations.

Problems and Exercises

5.5.24 Show that

$$1 + \frac{1}{1!} + \frac{1}{2!} + \frac{1}{3!} + \cdots < 3.$$

(You have probably learned in a calculus course that this sum is equal to $e \approx 2.718$. But that's cheating! Use "low-tech" methods.)

5.5.25 Reread the argument about the monotonicity of the function in Example 5.5.10 on page 191. How does this relate to the symmetry-product principle?

5.5.26 Rediscover Cauchy's ingenious proof of the AM-GM inequality for n variables with the following hints.

(a) First use induction to prove that AM-GM is true as long as the number of variables is a power of 2.

(b) Next, suppose that you knew that AM-GM was true for 4 variables, and you wanted to prove that it was true for 3 variables. In other words, you wish to prove that

$$\frac{x + y + z}{3} \geq \sqrt[3]{xyz} \tag{12}$$

for all positive x, y, z, and you are allowed to use the fact that

$$\frac{a + b + c + d}{4} \geq \sqrt[4]{abcd} \tag{13}$$

for all positive a, b, c, d. Can you think of some clever things to substitute for a, b, c, d in (13) which will transform it into (12).

(c) Generalize this method to devise a way to go *backward* from the 2^r-variable AM-GM case down to AM-GM with any smaller number of variables (down to 2^{r-1}).

5.5.27 For which integer n is $1/n$ closest to

$$\sqrt{1,000,000} - \sqrt{999,999}?$$

No calculators, please!

5.5.28 Prove that $n! < \left(\dfrac{n+1}{2}\right)^n$, for $n = 2, 3, 4, \ldots$.

5.5.29 (IMO 1976) Determine, with proof, the largest number which is the product of positive integers whose sum is 1976. Hint: try an "algorithmic" approach.

5.5.30 Example 3.1.12 on page 78 demonstrated the factorization

$$a^3 + b^3 + c^3 - 3abc = (a+b+c)(a^2 + b^2 + c^2 - ab - bc - ac).$$

Use this to prove the AM-GM inequality for 3 variables. Does this method generalize?

5.5.31 Show that

$$\frac{1}{\sqrt{4n}} \leq \left(\frac{1}{2}\right)\left(\frac{3}{4}\right)\cdots\left(\frac{2n-1}{2n}\right) < \frac{1}{\sqrt{2n}}.$$

5.5.32 Let $a_1, a_2, \ldots, a_n$ be a sequence of positive numbers. Show that for all positive x,

$$(x+a_1)(x+a_2)\cdots(x+a_n) \leq \left(x + \frac{a_1 + a_2 + \cdots + a_n}{n}\right)^n.$$

5.5.33 Find all ordered pairs of positive real numbers (x, y) such that $x^y = y^x$. Notice that the set of pairs of the form (t, t) where t is any positive number is *not* the full solution, since $2^4 = 4^2$.

5.5.34 Show that

$$\sqrt{a_1^2 + b_1^2} + \sqrt{a_2^2 + b_2^2} + \cdots + \sqrt{a_n^2 + b_n^2} \geq$$
$$\sqrt{(a_1 + a_2 + \cdots + a_n)^2 + (b_1 + b_2 + \cdots + b_n)^2}$$

for all real values of the variables, and give a condition for equality to hold. Algebraic methods will certainly work, but there must be a better way ...

5.5.35 If you have studied vector dot-product, you should be able to give a geometric interpretation of the Cauchy-Schwarz inequality. Think about magnitudes, cosines, etc.

5.5.36 Here's another way to prove Cauchy-Schwarz, which employs several useful ideas: Define

$$f(t) := (a_1 t + b_1)^2 + (a_2 t + b_2)^2 + \cdots + (a_n t + b_n)^2.$$

Observe that f is a quadratic polynomial in t. It is possible that $f(t)$ has zeros, but only if $a_1/b_1 = a_2/b_2 = \cdots = a_n/b_n$. Otherwise, $f(t)$ is strictly positive. Now use the quadratic formula, and look at the discriminant. It must be negative; why? Show that the negativity of the discriminant implies Cauchy-Schwarz.

5.5.37 Give a quick Cauchy-Schwarz proof of the inequality in Example 5.5.18.

5.5.38 Let $a_1, a_2, \ldots, a_n$ be positive, with a sum of 1. Show that $\sum_{i=1}^{n} a_i^2 \geq 1/n$.

5.5.39 If $a, b, c > 0$, prove that

$$(a^2b + b^2c + c^2a)(ab^2 + bc^2 + ca^2) \geq 9a^2b^2c^2.$$

5.5.40 Let $a, b, c \geq 0$. Prove that

$$\sqrt{3(a+b+c)} \geq \sqrt{a} + \sqrt{b} + \sqrt{c}.$$

5.5.41 Let $a, b, c, d \geq 0$. Prove that

$$\frac{1}{a} + \frac{1}{b} + \frac{4}{c} + \frac{16}{d} \geq \frac{64}{a+b+c+d}.$$

5.5.42 (USAMO 1983) Prove that the zeros of

$$x^5 + ax^4 + bx^3 + cx^2 + dx + e = 0$$

cannot *all* be real if $2a^2 < 5b$.

5.5.43 Let $x, y, z > 0$ with $xyz = 1$. Prove that

$$x + y + z \leq x^2 + y^2 + z^2.$$

5.5.44 Let $x, y, z \geq 0$ with $xyz = 1$. Find the minimum value of

$$\frac{x}{y+z} + \frac{y}{x+z} + \frac{z}{x+y}.$$

5.5.45 (IMO 1984) Prove that

$$0 \leq yz + zx + xy - 2xyz \leq 7/27,$$

where x, y and z are non-negative real numbers for which $x + y + z = 1$.

5.5.46 (Putnam 1968) Determine all polynomials that have only real roots and all coefficients are equal to ± 1.

5.5.47 Let $a_1, a_2, \ldots, a_n$ be a sequence of positive numbers, and let $b_1, b_2, \ldots, b_n$ be any permutation of the first sequence. Show that

$$\frac{a_1}{b_1} + \frac{a_2}{b_2} + \cdots + \frac{a_n}{b_n} \geq n.$$

5.5.48 Let $a_1, a_2, \ldots, a_n$ and $b_1, b_2, \ldots, b_n$ be increasing sequences of real numbers and let $x_1, x_2, \ldots, x_n$ be any permutation of $b_1, b_2, \ldots, b_n$. Show that

$$\sum a_i b_i \geq \sum a_i x_i.$$

Chapter 6

Combinatorics

6.1 Introduction to Counting

Combinatorics is the study of counting. That sounds rather babyish, but in fact counting problems can be quite deep and interesting and have many connections to other branches of mathematics. For example, consider the following problem.

Example 6.1.1 (Czech and Slovak, 1995) Decide whether there exist 10,000 ten-digit numbers divisible by seven, all of which can be obtained from one another by a reordering of their digits.

On the surface, it looks like a number theory problem. But it is actually just a question of carefully counting the correct things. We will solve this problem soon, on page 221, but first we need to develop some basic skills.

Our first goal is a good understanding of the ideas leading up to **binomial theorem**. We assume that you have studied this subject a little bit before, but intend to review it and expand upon it now. Many of the concepts will be presented as a sequence of statements for you to verify before moving on. Please do not rush; make sure that you really understand each statement! In particular, pay attention to the tiniest of arithmetical details: good combinatorial reasoning is largely a matter of knowing exactly when to add, multiply, subtract, or divide.

Permutations and Combinations

Items 6.1.2–6.1.12 introduce the concepts of permutations and combinations, and use only addition, multiplication, and division.

6.1.2 *Simple Addition.* If there are a varieties of soup and b varieties of salad, then there are $a + b$ possible ways to order a meal of soup *or* salad (but not both soup and salad).

6.1.3 *Simple Multiplication.* If there are a varieties of soup and b varieties of salad, then there are ab possible ways to order a meal of soup *and* salad.

6.1.4 Let A and B be finite sets that are disjoint ($A \cap B = \emptyset$). Then 6.1.2 is equivalent to the statement

$$|A \cup B| = |A| + |B|.$$

6.1.5 Notice that 6.1.3 is equivalent to the statement

$$|A \times B| = |A| \cdot |B|,$$

for any two finite sets A and B (not necessarily disjoint).

6.1.6 A **permutation** of a collection of objects is a reordering of them. For example, there are 6 different permutations of the letters ABC, namely ABC, ACB, BAC, BCA, CAB, CBA. There are $5 \cdot 4 \cdot 3 \cdot 2 \cdot 1 = 120$ different ways to permute the letters in "HARDY." You probably know that $n \cdot (n-1) \cdots 1$ is denoted by the symbol $n!$, called "n factorial." You shouldlearn (at least passively) the values of $n!$ for $n \leq 10$. (While you are at it, learn the first dozen or so of all the common sequences: squares, cubes, Fibonacci, etc.)

6.1.7 *Permutations of n things taken r at a time.* The number of different 3-letter words we could make using the 9 letters in "CHERNOBYL" is $9 \cdot 8 \cdot 7 = 504$. In general, the number of distinct ways of permuting r things chosen from n things is

$$n \cdot (n-1) \cdots (n-r+1).^{1}$$

This product is also equal to $\dfrac{n!}{(n-r)!}$ and is denoted by $P(n, r)$.

6.1.8 But what about permuting the letters in "GAUSS"? At first you might think the answer is $6!$, but it is actually $\frac{6!}{2}$. Why?

6.1.9 Furthermore, explain that the number of different permutations of "PARADOX-ICAL" is $\frac{11!}{6}$, not $\frac{11!}{3}$. Why?

6.1.10 Likewise, verify that "RAMANUJAN" has $\frac{9!}{6 \cdot 2}$ different permutations. State a general formula. Check with small numbers to make sure that your formula works.

We shall call the formula you got in problem 6.1.10 the **Mississippi formula**, since an amusing example of it is computing the number of permutations in "MISSISSIPPI," which is, of course, $\frac{11!}{4!4!2!}$. The Mississippi formula is easy to remember by doing examples, but let's write it out formally. This is a good exercise in notation, which also helps clarify just what we are counting. Here goes:

> *We are given a collection of balls that are indistinguishable except for color. If there are a_i balls of color i, for $i = 1, 2, \ldots, n$, then the number of different ways that these balls can be arranged in a row is*
>
> $$\frac{(a_1 + a_2 + \cdots + a_n)!}{a_1! a_2! \cdots a_n!}.$$

---

[1] Notice that the product ends with $(n-r+1)$ and not $(n-r)$. This is a frequent source of minor errors, known as "OBOB," the "Off-by-one Bug."

The Mississippi formula involves both multiplication and division. Let's examine the role of each operation carefully, with the example of "PARADOXICAL." If all of the letters were different, then there would be 9 possible choices for the first letter, 8 for the second, 7 for the third, etc., yielding a total of

$$9 \times 8 \times 7 \times \cdots \times 1 = 9!$$

different possible permutations. We multiplied the numbers because we were counting a *joint* event composed of 9 sub-events, respectively, with $9, 8, 7, \ldots, 1$ choices. But the letters are *not* all different; there are 3 indistinguishable A's. Hence the value 9! has *overcounted* what we want. Pretend for a moment that we can distinguish the A's by labeling them A_1, A_2, A_3. Then, for example, we would count $CA_1LPORA_2A_3XID$ and $CA_3LPORA_1A_2XID$ as two different words, when they are really indistinguishable. We want to count the word CALPORAAXID just once, but we will count it $3! = 6$ times since there are 3! ways of permuting those 3 A's. In other words, we are *uniformly* overcounting by a factor of 3!, and can rectify this by dividing by 3!. That is why the correct answer is $\frac{11!}{3!}$. In general,

> *To count the number of ways a joint event occurs, multiply together the number of choices for each sub-event. To rectify **uniform** overcounting, divide by the overcounting factor.*

6.1.11 Let's apply the Mississippi formula to an n-letter string consisting of r 0's and $(n - r)$ 1's (also known as an **n-bit string**). Verify that the number of permutations would be

$$\frac{n!}{r!(n - r)!}.$$

This is often denoted by $\binom{n}{r}$ ("n choose r") and is also called a **binomial coefficient**. For example, the number of distinct permutations of "0000111" is $\binom{7}{3} = \frac{7!}{3!4!} = 35$.

6.1.12 *Combinations of n things taken r at a time.* Here is a seemingly different problem: We are ordering a pizza, and there are 8 different toppings available (anchovies, garlic, pineapple, sausage, pepperoni, mushroom, olive, and green pepper). We would like to know how many different pizzas can be ordered with exactly 3 toppings. In contrast to permutations, *the order that we choose the toppings does not matter.* For example, a "sausage-mushroom- garlic" pizza is indistinguishable from a "mushroom-garlic-sausage" pizza.

To handle this difficulty, we proceed as we did with the Mississippi problem. If the order did matter, then the number of different pizzas would be the simple permutation $P(8, 3)$, but then we are uniformly overcounting by a factor of 3!. So the correct answer is

$$\frac{P(8, 3)}{3!}.$$

Notice that

$$\frac{P(8, 3)}{3!} = \frac{8!}{5!3!} = \binom{8}{3}.$$

In general, the number of ways you can select a subset of r distinct elements from a set of n distinct elements, *where the order of selection doesn't matter*, is

$$\frac{P(n, r)}{r!} = \binom{n}{r}.$$

This is called a **combination**. If the order does matter, then the number of ways is $P(n, r)$ and it is called a **permutation**.

For example, the number of different ways to pick a 3-person committee out of a 30-person class is $\binom{30}{3}$. However, if the committee members have specific roles, such as president, vice-president, and secretary, then there are $P(30, 3)$ different committees (since the committee where Joe is president, Karen is VP, Tina is secretary is not the same as the committee where Joe is secretary, Tina is VP, Karen is president).

Incidentally, it makes sense to define binomial coefficients involving the number zero. For example, $\binom{10}{0} = 1$, because there is only one way to choose a committee with no people. This interpretation will be consistent with the formula if we define $0!$ to equal 1.

The interpretation of the binomial coefficient as a permutation divided by a factorial leads to a nice computational shortcut. Notice that it is much easier to compute $\binom{11}{4}$ as

$$\frac{11 \cdot 10 \cdot 9 \cdot 8}{1 \cdot 2 \cdot 3 \cdot 4} = 11 \cdot 5 \cdot 3 \cdot 2 = 330$$

than it would be to compute $\frac{11!}{4!7!}$.

Combinatorial Arguments

Keeping a flexible point of view is a powerful strategy. This is especially true with counting problems where often the crux move is to count the same thing in two different ways. To help develop this flexibility, you should practice creating "combinatorial arguments." This is just fancy language for a story that rigorously describes *in English* how you count something. Here are a few examples that illustrate the method. First we state something algebraically, then the combinatorial story that corresponds to it. Pay attention to the building blocks of "algebra to English" translation, and in particular, make sure you understand when and why multiplication rather than addition happens, and vice versa.

7×6	If there are 7 choices of pasta and 6 choices of sauces, there are 7×6 ways of choosing a pasta dinner with a sauce.
$12 + 6 + 5$	If the things you are counting fall into three mutually exclusive classes, with 12 in the first, 6 in the second, 5 in the third class, then the total number of things you are counting is $12 + 6 + 5$.
7^5	If you are eating a 5-course dinner, with 7 choices per course, you can order 7^5 different dinners.

$\binom{10}{4}$	The number of ways of choosing a team of 4 people out of a room of 10 people (where the order that we pick the people does not matter).
$P(10, 4)$	The number of ways of choosing a team of 4 people out a room of 10 people, where the order *does* matter (for example, we choose the 4 people, and then designate a team captain, co-captain, mascot, and manager).
$5 \cdot \binom{13}{5}$	The number of ways you can choose a team of 5 people chosen from a room of 13 people, where you are designating the captain.
$\binom{17}{8}\binom{10}{2}$	The number of different teams of 8 girls and 2 boys, chosen from a pool of 17 girls and 10 boys.
$\binom{17}{4} + \binom{17}{3}$	The number of ways of picking a team of 3 *or* 4 people, chosen from a room of 17 people.
$\binom{17}{10} = \binom{17}{7}$	Each selection of 10 winners from a group of 17 is simultaneously a selection of 7 losers from this group.

6.1.13 *The Symmetry Identity.* Generalizing the last example, observe that for all integers n, r with $n \geq r \geq 0$, we have

$$\binom{n}{r} = \binom{n}{n-r}.$$

This can also be verified with algebra, using the formula from 6.1.11, but the combinatorial argument used above is much better. The combinatorial argument shows us *why* it is true, while algebra merely shows us *how* it is true.

Pascal's Triangle and the Binomial Theorem

6.1.14 *The Summation Identity.* Here is a more sophisticated identity with binomial coefficients: For all integers n, r with $n \geq r \geq 0$,

$$\binom{n}{r} + \binom{n}{r+1} = \binom{n+1}{r+1}.$$

Algebra can easily verify this, but consider the following combinatorial argument: Without loss of generality, let $n = 17, r = 10$. Then we need to show *why*

$$\binom{17}{10} + \binom{17}{11} = \binom{18}{11}.$$

Let us count all 11-member committees formed from a group of 18 people. Fix 1 of the 18 people, say, "Erika." The 11-member committees can be broken down into two mutually exclusive types: those with Erika, and those without. How many include Erika? Having already chosen Erika, we are free to chose 10 more people from the remaining pool of 17. Hence there are $\binom{17}{10}$ committees that include Erika. To count the committees without Erika, we must choose 11 people, but again out of 17, since we need to remove Erika from the original pool of 18. Thus $\binom{17}{11}$ committees exclude Erika. The total number of 11-member committees is the sum of the number of committees with Erika plus the number without Erika, which establishes the equality. The argument certainly works if we replace 17 and 10 with n and r (but it is easier to follow the reasoning with concrete numbers). ■

6.1.15 Recall **Pascal's Triangle**, which you first encountered in Problem 1.3.17 on page 12. Here are the first few rows. The elements of each row are the sums of pairs of adjacent elements of the prior row (for example, $10 = 4 + 6$).

$$
\begin{array}{ccccccccccc}
 & & & & & 1 & & & & & \\
 & & & & 1 & & 1 & & & & \\
 & & & 1 & & 2 & & 1 & & & \\
 & & 1 & & 3 & & 3 & & 1 & & \\
 & 1 & & 4 & & 6 & & 4 & & 1 & \\
1 & & 5 & & 10 & & 10 & & 5 & & 1 \\
\end{array}
$$

Pascal's Triangle contains all of the binomial coefficients: Label the rows and columns, starting with zero, so that, for example, the element in row 5, column 2 is 10. In general, the element in row n, column r will be equal to $\binom{n}{r}$. This is a consequence of the summation identity (6.1.14) and the fact that

$$\binom{n}{0} = \binom{n}{n} = 1$$

for all n. Ponder this carefully. It is very important.

6.1.16 When we expand $(x+y)^n$, for $n = 0, 1, 2, 3, 4, 5$, we get

$$
\begin{aligned}
(x+y)^0 &= 1, \\
(x+y)^1 &= x + y, \\
(x+y)^2 &= x^2 + 2xy + y^2, \\
(x+y)^3 &= x^3 + 3x^2y + 3xy^2 + y^3, \\
(x+y)^4 &= x^4 + 4x^3y + 6x^2y^2 + 4xy^3 + y^4, \\
(x+y)^5 &= x^5 + 5x^4y + 10x^3y^2 + 10x^2y^3 + 5xy^4 + y^5.
\end{aligned}
$$

Certainly it is no coincidence that the coefficients are exactly the elements of Pascal's Triangle. Indeed, in general it is true that the coefficient of $x^r y^{n-r}$ in $(x+y)^n$ is equal to $\binom{n}{r}$. You should be able to explain why by thinking about what happens when you multiply $(x+y)^k$ by $(x+y)$ to get $(x+y)^{k+1}$. You should see the summation identity in action and essentially come up with an induction proof.

6.1.17 *The Binomial Theorem.* Formally, the **binomial theorem** states that, for all positive integers n,

$$(x+y)^n = \sum_{r=0}^{n} \binom{n}{r} x^{n-r} y^r.$$

Expanding the sum out gives the easier-to-read formula

$$(x+y)^n = \binom{n}{0}x^n + \binom{n}{1}x^{n-1}y + \binom{n}{2}x^{n-2}y^2 + \cdots + \binom{n}{n-1}xy^{n-1} + \binom{n}{n}y^n.$$

(Finally we see why $\binom{n}{r}$ is called a binomial coefficient!)

6.1.18 *A Combinatorial Proof of the Binomial Theorem.* We derived the binomial theorem above by observing that the coefficients in the multiplication of the polynomial $(x+y)^k$ by $(x+y)$ obeyed the summation formula. Here is a more direct "combinatorial" approach, one where we think about how multiplication takes place in order to understand *why* the coefficients are what they are. Consider the expansion of, say,

$$(x+y)^7 = \underbrace{(x+y)(x+y)\cdots(x+y)}_{7 \text{ factors}}.$$

When we begin to multiply all this out, we perform "FOIL" with the first two factors, getting (before performing any simplifications)

$$(x^2 + yx + xy + y^2)(x+y)^5.$$

To perform the next step, we multiply each term of the first factor by x, and then multiply each by y, and then add them up, and then all that would now be multiplied by $(x+y)^4$. In other words, we get (without simplifying)

$$(x^3 + yx^2 + xyx + y^2x + x^2y + yxy + xy^2 + y^3)(x+y)^4.$$

Notice how we can read off the "history" of each term in the first factor. For example, the term xy^2 came from multiplying x by y and then y again in the product $(x+y)(x+y)(x+y)$. There are a total of $2 \times 2 \times 2 = 8$ terms, since there are 2 "choices" for each of the 3 factors. Certainly this phenomenon will continue as we multiply out all 7 factors and we will end up with a total of 2^7 terms.

Now let us think about combining like terms. For example, what will be the coefficient of $x^3 y^4$? This is equivalent to determining how many of the 2^7 unsimplified terms contained 3 x's and 4 y's. As we start to list these terms,

$$xxxyyyy, xxyxyyy, xxyyxyy, \ldots$$

we realize that counting them is just the "Mississippi" problem of counting the permutations of the word "XXXYYYY." The answer is $\frac{7!}{3!4!}$, and this is also equal to $\binom{7}{3}$.

6.1.19 Ponder 6.1.18 carefully, and come up with a general argument. Also, work out the complete multiplications for $(x+y)^n$ for n up to 10. If you have access to a computer, try to print out Pascal's Triangle for as many rows as possible. Whatever you do, become very comfortable with Pascal's Triangle and the binomial theorem.

Strategies and Tactics of Counting

When it comes to strategy, combinatorial problems are no different from other mathematical problems. The basic principles of wishful thinking, penultimate step, make it easier, etc. are all helpful investigative aids. In particular, careful experimentation with small numbers is often a crucial step. For example, many problems succumb to a three-step attack: experimentation, conjecture, proof by induction. The strategy of recasting is especially fruitful: to counteract the inherent dryness of counting, it helps to creatively *visualize* problems (for example, devise interesting "combinatorial arguments") and look for hidden symmetries. Many interesting counting problems involve very imaginative multiple viewpoints, as you will see below.

But mostly, combinatorics is a tactical game. You have already learned the fundamental tactics of multiplication, division, addition, permutations, and combinations. In the sections below, we will elaborate on these and develop more sophisticated tactics and tools.

Problems and Exercises

6.1.20 Find a combinatorial explanation for the following facts or identities.

(a) $\dbinom{2n}{2} = 2\dbinom{n}{2} + n^2$.

(b) $\dbinom{2n+2}{n+1} = \dbinom{2n}{n+1} + 2\dbinom{2n}{n}. + \dbinom{2n}{n-1}$

6.1.21 Define $d(n)$ to be the number of divisors of a positive integer n (including 1 and n).

(a) Show that if

$$n = p_1^{e_1} p_2^{e_2} \cdots p_t^{e_t}$$

is the prime factorization of n, then

$$d(n) = (e_1 + 1)(e_2 + 1) \cdots (e_t + 1).$$

For example, $360 = 2^3 3^2 5^1$ has $(3+1)(2+1)(1+1) = 24$ distinct divisors.

(b) Complete the solution of the locker problem (problem 2.2.3), which we began on page 32.

6.1.22 Use the binomial theorem and the algebraic tactic of substituting convenient values to prove the following identities (for all positive integers n):

(a) $\dbinom{n}{0} + \dbinom{n}{1} + \dbinom{n}{2} + \cdots + \dbinom{n}{n} = 2^n$.

(b) $\dbinom{n}{0} - \dbinom{n}{1} + \dbinom{n}{2} - \cdots + (-1)^n \dbinom{n}{n} = 0$.

6.1.23 Show that the total number of subsets of a set with n elements is 2^n. Include the set itself and the empty set.

6.1.24 Prove the identities in 6.1.22 again, but this time using combinatorial arguments.

6.1.25 (Russia 1996) Which are there more of among the natural numbers between 1 and 1,000,000: numbers that can be represented as a sum of a perfect square and a (positive) perfect cube, or numbers that cannot be?

6.1.26 (AIME 1996) Two of the squares of a 7×7 checkerboard are painted yellow, and the rest are painted green. Two color schemes are equivalent if one can be obtained from the other by applying a rotation in the plane of the board. How many inequivalent color schemes are possible?

6.1.27 If you understood the binomial theorem, you should have no trouble coming up with a **multinomial** theorem. As a warm-up, expand $(x + y + z)^2$ and $(x + y + z)^3$. Think about what make the coefficients what they are. Then come up with a general formula for $(x_1 + x_2 + \cdots + x_n)^r$.

6.2 Partitions and Bijections

We stated earlier that combinatorial reasoning is largely a matter of knowing exactly when to add, multiply, subtract, or divide. We shall now look at two tactics, often used in tandem. **Partitioning** is a tactic that deliberately focuses our attention on addition, by breaking a complex problem into several smaller and simpler pieces. In contrast, the **encoding** tactic attempts to count something in one step, by first producing a **bijection** (a fancy term for a 1-1 correspondence) between each thing we want to count and the individual "words" in a simple "code." The theory behind these tactics is quite simple, but mastery of them requires practice and familiarity with a number of classic examples.

Counting Subsets

A **partition** of a set S is a division of S into a union of *mutually exclusive* (pairwise disjoint) sets. We write

$$S = A_1 \cup A_2 \cup \cdots \cup A_r, \quad A_i \cap A_j = \emptyset, i \neq j.$$

Another notation that is sometimes used is the symbol $\sqcup$ to indicate "union of pairwise disjoint sets." Thus we could write

$$S = A_1 \sqcup A_2 \sqcup \cdots \sqcup A_r = \bigsqcup_{i=1}^{r} A_i$$

to indicate that the set S has been partitioned by the A_i.

Recall that $|S|$ denotes the cardinality (number of elements) of the set S. Obviously if S has been partitioned by the A_i, we must have

$$|S| = |A_1| + |A_2| + \cdots + |A_r|,$$

since there is no overlapping.

This leads to a natural combinatorial tactic: Divide the thing that we want to count into mutually exclusive and easy-to-count pieces. We call this tactic **partitioning**. For example, let us apply this method to problem 6.1.23, which asked us to prove that a set of n elements has 2^n subsets. Denote the set of subsets of the set S by sub(S). For example, if $S = \{a, b, c\}$, then

$$\text{sub}(S) = \{\emptyset, \{a\}, \{b\}, \{c\}, \{a, b\}, \{a, c\}, \{b, c\}, \{a, b, c\}\}.$$

Let $\mathcal{E}_k$ denote the collection of subsets of S which have k elements.[2] The $\mathcal{E}_i$ naturally partition sub(S), since they are mutually disjoint. In other words,

$$\text{sub}(S) = \mathcal{E}_0 \sqcup \mathcal{E}_1 \sqcup \mathcal{E}_2 \sqcup \mathcal{E}_3.$$

The cardinality of $\mathcal{E}_i$ is just $\binom{3}{i}$, since counting the number of i-element subsets is exactly the same as counting the number of ways we can choose i elements from the original set of 3 elements. This implies that

$$|\text{sub}(S)| = |\mathcal{E}_0| + |\mathcal{E}_1| + |\mathcal{E}_2| + |\mathcal{E}_3| = \binom{3}{0} + \binom{3}{1} + \binom{3}{2} + \binom{3}{3}.$$

In general, then, if $|S| = n$, the number of subsets of S must be

$$\binom{n}{0} + \binom{n}{1} + \binom{n}{2} + \cdots + \binom{n}{n}.$$

Now we must prove that this sum is equal to 2^n. This can be done by induction, but we will try another approach, the **encoding** tactic. Instead of partitioning the collection of subsets into many classes, look at this collection as a whole and encode each of its elements (which are subsets) as a string of symbols. Imagine storing information in a computer. How can you indicate a particular subset of $S = \{a, b, c\}$? There are many possibilities, but what we want is a uniform coding method that is simple to describe and works essentially the same way for all cases. That way it will be easy to count. For example, any subset of S is uniquely determined by the answers to the following yes/no questions.

- Does the subset include a?
- Does the subset include b?
- Does the subset include c?

We can encode the answers to these questions by a three-letter string which uses only the letters y and n. For example, the string yyn would indicate the subset $\{a, b\}$. Likewise, the string nnn indicates $\emptyset$ and yyy indicates the entire set S. Thus

There is a bijection between strings and subsets.

---

[2]We often use the word "collection" or "class" for sets of sets, and conventionally denote them with script letters.

In other words, to every three-letter string there corresponds exactly one subset, and to every subset there corresponds exactly one string. And it is easy to count the number of strings; two choices for each letter and three letters per string means 2^3 different strings in all.

This method certainly generalizes to sets with n elements, so we have proven that the number of subsets of a set with n elements is 2^n. Combining this with our previous partitioning argument, we have the combinatorial identity

$$|\text{sub}(S)| = \binom{n}{0} + \binom{n}{1} + \binom{n}{2} + \cdots + \binom{n}{n} = 2^n.$$

Let us look at a few more examples which explore the interplay between encoding, partitioning, and the use of combinatorial argument.

Example 6.2.1 Prove that for all positive integers n,

$$1 \cdot \binom{n}{1} + 2 \cdot \binom{n}{2} + \cdots + n \cdot \binom{n}{n} = n2^{n-1}. \tag{1}$$

Solution: One approach, of course, is to ignore the combinatorial aspect of the identity and attack it algebraically. We must compute $\sum_r r\binom{n}{r}$. Since we understand the simpler sum $\sum_r \binom{n}{r}$, it might be profitable to try to remove the difficult r factor from each term of (1):

$$\sum_{r=1}^{n} r \binom{n}{r} = \sum_{r=1}^{n} \frac{rn!}{r!(n-r)!}$$

$$= \sum_{r=1}^{n} \frac{rn!}{r(r-1)!(n-r)!}$$

$$= \sum_{r=1}^{n} \frac{n!}{(r-1)!(n-r)!}$$

$$= \sum_{r=1}^{n} \frac{n(n-1)!}{(r-1)!(n-r)!}$$

$$= n \sum_{r=1}^{n} \binom{n-1}{r-1}$$

$$= n \left(\binom{n-1}{0} + \binom{n-1}{1} + \cdots + \binom{n-1}{n-1} \right)$$

$$= n2^{n-1}.$$

The algebraic manipulations used here may seem to come out of thin air, but they are motivated by strategy. We start with a complicated sum, and deliberately strive to make its terms look like a simpler sum that you already know. Since the right- hand side of (1) is $n2^{n-1}$, it is plausible to see if the left-hand side can merely be turned into

n times a simpler sum which equals 2^{n-1}. This is just another example of using the wishful thinking and penultimate step strategies, perhaps the most useful combination in problem solving. And do note the careful use of factoring to extract, for example, $(n-1)!$ from $n!$, as well as the use of the interesting identity $\binom{n}{k} = \frac{n}{k}\binom{n-1}{k-1}$.

While this algebraic proof is elegant and instructive, in a sense it is an empty argument, for the terms in (1) obviously have combinatorical meaning which we have ignored. Let us now prove it with combinatorial reasoning.

The left-hand side of (1) is a sum, so we interpret it as a partitioning of a large set. Each term has the form $r\binom{n}{r}$, which can be interpreted as the number of ways you choose an r-person team from a pool of n people, *and designate a team leader*. The entire left-hand side is the total number of ways of picking a team of any size (between 1 and n) with a designated leader.

The right-hand side counts the same thing, but with an encoding interpretation: Suppose we number the n people from 1 to n in alphabetical order. First we pick a leader, which we encode by his or her number (ranging from 1 to n). Then we place the remaining $n-1$ people in alphabetical order, and either include them on the team or not, encoding our action with the letters y or n. For example, suppose $n = 13$. The string 11nynnnnnnyyy indicates a team with 5 people, led by #11 and also including #2,#10, #12 and #13. This method codes every possible team-with-leader uniquely, and the number of such strings is equal to $n2^{n-1}$, since there are n choices for the first place in the string (notice that this spot may be occupied with a multi-digit number, but what we are counting is choices, not digits), and 2 choices for each of the remaining $n-1$ places. ∎

Information Management

Proper encoding demands precise information management. The model of storing information in a computer compels one not to waste "memory" with redundant information. Here are a few examples of how and *how not* to organize information.

Example 6.2.2 Suppose that you wanted to count the number of permutations of the word "BOOBOO." We already know that the answer is $\frac{6!}{2!4!}$ from the Mississippi formula. How to encode this? There are 6 "slots" in which to place letters. Each permutation is *uniquely* determined by the knowledge of which 2 slots one places the letter B, since it is understood that the remaining 4 slots will be occupied by O's. In other words, there is a bijection between choices of 2 slots from the 6 slots and permutation of the letters. Thus our answer is $\binom{6}{2}$, which of course equals $\frac{6!}{2!4!}$. A common error is to count choosing slots for B's and choosing slots for O's as independent choices, yielding the erroneous answer of $\binom{6}{2}\binom{6}{4}$.

It is hard to avoid all errors of this kind, but try to think carefully about "freedom of choice": ask yourself what has already been *completely determined* from previous choices. In this case, once we choose the slots for the B's, the locations of the O's are determined and we have no freedom of choice.

Example 6.2.3 Suppose we have three different toys and we want to give them away to 2 girls and 1 boy (one toy per child). The children will be selected from 4 boys and 6 girls. In how many ways can this be done?

Solution: The numerical answer is 360. Here are two different approaches.

1. Ignore order, but don't ignore sex: Then there are $\binom{4}{1}$ ways to pick the single boy and $\binom{6}{2}$ ways to pick the two girls. Thus there are $\binom{4}{1} \cdot \binom{6}{2}$ ways of picking one boy and two girls. But this does not distinguish between, say, "Joe, Sue, Jane" and "Jane, Joe, Sue." In other words, we need to correct for order by multiplying by 3! (since we give the toys, which are different, away in order). So the answer is $\binom{4}{1} \cdot \binom{6}{2} \cdot 3!$. ■

2. Include order from the start: First, pick a boy (we can do this in $\binom{4}{1}$ ways). Then, pick a toy for this boy (3 ways). Then, pick the two girls, but counting order ($P(6, 2) = 6 \cdot 5$ ways). The answer is then $P(6, 2) \cdot \binom{4}{1} \cdot 3$. ■

Example 6.2.4 Suppose again that we have three different toys and we want to give them away (one toy per kid) to 3 children selected from a pool of 4 boys and 6 girls, but now we require that at least 2 boys get a toy. In how many ways can this be done?

Solution: Beginners are often seduced by the quick answers provided by encoding and attempt to convert just about any counting problem into a simple multiplication or binomial coefficient.

The following argument is wrong, but not obviously wrong. Imagine that the toys are ordered (for example, in order, we give away a video game, a doll, a puzzle). We first pick a boy to get the video game (4 choices). Then another boy gets the doll (3 choices). Then we give the puzzle to one of the 8 remaining kids (8 choices). The number of ways we can do this is just $4 \cdot 3 \cdot 8$. Of course, we need to correct for sex bias. With this method, we are guaranteeing that only boys get the video game and the doll. The puzzle is the "leftover" toy. So, to make the count fair, and symmetrical, we multiply by 3, for the three different leftover toy possibilities. (We don't multiply by $3! = 6$, since the order of the pick of the two has already been incorporated into our count). Anyway, this yields $4 \cdot 3 \cdot 8 \cdot 3 = 288$, which is too large! Why?

The problem is the "leftover" toy. Sometimes a boy gets it, sometimes a girl. If a girl gets the leftover toy, the count is OK. But if a boy gets it, we will be overcounting by a factor of three. To see this, imagine that the puzzle is the leftover toy and we picked (in order) "Joe, Bill, Fred." Then Joe gets the video game, Bill gets the doll, and Fred gets the puzzle. Now, let the video game be the overflow toy. We will still end up counting a choice where Fred gets the puzzle, Bill gets the doll, and Joe gets the video game as a different choice! If a girl ended up with the overflow toy, there wouldn't be an overcount. For example, if Joe gets the video game, Bill gets the doll, and Sue gets the puzzle, and then later made the video game the overflow toy, the new count would give Sue the video game, which is certainly a different choice!

This is confusing! How do you guard against such subtle errors? There is no perfect solution, but it is crucial that you *check* your counting method by looking at smaller numbers, where you can directly count the choices. And also, in this case, let the language ("at least") clue you in to other methods. Whenever you see "at least," you should investigate partitioning.[3] This gives an easy and reliable solution, for there are two cases:

1. *There are exactly two boys, and one girl chosen.* Arguing as we did in 6.2.3 above, there are $\binom{4}{2}$ ways to pick the boys, and $\binom{6}{1}$ ways to pick a girl, and then we must correct for order by multiplying by 3!.
2. *There are exactly three boys.* There will be just $P(4, 3)$ choices (since order matters).

The final answer is just the sum of these two, or

$$\binom{4}{2}\binom{6}{1}3! + P(4, 3) = 6 \cdot 6 \cdot 6 + 24 = 240,$$

which indeed is smaller than the incorrect answer of 288.   ∎

Example 6.2.5 *The Hockey Stick Identity.* Consider the "hockey stick" outlined in Pascal's Triangle below.

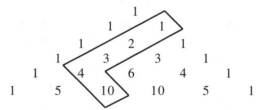

The sum of the elements in the handle is equal to the element in the blade; i.e.,

$$1 + 2 + 3 + 4 = 10.$$

This works for any parallel "hockey stick" of any length or location that begins at the border of the triangle. For example, $1 + 3 + 6 = 10$ and $1 + 4 + 10 + 20 = 35$ (check that this one works by writing in the next few rows of the triangle). In general, if we start at row s and continue until row f, we have

$$\binom{r}{r} + \binom{r+1}{r} + \binom{r+2}{r} + \cdots + \binom{f}{r} = \binom{f+1}{r+1}.$$

Why is this true?

Solution: Consider a specific example, say $1 + 3 + 6 + 10 = 20$. We need to explain why

$$\binom{2}{2} + \binom{3}{2} + \binom{4}{2} + \binom{5}{2} = \binom{6}{3}.$$

[3] Also, you should consider the tactic of **counting the complement** (page 225).

The simple sum suggests partitioning: Let us look at all 3-element subsets (which we will call **3-sets**) chosen from a set of 6 elements, and see if we can break them up into 4 "natural" classes.

Write the 6-element set as $\{a, b, c, d, e, f\}$ and follow the natural convention that subsets are written in alphabetic order. This tactic is a simple and natural way to get "free" information and is just another example of monotonizing (see page 83). Since we are counting combinations and not permutations, order is irrelevant as far as counting is concerned, but that doesn't mean that we cannot impose useful order on our own.

Following this convention, let us list all 20 different 3-sets, as 3-letter "words." And of course, we will list them in alphabetical order!

$$abc, abd, abe, abf, acd, ace, acf, ade, adf, aef;$$
$$bcd, bce, bcf, bde, bdf, bef;$$
$$cde, cdf, cef;$$
$$def.$$

The solution practically cries out: *Classify subsets by their initial letter!*

Of the 20 different 3-sets, $\binom{5}{2} = 10$ of them begin with the letter a, since we are fixing the first letter and still have to choose 2 more from the remaining 5 letters available (b through f). If we start with the letter b, we must still chose 2 more, but have only 4 choices available (c through f), so we have $\binom{4}{2}$ possibilities. Similarly, $\binom{3}{2} = 3$ of the 3-sets start with c and just one $(\binom{2}{2})$ starts with d. (No 3-sets can start with e or f since the letters in each word are in alphabetical order.)

By now you should have a good understanding of *why* the hockey stick identity is true. But let us write a formal argument, just for practice. Note the careful use of notation and avoidance of "OBOB."

We shall show that

$$\binom{r}{r} + \binom{r+1}{r} + \binom{r+2}{r} + \cdots + \binom{f}{r} = \binom{f+1}{r+1} \tag{2}$$

holds for all positive integers $f \geq r$. Let us consider all $(r+1)$-element subsets of the set $\{1, 2, \ldots, f+1\}$. For $j = 1, 2, \ldots, f-r+1$, let $\mathcal{E}_j$ denote the collection of these subsets whose minimal element is j. The $\mathcal{E}_j$ partition the collection of $(r+1)$-element subsets, so

$$|\mathcal{E}_1| + |\mathcal{E}_2| + \cdots + |\mathcal{E}_{f-r+1}| = \binom{f+1}{r+1}.$$

If the minimal element of an $(r+1)$-element subset is j, then the remaining r elements will be chosen from the range $j+1, j+2, \ldots, f+1$. There are $(f+1) - (j+1) + 1 = f - j + 1$ integers in this range, so consequently

$$|\mathcal{E}_j| = \binom{f-j+1}{r}.$$

Summing this as j ranges from 1 to $f - r + 1$ yields (2). ∎

Balls in Urns and Other Classic Encodings

The following well-known problem crops up in many different forms.

Example 6.2.6 Imagine a piece of graph paper. Starting at the origin draw a path to the point $(10, 10)$ that stays on the grid lines (which are one unit apart) and has a total length of 20. For example, one path is to go from $(0, 0)$ to $(0, 7)$ to $(4, 7)$ to $(4, 10)$ to $(10, 10)$. Another path goes from $(0, 0)$ to $(10, 0)$ to $(10, 10)$. How many possible different paths are there?

Solution: Each path can be completely described by a 20-letter sequence of U's and R's, where "U" means "move one unit up" and "R" means "move one unit to the right." For example, the path that goes from $(0, 0)$ to $(0, 7)$ to $(4, 7)$ to $(4, 10)$ to $(10, 10)$ would be described by the string

$$\underbrace{UUUUUUU}_{7}\underbrace{RRRR}_{4}\underbrace{UUU}_{3}\underbrace{RRRRRR}_{6}.$$

Since the path starts at $(0, 0)$, has length 20, and ends at $(10, 10)$, each path must have exactly 10 U's and 10 R's. Hence the total number of paths is just a simple Mississippi problem, whose answer is

$$\frac{20!}{10!10!} = \binom{20}{10}.$$ ∎

The next example can be done by partitioning as well as encoding, although the encoding solution is ultimately more productive.

Example 6.2.7 How many different ordered triples (a, b, c) of non-negative integers are there such that $a + b + c = 50$?

Solution: We will present two completely different solutions, the first using partitioning, the second using encoding. First, it is easy to see that for each non-negative integer n, the equation $a + b = n$ is satisfied by $n + 1$ different ordered pairs (a, b), namely

$$(0, n), (1, n - 1), (2, n - 2), \dots, (n, 0).$$

So we can partition the solutions of $a + b + c = 50$ into the disjoint cases where $c = 0, c = 1, c = 2, \dots, c = 50$. For example, if $c = 17$, then $a + b = 33$ so there will be 34 different ordered triples $(a, b, 17)$ that satisfy $a + b + c = 50$. Our answer is therefore

$$1 + 2 + 3 + \cdots + 51 = \frac{51 \cdot 52}{2}.$$

The problem was solved easily by partitioning, but encoding also works. To see how, let us replace 50 with a smaller number, say, 11, and recall how we first learned about addition in first or second grade. You practiced sums like "$3 + 6 + 2 = 11$" by drawing sequences of dots:

$$\bullet\bullet\bullet + \bullet\bullet\bullet\bullet\bullet\bullet + \bullet\bullet \;\; = \;\; \bullet\bullet\bullet\bullet\bullet\bullet\bullet\bullet\bullet\bullet\bullet.$$

Each ordered triple (a, b, c) of non-negative integers which sum to 11 can be thought of as an allocation of 11 dots, separated by two plus signs. The triple $(3, 6, 2)$ can be encoded by the string $\bullet\bullet\bullet + \bullet\bullet\bullet\bullet\bullet\bullet + \bullet\bullet$. Zero is not a problem; we encode it with no dots. Hence $+ + \bullet\bullet\bullet\bullet\bullet\bullet\bullet\bullet\bullet\bullet\bullet$ corresponds to $(0, 0, 11)$ and $\bullet\bullet\bullet\bullet\bullet + {} + \bullet\bullet\bullet\bullet\bullet\bullet$ corresponds to $(5, 0, 6)$. This method uniquely encodes each triple with a string of 13 symbols containing 2 $+$ symbols and 11 $\bullet$ symbols. There are $\binom{13}{2}$ ways of counting these strings (for each string is uniquely determined by which 2 of the 13 possible slots we place the $+$ symbol).

Finally, replacing 11 with 50, we see that the answer is

$$\binom{52}{2} = \frac{52 \cdot 51}{2}. \qquad\blacksquare$$

This is of course just one example of a general tool, the **balls in urns** formula:

*The number of different ways we can place b **indistinguishable** balls into u **distinguishable** urns is*

$$\binom{b + u - 1}{b} = \binom{b + u - 1}{u - 1}.$$

The balls in urns formula is amazingly useful, and turns up in many unexpected places. Let us use it, along with the pigeonhole principle (Section 3.3), to solve Example 6.1.1 on page 205. Recall that this problem asked whether there exist 10,000 ten-digit numbers divisible by seven, all of which can be obtained from one another by a reordering of their digits.

Solution:

Ignore the issue of divisibility by 7 for a moment, and concentrate on understanding sets of numbers "which can be obtained from one another by a reordering of their digits." Let us call two numbers which can be obtained from one another in this way "sisters," and let us call a set that is as large as possible whose elements are all sisters a "sorority." For example, 1,111,233,999 and 9,929,313,111 are sisters who belong to a sorority with $\frac{10!}{4!3!2!}$ members, since the membership of the sorority is just the number of ways of permuting the digits. The sororities have vastly different sizes. The most "exclusive" sororities have only 1 member (for example, the sorority consisting entirely of 6,666,666,666) yet one sorority has 10! members (the one containing 1,234,567,890).

In order to solve this problem, we need to show that there is a sorority with at least 10,000 members that are divisible by 7. One approach is to look for big sororities (like the one with 10! members), but it is possible (even though is seems unlikely) that somehow most of the members will not be multiples of 7.

Instead, the crux idea is that

> *It is not the size of the sororities that really matters, but how many*
> *sororities there are.*

If the number of sororities is fairly small, then even if the multiples of 7 were dispersed very evenly, "enough" of them would land in some sorority.

Let's make this more precise: Suppose it turned out that there were only 100 sororities (of course there are more than that). There are $\lfloor 10^{10}/7 \rfloor = 1,428,571,428$ multiples of seven. By the pigeonhole principle, at least one sorority will contain $\lceil 1,428,571,428/100 \rceil$ multiples of seven, which is way more than we need. In any event, we have the penultimate step to work toward: compute (or at least estimate) the number of sororities.

We can compute the exact number. Each sorority is uniquely determined by its collection of 10 digits *where repetition is allowed.* For example, one (highly exclusive) sorority can be named "ten 6's," while another is called "three 4's, a 7, two 8's, and four 9's." So now the question becomes, in how many different ways can we choose 10 digits, with repetition allowed? Crux move #2: this is equivalent to putting 10 balls into 10 urns that are labeled $0, 1, 2, \ldots, 9$. By the balls in urns formula, this is $\binom{19}{9} = 92,378$.

Finally, we conclude that there will be a sorority with at least

$$\lceil 1,428,571,428/92,378 \rceil = 15,465$$

members. This is greater than 10,000, so the answer is "yes." ∎

Problems and Exercises

6.2.8 (Jim Propp) Sal the Magician asks you to pick any five cards from a standard deck.[4] You do so, and then show them to Sal's assistant Pat, who places one of the five cards back in the deck and then puts the remaining four cards into a pile. Sal is blindfolded, and does not witness any of this. Then Sal takes off the blindfold, takes the pile of four cards, reads the four cards that Pat has arranged, and is able to find the fifth card in the deck (even if you shuffle the deck after Pat puts the card in the deck). Assume that neither Sal nor Pat has supernatural powers, and that the deck of cards is not marked. How is the trick done? Harder version: you pick which of the five cards goes back into the deck (instead of Pat).

6.2.9 Eight people are in a room. One or more of them get an ice-cream cone. One or more of them get a chocolate-chip cookie. In how many different ways can this happen, given that at least one person gets both an ice-cream cone and a chocolate-chip cookie?

6.2.10 How many subsets of the set $\{1, 2, 3, 4, \ldots, 30\}$ have the property that the sum of the elements of the subset is greater than 232?

---

[4]A standard deck contains 52 cards, 13 denominations $(2, 3, \ldots, 10, J, Q, K, A)$ in each of 4 suits $(\diamondsuit, \heartsuit, \clubsuit, \spadesuit)$.

6.2.11 How many *strictly increasing* sequences of positive integers begin with 1 and end with 1000?

6.2.12 Find another way to prove the Hockey Stick Identity (Example 6.2.5 on page 218), using the summation identity (6.1.14).

6.2.13 For any set, prove that the number of its subsets with an even number of elements is equal to the number of subsets with an odd number of elements. For example, the set $\{a, b, c\}$ in the problem above has 4 subsets with an even number of elements (the empty set has 0 elements, which is even), and 4 with an odd number of elements.

6.2.14 In how many ways can two squares be selected from an 8-by-8 chessboard so that they are not in the same row or the same column?

6.2.15 In how many ways can four squares, not all in the same row or column, be selected from an 8-by-8 chessboard to form a rectangle?

6.2.16 In how many ways can we place r red balls and w white balls in n boxes so that each box contains at least one ball of each color?

6.2.17 A parking lot for compact cars has 12 adjacent spaces, and 8 are occupied. A large sport-utility vehicle arrives, needing 2 adjacent open spaces. What is the probability that it will be able to park? Generalize!

6.2.18 Find a nice formula for the sum

$$\binom{n}{0}^2 + \binom{n}{1}^2 + \binom{n}{2}^2 + \cdots + \binom{n}{n}^2.$$

Can you explain why your formula is true?

6.2.19 How many ways can the positive integer n can be written as an ordered sum of at least one positive integer? For example,

$$4 = 1 + 3 = 3 + 1 = 2 + 2 = 1 + 1 + 2 = 1 + 2 + 1 = 2 + 1 + 1 = 1 + 1 + 1 + 1,$$

so when $n = 4$, there are 8 such *ordered partitions*.

6.2.20 Ten dogs encounter 8 biscuits. Dogs do not share biscuits! Verify that the number of different ways that the biscuits can be consumed will equal

(a) $\binom{17}{8}$ if we assume that the dogs are distinguishable, but the biscuits are not;

(b) 10^8 if we assume that the dogs and the biscuits are distinguishable (for example, each biscuit is a different flavor).

6.2.21 In problem 6.2.20 above, what would the answer be if we assume that neither dogs nor biscuits are distinguishable? (The answer is *not* 1!)

6.2.22 When $(x + y + z)^{1999}$ is expanded and like terms are collected, how many terms will there be? [For example, $(x + y)^2$ has 3 terms after expansion and simplification.]

6.2.23 Find a formula for the number of different ordered triples (a, b, c) of *positive* integers which satisfy $a + b + c = n$.

6.2.24 Let S be a set with n elements. In how many different ways can one select two not necessarily distinct or disjoint subsets of S so that the union of the two subsets is S? The order of selection does not matter; for example, the pair of subsets $\{a, c\}, \{b, c, d, e, f\}$ represents the same selection as the pair $\{b, c, d, e, f\}, \{a, c\}$.

6.2.25 (AIME 1988) In an office, at various times during the day, the boss gives the secretary a letter to type, each time putting the letter on top of the pile in the secretary's in-box. When there is time, the secretary takes the top letter off the pile and types it. There are nine letters to be typed during the day, and the boss delivers them in the order $1, 2, 3, 4, 5, 6, 7, 8, 9$.

While leaving for lunch, the secretary tells a colleague that letter 8 has already been typed, but says nothing else about the morning's typing. The colleague wonders which of the nine letters remain to be typed after lunch and in what order they will be typed. Based upon the above information, how many such *after-lunch typing orders* are possible? (That there are no letters left to be typed is one of the possibilities.)

6.2.26 (AIME 1983)A gardener plants three maple trees, four oak trees and five birch trees in a row. He plants them in random order, each arrangement being equally likely. Find the probability that no two birch trees are next to one another.

6.2.27 (AIME 1984)Twenty-five people sit around a circular table. Three of them are chosen randomly. What is the probability that at least two of the three had been sitting next to one another?

6.2.28 We are given n points arranged around a circle and the chords connecting each pair of points are drawn. If no three chords meet in a point, how many points of intersection are there? For example, when $n = 6$, there are 15 intersections.

6.2.29 Given g girls and b boys, how many different ways can you seat these people in a row of $g + b$ seats so that no two boys sit together? There are two different interpretations: one where you do not distinguish between individual girls and between individual boys (i.e., you would not consider the seating arrangements "Becky, Sam, Amy, Tony" and "Amy, Tony, Becky, Sam" as different), and one where you do distinguish between individuals. Answer the question for each interpretation. Find a general formula, and prove why your formula is correct.

6.2.30 The balls in urns formula says that if you are order h hats from a store which sells k different kinds of hats, then there are $\binom{k + h - 1}{h}$ different possible orders.

Use a partitioning argument to show that this is also equal to $\displaystyle\sum_{r=1}^{k} \binom{k}{r}\binom{h-1}{r-1}$.

6.2.31 Use a combinatorial argument to show that for all positive integers n, m, k with n and m greater than or equal to k,

$$\sum_{j=0}^{k} \binom{n}{j}\binom{m}{k-j} = \binom{n+m}{k}.$$

This is known as the **Vandermonde convolution formula**.

6.2.32 In Example 6.2.4 on page 217, the incorrect argument overcounted by 48. Account *exactly* for this discrepancy.

6.2.33 As defined in Problem 4.3.17 on page 153, let $p(n)$ denote the number of unrestricted partitions of n. Show that $p(n) \geq 2^{\lfloor \sqrt{n} \rfloor}$ for all $n \geq 2$.

6.2.34 (Putnam 1993) Let $\mathcal{P}_n$ be the set of subsets of $\{1, 2, \ldots, n\}$. Let $c(n, m)$ be the number of functions $f : \mathcal{P}_n \to \{1, 2, \ldots, m\}$ such that $f(A \cap B) = \min\{f(A), f(B)\}$. Prove that

$$c(n, m) = \sum_{j=1}^{m} j^n.$$

6.3   The Principle of Inclusion-Exclusion

The partitioning tactic makes a pretty utopian assumption, namely that the thing we wish to count can be nicely broken down into *pairwise disjoint* subsets. Reality is often messier, and new tactics are needed. We shall explore a number of ways of dealing with situations where sets overlap and overcounting is *not* uniform.

Count the Complement

By now you are used to the strategy of changing your point of view. A particular application of this in counting problems is

> *If the thing you wish to count is confusing, try looking at its complement instead.*

Example 6.3.1 How many n-bit strings contain at least 1 zero?

Solution: Counting this directly is hard, because there are n cases, namely, exactly 1 zero, exactly 2 zeros,.... This is certainly not impossible to do, but an easier approach is to first note that there are 2^n possible n-bit strings, and then count how many of them contain *no* zeros. There is only one such string (the string containing n 1's). So our answer is $2^n - 1$. ∎

Example 6.3.2 Ten children order ice-cream cones at a store featuring 31 flavors. How many orders are possible in which at least two children get the same flavor?

Solution: We shall make the humanistic assumption that the children are distinguishable. Then we are counting a pretty complex thing. For example, one order might be that all the children order flavor #6. Another order might specify that child #7 and child #9 each order flavor #12 and children #1–4 order flavor #29, etc. Let us first count all possible orders with no restrictions; that is just 31^{10}, since each of the 10 kids has 31 choices. Now we count orders where there is no duplication of flavor; that is just $31 \times 30 \times 29 \times \cdots \times 22 = P(31, 10)$. The answer is then the difference $31^{10} - P(31, 10)$. ∎

PIE with Sets

Sometimes the complement counting tactic fails us, because the complement is just as complicated as the original set. The principle of inclusion-exclusion (PIE) is a systematic way to handle these situations.

In simplest form, PIE states that the number of elements in the union of two sets is equal to the sum of the number of elements in the sets minus the number of elements in their intersection. Symbolically, we can write this as

$$|A \cup B| = |A| + |B| - |A \cap B|.$$

It is easy to see why this is true: Adding $|A|$ and $|B|$ overcounts the value of $|A \cup B|$. This overcounting is not uniform; we did not count everything twice, just the elements of $A \cap B$. Consequently, we can correct for overcounting by subtracting $|A \cap B|$.

A bit of experimenting quickly leads to a conjecture for the general case of PIE with n sets.

6.3.3 Verify that for three sets, PIE is

$$|A \cup B \cup C| = |A| + |B| + |C| - (|A \cap B| + |A \cap C| + |B \cap C|) + |A \cap B \cap C|.$$

6.3.4 Verify that for four sets, PIE is

$$\begin{aligned}
|A \cup B \cup C \cup D| = \\
+ (|A| + |B| + |C| + |D|) \\
- (|A \cap B| + |A \cap C| + |A \cap D| + |B \cap C| + |B \cap D| + |C \cap D|) \\
+ (|A \cap B \cap C| + |A \cap C \cap D| + |A \cap B \cap D| + |B \cap C \cap D|) \\
- |A \cap B \cap C \cap D|.
\end{aligned}$$

In general, we conjecture

The cardinality of the union of n sets $=$

$+$ (sum of the cardinalities of the sets)

$-$ (sum of the cardinalities of all possible intersections of two sets)

+ (sum of the cardinalities of all possible intersections of three sets)

$$\vdots$$

± (the cardinality of the intersection of all n sets),

where the last term is negative if n is even, and positive is n is odd.

It is pretty easy to see why the formula is true in an informal way. For example, consider the following diagram, which illustrates the three-set situation.

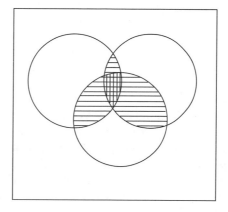

Visualize each set as a rubber disk. The cardinality of the union of the sets corresponds to the "map area" of the entire shaded region. However, the shading is in three intensities. The lightest shading corresponds to a single thickness of rubber, the intermediate shading means double thickness, and the darkest shading indicates triple thickness. The map area that we desire is just the single-thickness area. If we merely add up the areas of the three disks, we have "overcounted"; the light area is OK but the medium-dark area has been counted twice and the darkest shading area was counted three times. To rectify this, we subtract the areas of the intersections $A \cap B$, $A \cap C$, $B \cap C$. But now we have undercounted the darkest area $A \cap B \cap C$ — it was originally counted three times, but now has been subtracted three times. So we add it back, and we are done. It makes sense that in the general case, we would alternate between adding and subtracting the different thicknesses.

This argument is attractive, but hard to generalize rigorously to n sets. Let us attempt a rigorous proof of PIE, one that illustrates a nice counting idea and useful notation. Let our sets be $A_1, A_2, \ldots, A_n$ and let S_k denote the sum of the cardinalities of all possible intersections of k of these sets. For example,

$$S_1 = |A_1| + |A_2| + \cdots + |A_n| = \sum_{1 \leq i \leq n} |A_i|$$

and

$$S_2 = |A_1 \cap A_2| + |A_1 \cap A_3| + \cdots + |A_{n-1} \cap A_n| = \sum_{1 \leq i < j \leq n} |A_i \cap A_j|.$$

Notice the subscript notation. (Take some time to study it carefully, perhaps by writing out several examples.) The condition $1 \leq i < j \leq n$ gives us all $\binom{n}{2}$ possible combinations of two different indices, with no repeats. For example, $|A_3 \cap A_7|$ appears just once in the sum, since $|A_7 \cap A_3|$ is not allowed. In general,

$$S_k = \sum_{1 \leq i_1 < i_2 < \cdots < i_k \leq n} |A_{i_1} \cap A_{i_2} \cap \cdots \cap A_{i_k}|.$$

With this notation, PIE becomes the statement

$$|A_1 \cup A_2 \cup \cdots \cup A_n| = S_1 - S_2 + S_3 - \cdots + (-1)^{n-1} S_n. \tag{3}$$

To prove this, let $x \in A_1 \cup A_2 \cup \cdots \cup A_n$ be an arbitrary element of the union of the n sets. This element x is counted exactly once by the left-hand side of (3), since the left-hand side means "the number of elements in the union of the n sets." Thus, if we can show that the right-hand side of (3) also counts the element x exactly once, we will be done.[5]

Let r be the number of sets that x is a member of. For example, if $x \in A_3$ and no other set, then $r = 1$. Certainly, r can range between 1 and n, inclusive. Let us see how many times each S_k counts the element x. When S_1 is computed, each element in each set A_i is counted. Thus the element x is counted exactly r times, once for each set it is a member of. To compute S_2, we count the elements in each set of the form $A_i \cap A_j$, $i \leq j$. The only sets that are relevant to us are the r sets that x is a member of. There will be $\binom{r}{2}$ intersections of pairs of these sets, and for each of them, we will count x once. Hence S_2 counts x exactly $\binom{r}{2}$ times. In general, S_k counts the element x exactly $\binom{r}{k}$ times. If $k > r$, then S_k is a sum of cardinalities of intersections of $k > r$ sets, and none of these sets can contain x, which only lies in r sets! And of course, S_r counts x exactly once (i.e., $\binom{r}{r}$ times), since there is only one intersection of r sets which actually contains x.

We have reduced PIE, then, to proving the identity

$$1 = r - \binom{r}{2} + \binom{r}{3} + \cdots + (-1)^{r-1} \binom{r}{r}.$$

Recall that $r = \binom{r}{1}$ and $1 = \binom{r}{0}$, so the above equality has the equivalent formulation

$$\binom{r}{0} - \binom{r}{1} + \binom{r}{2} + \cdots + (-1)^r \binom{r}{r} = 0, \tag{4}$$

which was part of problem 6.1.22. One can prove equation (4) easily in at least three different ways. Try them all!

- Induction plus the summation identity (6.1.14) of Pascal's Triangle;
- A direct examination of the elements in Pascal's Triangle, using the symmetry identity (6.1.13) and the summation identity;

---

[5]Notice the new counting idea: to see if a combinatorial identity is true, examine how each side of the equation counts a representative element.

- The slickest way, perhaps: Just expand $0 = (1-1)^r$ with the binomial theorem and you *immediately* get (4)!

In any event, now that we know that (4) is true, we have proven PIE. ■

A few examples will convince you of the power in PIE. The key to approaching these problems is a flexible attitude about whether to count something or its complement.

Example 6.3.5 How many five-card hands from a standard deckcards of cards contain at least one card in each suit?

Solution: First note that there are $\binom{52}{5}$ possible hands, since the order of the cards in a hand is immaterial. Whenever you the words "at least," you should be alerted to the possibility of counting complements. Which is easier to compute: the number of suits containing no diamonds, or the number of suits containing at least one diamond? Certainly the former is easier to calculate; if there are no diamonds, then there are only $52 - 13 = 39$ cards to choose from, so the number of hands with no diamonds is $\binom{39}{5}$. This suggests that we define as our "foundation" sets C, D, H, S to be the sets of hands containing *no* clubs, diamonds, hearts, spades, respectively. Not only do we have

$$|C| = |D| = |H| = |S| = \binom{39}{5},$$

but the intersections are easy to compute as well. For example, $D \cap H$ is the set of hands that contain neither diamonds nor hearts. There are $52 - 2 \cdot 13 = 26$ cards to choose from, so

$$|D \cap H| = \binom{26}{5}.$$

By similar reasoning, $|D \cap H \cap S| = \binom{13}{5}$. Notice that $C \cap D \cap H \cap S = \emptyset$, since it is impossible to omit all 4 suits!.

These sets are not just easy to compute with, they are useful as well, because $C \cup D \cup H \cup S$ consists of all hands for which at least one suit is absent. That is exactly the complement of what we want!

Thus we will use PIE to compute $|C \cup D \cup H \cup S|$, and subtract this result from $\binom{52}{5}$. We have

$$|C \cup D \cup H \cup S| = S_1 - S_2 + S_3 - S_4,$$

where

$$S_1 = |C| + |D| + |H| + |S|,$$
$$S_2 = |C \cap D| + |C \cap H| + |C \cap S| + |D \cap H| + |D \cap S| + |H \cap S|,$$
$$S_3 = |C \cap D \cap H| + |C \cap D \cap S| + |C \cap H \cap S| + |D \cap H \cap S|,$$
$$S_4 = |C \cap D \cap H \cap S|.$$

In other words,

$$|C \cup D \cup H \cup S| = \binom{39}{5} - \binom{4}{2}\binom{26}{5} + \binom{4}{3}\binom{13}{5} - \binom{4}{4} \cdot 0. \qquad \blacksquare$$

The combination of PIE with counting the complement is so common, it is worth noting (and verifying) the following alternative formulation of PIE.

6.3.6 Complement PIE. Given N items, and k sets $A_1, A_2, \ldots, A_k$, the number of these items which lie in *none* of the A_j is equal to

$$N - S_1 + S_2 - S_3 + \cdots \pm S_k,$$

where S_i is the sum of the cardinalities of the intersections of the A_j's taken i at at time, and the final sign is plus or minus, depending on whether k is even or odd.

The next example combines "Complement PIE" with other ideas, including the useful encoding tool, **invent a font**, whereby we temporarily "freeze" several symbols together to define a single new symbol.

Example 6.3.6 Four young couples are sitting in a row. In how many ways can we seat them so that no person sits next to his or her "significant other?"

Solution: Clearly, there are 8! different possible seatings. Without loss of generality, let the people be boys and girls denoted by $b_1, b_2, b_3, b_4, g_1, g_2, g_3, g_4$, where we assume that the couples are b_i and g_i, for $i = 1, 2, 3, 4$. Define A_i to be the set of all seatings for which b_i and g_i sit together. To compute $|A_i|$, we have two cases: either b_i is sitting to the left of g_i or vice versa. For each case, there will be 7! possibilities, since we are permuting 7 symbols: the *single* symbol $b_i g_i$ (or $g_i b_i$), plus the other 6 people. Hence $|A_i| = 2 \cdot 7!$ for each i. Next, let us compute $|A_i \cap A_j|$. Now we are fixing couple i and couple j, and letting the other 4 people permute freely. This is the same as permuting 6 symbols, so we get 6!. However, there are $2^2 = 4$ cases, since couple i can be seated either boy-girl or girl-boy, and the same is true for couple j. Hence $|A_i \cap A_j| = 4 \cdot 6!$. By the same reasoning, $|A_i \cap A_j \cap A_k| = 2^3 \cdot 5!$ and $|A_1 \cap A_2 \cap A_3 \cap A_4| = 2^4 \cdot 4!$. Finally, PIE yields

$$|A_1 \cup A_2 \cup A_3 \cup A_4| = 4 \cdot 2 \cdot 7! - \binom{4}{2} 4 \cdot 6! + \binom{4}{3} 2^3 \cdot 5! - 2^4 \cdot 4!,$$

where the binomial coefficients were used because there were $\binom{4}{2}$ ways of intersecting two of the sets, etc. Anyway, when we subtract this from 8!, we get the number of permutations in which no boy sits next to his girl, which is 13,824.

PIE with Indicator Functions

We're not done with PIE yet. By now you should feel comfortable with the truth of PIE, and you may understand the proof given above, but you should feel a bit baffled by the peculiar equivalence of PIE and the fact that $(1 - 1)^r = 0$. We shall now present a proof of the complement formulation of PIE (6.3.6), using the "binary" language of indicator functions.

Recall that the **indicator function of** A (see page 160) is denoted by $\mathbf{1}_A$ and is a function with domain U (where U is a "universal set" containing A) and range $\{0, 1\}$ defined by

$$\mathbf{1}_A(x) = \begin{cases} 0 & \text{if } x \notin A, \\ 1 & \text{if } x \in A, \end{cases}$$

for each $x \in U$. For example, if $U = \mathbf{N}$ and $A = \{1, 2, 3, 4, 5\}$, then $\mathbf{1}_A(3) = 1$ and $\mathbf{1}_A(17) = 0$.

Also recall that for any two sets A, B, the following are true (this was Problem 5.1.2 on page 160):

$$\mathbf{1}_A(x)\mathbf{1}_B(x) = \mathbf{1}_{A \cap B}(x). \tag{5}$$

$$1 - \mathbf{1}_A(x) = \mathbf{1}_{\overline{A}}(x). \tag{6}$$

In other words, the product of two indicator functions is the indicator function of the intersection of the two sets and the indicator function of a set's complement is just one subtracted from the indicator function of that set.

Another easy thing to check (Problem 5.3.11 on page 178) is that for any finite set A,

$$\sum_{x \in U} \mathbf{1}_A(x) = |A|. \tag{7}$$

This is just a fancy way of saying that if you consider each x in U and write down a "1" whenever x lies in A, then the sum of these "1"s will of course be the number of elements in A.

Let us apply these simple concepts to get a new proof of the complement form of PIE (6.3.6). Let the universal set U contain N elements, and without loss of generality, suppose that we have just 4 sets A_1, A_2, A_3, A_4. Define N_0 to be the number of elements in U which have none of these properties. In other words, N_0 is counting the number of elements of U which are not in any of A_1, A_2, A_3, A_4. Define the function $g(x)$ by

$$g(x) := (1 - \mathbf{1}_{A_1}(x))(1 - \mathbf{1}_{A_2}(x))(1 - \mathbf{1}_{A_3}(x))(1 - \mathbf{1}_{A_4}(x)).$$

Then, by applying equations (5) and (6) we see (verify!) that $g(x) = \mathbf{1}_{N_0}(x)$ and thus (7) implies that (verify!)

$$N_0 = \sum_{x \in U} g(x).$$

In other words,

$$N_0 = \sum_{x \in U} (1 - \mathbf{1}_{A_1}(x))(1 - \mathbf{1}_{A_2}(x))(1 - \mathbf{1}_{A_3}(x))(1 - \mathbf{1}_{A_4}(x)).$$

When we multiply out the four factors in the right-hand side, we get

$$N_0 = + \sum_{x \in U} 1$$

$$-\sum_{x\in U}\left(\mathbf{1}_{A_1}(x)+\mathbf{1}_{A_2}(x)+\mathbf{1}_{A_3}(x)+\mathbf{1}_{A_4}(x)\right)$$

$$+\sum_{x\in U}\left(\mathbf{1}_{A_1}(x)\mathbf{1}_{A_2}(x)+\mathbf{1}_{A_1}(x)\mathbf{1}_{A_3}(x)+\mathbf{1}_{A_1}(x)\mathbf{1}_{A_4}(x)\right.$$

$$\left.+\mathbf{1}_{A_2}(x)\mathbf{1}_{A_3}(x)+\mathbf{1}_{A_2}(x)\mathbf{1}_{A_4}(x)+\mathbf{1}_{A_3}(x)\mathbf{1}_{A_4}(x)\right)$$

$$-\sum_{x\in U}\left(\mathbf{1}_{A_1}(x)\mathbf{1}_{A_2}(x)\mathbf{1}_{A_3}(x)+\mathbf{1}_{A_1}(x)\mathbf{1}_{A_2}(x)\mathbf{1}_{A_4}(x)\right.$$

$$\left.+\mathbf{1}_{A_1}(x)\mathbf{1}_{A_3}(x)\mathbf{1}_{A_4}(x)+\mathbf{1}_{A_2}(x)\mathbf{1}_{A_3}(x)\mathbf{1}_{A_4}(x)\right)$$

$$+\sum_{x\in U}\left(\mathbf{1}_{A_1}(x)\mathbf{1}_{A_2}(x)\mathbf{1}_{A_3}(x)\mathbf{1}_{A_4}(x)\right).$$

If we apply equations (5), (6) and (7), we see that this ugly sum is exactly the same as

$$\begin{aligned} N_0 \;=\; & N \\ & -\; S_1 \\ & +\; S_2\;, \\ & -\; S_3 \\ & +\; S_4 \end{aligned}$$

using the notation of (6.3.6). In other words, we just demonstrated the truth of PIE for 4 sets. The argument certainly generalizes, for it uses only the algebraic fact that the expansion of

$$(1-a)(1-b)(1-c)\cdots$$

is equal to the alternating sum

$$1-(a+b+\cdots)-(ab+ac+\cdots)+\cdots. \qquad\blacksquare$$

Problems and Exercises

6.3.7 In Example 6.3.2 on page 226, we assumed that the children are distinguishable. But if we are just counting *orders*, then the children are not. For example, one order could be "Three cones of flavor #16, seven of flavor #28." How many such orders are there in this case?

6.3.8 What is wrong with the following "solution" to Example 6.3.6?

The first person can of course be chosen freely, so there are 8 choices. The next person must not be that person's partner, so there are 6 available. The third person cannot be the second person's partner, so there are 5 choices. Thus the product is

$$8\cdot6\cdot5\cdot4\cdot3\cdot2\cdot1\cdot1,$$

since the last two slots have no freedom of choice.

6.3.9 How many integers between 1 and 1000, inclusive, are divisible by neither 2, 3, or 5?

6.3.10 (USAMO 72) A random number generator randomly generates the integers $1, 2, \ldots, 9$ with equal probability. Find the probability that after n numbers are generated, the product is a multiple of 10.

6.3.11 How many nonnegative integer solutions are there to $a + b + c + d = 17$, provided that $d \le 12$?

6.3.12 Let $a_1, a_2, \ldots, a_n$ be an ordered sequence of n distinct objects. A **derangement** of this sequence is a permutation that leaves no object in its original place. For example, if the original sequence is $\{1, 2, 3, 4\}$, then $\{2, 4, 3, 1\}$ is not a derangement, but $\{2, 1, 4, 3\}$ is. Let D_n denote the number of an n-element sequence. Show that

$$D_n = n! \left(1 - \frac{1}{1!} + \frac{1}{2!} - \cdots + (1-)^n \frac{1}{n!} \right).$$

6.3.13 Use a combinatorial argument (no formulas!) to prove that

$$n! = \sum_{r=0}^{n} \binom{n}{r} D_{n-r},$$

where D_k is defined in the problem above.

6.3.14 Consider 10 people sitting around a circular table. How many different ways can they change seats so that each person has a different neighbor to the right?

6.3.15 Imagine that you are going to give n kids ice-cream cones, one cone per kid, and there are k different flavors available. Assuming that no flavors get mixed, show that the number of ways we can give out the cones *using all k flavors* is

$$k^n - \binom{k}{1}(k-1)^n + \binom{k}{2}(k-2)^n - \binom{k}{3}(k-3)^n + \cdots + (-1)^k \binom{k}{k} 0^n.$$

6.3.16 (IMO 89) Let a permutations π of $\{1, 2, \ldots, 2n\}$ have property P if

$$|\pi(i) - \pi(i+1)| = n$$

for at least one $i \in [2n - 1]$. Show that, for each n, there are more permutations with property P than without it.

6.4 Recurrence

Many problems involving the natural numbers require finding a formula or algorithm that is true for all natural numbers n. If we are lucky, a little experimenting suggests the general formula, and we then try to prove our conjecture. But sometimes the problem can be so complicated that at first it is difficult to "globally" comprehend it. The general formula may not be at all apparent. In this case, it is still possible to gain insight by focusing on the "local" situation, the transition from $n = 1$ to $n = 2$, and then, more generally, the transition from n to $n + 1$. Here is a very simple example.

Tiling and the Fibonacci Recurrence

Example 6.4.1 Define a **domino** to be a 1×2 rectangle. In how many ways can an $n \times 2$ rectangle be tiled by dominos?

Solution: Let t_n denote the number of tilings for an $n \times 2$ rectangle. Obviously $t_1 = 1$, and it is easy to check that $t_2 = 2$, since there are only the two possibilities below.

Consider t_7. Let us partition all the tilings of the 7×2 rectangle into two classes:

- Class V contains all tilings with a single vertical domino at the right end.

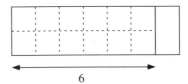

6

- Class H contains all other tilings. If the right end isn't a vertical domino, then it has to be two horizontal dominos.

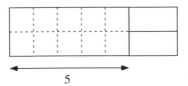

5

This is definitely a partition, since each and every tiling must be in one of these classes, and they do not share any elements. Class V contains t_6 members: Take any tiling of a 6×2 rectangle, append a vertical domino on the right, and you get a class V tiling of a 7×2 rectangle. Likewise, there are t_5 tilings in class H. In other words, we have shown that $t_7 = t_6 + t_5$. This argument certainly generalizes, so we have the **recurrence** formula.

$$t_{n+1} = t_n + t_{n-1}, \quad n = 2, 3, \ldots. \tag{8}$$

Have we solved the problem? Yes and no. Formula (8), plus the **boundary values** $t_1 = 1, t_2 = 2$, completely determine t_n for any value of n, and we have a simple algorithm for computing the values: just start at the beginning and apply the recurrence formula! The first few values are contained in the following table.

n	1	2	3	4	5	6	7	8	9	10	11	12
t_n	1	2	3	5	8	13	21	34	55	89	144	233

These values are precisely those of the Fibonacci numbers (see Problem 1.3.18 on page 12). Recall that the Fibonacci numbers f_n are defined by $f_0 = 0$, $f_1 = 1$ and $f_n = f_{n-1} + f_{n-2}$ for $n > 1$. The Fibonacci recurrence formula is the same as (8); only the boundary values are different. But since $f_2 = 1 = t_1$ and $f_3 = 2 = t_2$, we are guaranteed that $t_n = f_{n+1}$ for all n. So the problem is "solved," in that we have recognized that the tiling numbers are just Fibonacci numbers.   ■

Of course you may argue that the problem is not completely solved, as we do not have a "simple" formula for t_n (or f_n). In fact, problem 1.3.18 did in fact state the remarkable formula

$$f_n = \frac{1}{\sqrt{5}} \left\{ \left(\frac{1 + \sqrt{5}}{2} \right)^n - \left(\frac{1 - \sqrt{5}}{2} \right)^n \right\}, \tag{9}$$

which holds for all $n \geq 0$.

Let's verify this formula, deferring for now the more important question of where it came from. In other words, let's figure out the *how* of it, ignoring for now the *why* of it. All that we need to do is show that (9) satisfies both the Fibonacci recurrence formula and agrees with the two boundary values $f_0 = 0$, $f_1 = 1$. The last two are easy to check. And verifying the recurrence formula is a fun algebra exercise: Define

$$\alpha := \frac{1 + \sqrt{5}}{2}, \quad \beta := \frac{1 - \sqrt{5}}{2}.$$

Note that

$$\alpha + \beta = 1, \quad \alpha\beta = -1,$$

so both α and β are roots of the quadratic equation (see page 183)

$$x^2 - x - 1 = 0.$$

In other words, both α and β satisfy

$$x^2 = x + 1.$$

This means that if we define a sequence by $g_n := \alpha^n$, it will satisfy the Fibonacci recurrence! For any $n \geq 0$, we have

$$g_{n+1} = \alpha^{n+1} = \alpha^{n-1}\alpha^2 = \alpha^{n-1}(\alpha + 1) = \alpha^n + \alpha^{n-1} = g_n + g_{n-1}.$$

Likewise, if we define $h_n := \beta^n$, this sequence will also satisfy the recurrence. Indeed, if A and B are any constants, then the sequence

$$u_n := A\alpha^n + B\beta^n$$

will satisfy the recurrence, since

$$u_{n+1} = A\alpha^{n+1} + B\beta^{n+1} = A(\alpha^n + \alpha^{n-1}) + B(\beta^n + \beta^{n-1}) = u_n + u_{n-1}.$$

Thus, in particular, if we define

$$f_n := \frac{1}{\sqrt{5}}(\alpha^n - \beta^n),$$

then $f_{n+1} = f_n + f_{n-1}$. Since this also satisfies the boundary values, it must generate the entire Fibonacci sequence. ∎

While it is nice to have a "simple" formula that generates the Fibonacci sequence, knowing the recurrence formula is almost as good, and sometimes it is impossible or extremely difficult to get a "closed-form" solution to a recurrence. A few problems at the end of this section discuss some methods for solving recurrences,[6] but for now, let us just concentrate on discovering some interesting recurrence formulas.

The Catalan Recurrence

In the example below, we will discover a complicated recurrence formula that turns up in surprisingly many situations.

Example 6.4.2 The idea of a triangulation of a polygon was introduced in Example 2.3.9 on page 53. Compute the number of different triangulations of a convex n-gon.

Partial Solution: Experimentation yields $t_3 = 1, t_4 = 2, t_5 = 5$. For example, here are the 5 different triangulations for a convex pentagon:

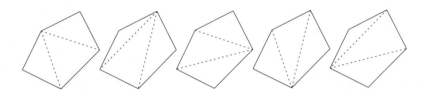

Let's move on to 6-gons, trying to discover a connection between them and smaller polygons. Fix a base, and consider the four possible vertices that we can draw a triangle from:

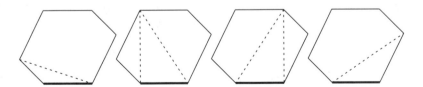

Notice that these four pictures partition the triangulations. The first picture yields t_5 new triangulations (corresponding to all the ways that the pentagon above the dotted

---

[6]Also see Section 4.3 for another general method for solving and analyzing recurrences.

line can be triangulated), while the second yields t_4 triangulations (the only choice involved is triangulating the quadrilateral). Continuing this reasoning, we deduce that

$$t_6 = t_5 + t_4 + t_4 + t_5 = 14,$$

which you can check by carefully drawing all the possibilities.

Before we rest, let's look at the 7-gon case. The triangulations are partitioned by the five cases below:

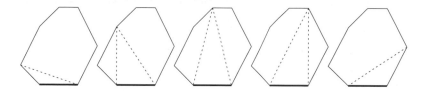

The first picture yields t_6 triangulations, and the second yields t_5, but the third picture gives rise to $t_4 \cdot t_4$ triangulations, since we are free to choose any triangulation in each of the two quadrilaterals on either side of the dotted triangle. The general situation is that each picture dissects the 7-gon into three polygons, one of which is the specified triangle with fixed base, and the other two may or may not involve choices. The other two polygons may include a "degenerate" 2-gon (for example, the first and last pictures) or a triangle, and in both cases this provides no new choices. But otherwise, we will be free to choose and will need to multiply to count the total number of triangulations for each picture. If we include the "degenerate" case and define t_2 to equal 1, we have the more consistent equation

$$t_7 = t_2 t_6 + t_3 t_5 + t_4 t_4 + t_5 t_3 + t_6 t_2,$$

and in general,

$$t_n = t_2 t_{n-1} + t_3 t_{n-2} + \cdots + t_{n-1} t_2 = \sum_{u+v=n+1} t_u t_v, \qquad (10)$$

where the indices of summation are all pairs u, v that add up to $n + 1$; we need no further restrictions as long as we accept the sensible convention that $t_u = 0$ for all $u \leq 1$.

Now that we have a recurrence plus boundary values, we have "solved" the problem, at least in a computational sense, since we can calculate as many values of t_n as we would like. ∎

Here is a table of the first few values.

n	2	3	4	5	6	7	8	9	10	11
t_n	1	1	2	5	14	42	132	429	1430	4862

The sequence $1, 1, 2, 5, 14, \ldots$ is known as the **Catalan** numbers (according to some conventions, the index starts at zero, so if C_r denotes the rth Catalan number, then $t_r = C_{r-2}$). Recurrence formulas such as (10) may seem rather complicated, but they are really straightforward applications of standard counting ideas (partitioning and simple encoding). Algebraically, the sum should remind you of the rule for multiplying polynomials (see page 180), which in turn should remind you of generating functions (Section 4.3).

Example 6.4.3 Use generating functions to find a formula for C_n.

Solution: Define the generating function

$$f(x) = C_0 + C_1 x + C_2 x^2 + \cdots.$$

Squaring, we have

$$f(x)^2 = C_0 C_0 + (C_1 C_0 + C_0 C_1)x + (C_2 C_0 + C_1 C_1 + C_0 C_2)x^2 + \cdots = C_1 + C_2 x + C_3 x^2 + \cdots$$

This implies that

$$x f(x)^2 = f(x) - C_0 = f(x) - 1.$$

Solving for $f(x)$ by the quadratic formula yields

$$f(x) = \frac{1 \pm \sqrt{1 - 4x}}{2x}.$$

A bit of thinking about the behavior as x approaches 0 should convince you that the correct sign to pick is negative, i.e.

$$f(x) = \frac{1 - \sqrt{1 - 4x}}{2x}.$$

Now all that remains is to find a formula for C_n by expanding $f(x)$ as a power series. We need the generalized binomial theorem (see Problem 8.4.11 on page 316), which states that

$$(1 + y)^\alpha = 1 + \sum_{n \geq 1} \frac{(\alpha)(\alpha - 1) \cdots (\alpha - n + 1)}{n!} y^n.$$

Thus,

$$\sqrt{1 - 4x} = 1 + \sum_{n \geq 1} \frac{\left(\frac{1}{2}\right)\left(\frac{1}{2} - 1\right) \cdots \left(\frac{1}{2} - n + 1\right)}{n!} (-4x)^n$$

$$= 1 + \sum_{n \geq 1} (-1)^{n-1} \frac{(1)(1)(3)(5) \cdots (2n - 3)}{2^n n!} (-4)^n x^n$$

$$= 1 - \left(\frac{2x}{n}\right) \sum_{n \geq 1} \frac{(1)(3)(5) \cdots (2n - 3) 2^{n-1} (n - 1)!}{(n - 1)!(n - 1)!} x^{n-1}$$

$$= 1 - \left(\frac{2x}{n}\right) \sum_{n \geq 1} \frac{(1)(3)(5) \cdots (2n - 3)(2)(4)(6) \cdots (2n - 2)}{(n - 1)!(n - 1)!} x^{n-1}$$

$$= 1 - \left(\frac{2x}{n}\right) \sum_{n \geq 1} \binom{2n - 2}{n - 1} x^{n-1}.$$

Finally, we have

$$f(x) = \frac{1 - \sqrt{1 - 4x}}{2x} = \frac{1}{n} \sum_{n \geq 1} \binom{2n - 2}{n - 1} x^{n-1},$$

and the coefficient of x^n is

$$\frac{1}{n + 1} \binom{2(n + 1) - 2}{(n + 1) - 1}.$$

Thus we can conclude that

$$C_n = \frac{1}{n + 1} \binom{2n}{n}.$$

∎

For a fascinating, purely combinatorial derivation of this formula, see [9], pp. 345–346.

Problems and Exercises

In the problems below, you should notice, among other things, the ubiquity of Fibonacci- and Catalan-style recurrences.

6.4.4 Consider a language whose alphabet has just one letter, but words of any length are allowed. Messages begin and end with words, and when you type a message, you hit the space bar once between words. How many different messages in this language can be typed using exactly n keystrokes?

6.4.5 For a set S of integers, define $S + 1$ to be $\{x + 1 : x \in S\}$. How many subsets S of $\{1, 2, \ldots, n\}$ satisfy $S \cup (S + 1) = \{1, 2, \ldots, n\}$? Can you generalize this?

6.4.6 Find the number of subsets of $\{1, 2, \ldots, n\}$ that contain no two consecutive elements of $\{1, 2, \ldots, n\}$.

6.4.7 Find the maximum number of regions in the plane that are determined by n "vee"s. A "vee" is two rays which meet at a point. The angle between them is any positive number.

6.4.8 Fix positive a, b, c. Define a sequence of real numbers by $x_0 = a$, $x_1 = b$, and $x_{n+1} = c x_n x_{n-1}$. Find a nice formula for x_n.

6.4.9 For each $n \geq 1$, we call a sequence of n ('s and n)'s "legal" if the parentheses match up, somehow. For example, if $n = 4$, the sequence "(()())" is legal, but "())(()" is not. Let ℓ_n denote the number of legal arrangements for the $2n$ parentheses. Find a recurrence relation for ℓ_n.

6.4.10 How many ways can you tile a $3 \times n$ rectangle with 2×1 dominos?

6.4.11 (Putnam 1996) Define a **selfish** set to be a set which has its own cardinality (number of elements) as an element. Find, with proof, the number of subsets of $\{1, 2, \ldots, n\}$ which are *minimal* selfish sets, that is, selfish sets none of whose proper subsets is selfish.

6.4.12 In Pascal's Triangle below, examine the sum along each dotted line.

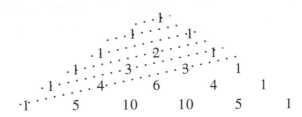

Make a conjecture and prove it.

6.4.13 Let $u(n)$ denote the number of ways in which a set of n elements can be partitioned. For example, $u(3) = 5$, corresponding to

$$\{a, b, c\}; \quad \{a, b\}, \{c\}; \quad \{a\}, \{b, c\}; \quad \{a, c\}, \{b\}; \quad \{a\}, \{b\}, \{c\}.$$

Find a recurrence relation for $u(n)$. You might hope that $u(4) = 14$, suggesting a Catalan-style recurrence, but unfortunately, $u(4) = 15$.

6.4.14 A movie theater charges \$5 for a ticket. The cashier starts out with no change, and each customer either pays with a 5-dollar bill or offers a 10-dollar bill (and gets change). Clearly, the cashier will be in trouble if there are too many customers with only 10-dollar bills. It turns out that there were $2n$ customers, and the cashier never had to turn anyone away, but after dealing with the last customer, there were no 5-dollar bills left in the cash register. Let w_n denote the number of different ways this could have happened. Find a recurrence for w_n.

6.4.15 The number of derangements D_n was defined in Problem 6.3.12. Show that

$$D_n = (n - 1)(D_{n-1} + D_{n-2}),$$

and use this to prove by induction the formula given in 6.3.12.

6.4.16 Denote by $B(n)$ the number of partitions of n into parts which are powers of two. For example, $B(6) = 6$, because

$$1+1+1+1+1+1 = 1+1+1+1+2 = 1+1+2+2 = 1+1+4 = 2+2+2 = 4+2.$$

Prove that

(a) $B(2n + 1) = B(2n)$ for all $n \geq 1$;

(b) $B(2n) = B(2n - 1) + B(n)$ for all $n \geq 1$;

(c) $B(n)$ is even for all $n \geq 2$.

Stirling Numbers

Problems 6.4.17– 6.4.23 develop a curious "dual" to the Binomial Theorem. For n, k positive integers with $n \geq k$, define $\left\{ {n \atop k} \right\}$ (called the **Stirling numbers of the second kind**)[7] to be the number of different partitions of a set with n elements into k non-empty subsets. For example, $\left\{ {4 \atop 3} \right\} = 6$, because there are 6 three- part partitions of the set $\{a, b, c, d\}$, namely

$$\{a\}, \{b\}, \{c, d\}; \quad \{a\}, \{c\}, \{b, d\}; \quad \{a\}, \{d\}, \{b, c\};$$

$$\{b\}, \{c\}, \{a, d\}; \quad \{b\}, \{d\}, \{a, c\}; \quad \{c\}, \{d\}, \{a, b\}.$$

6.4.17 Show that for all $n > 0$,

(a) $\left\{ {n \atop 1} \right\} = 1$.

(b) $\left\{ {n \atop 2} \right\} = 2^{n-1} - 1$.

(c) $\left\{ {n \atop n-1} \right\} = \binom{n}{2}$.

6.4.18 Find a combinatorial argument to explain the recurrence

$$\left\{ {n+1 \atop k} \right\} = \left\{ {n \atop k-1} \right\} + k \left\{ {n \atop k} \right\}.$$

6.4.19 Imagine that you are going to give n kids ice-cream cones, one cone per kid, and there are k different flavors available. Assuming that no flavors get mixed, show that the number of ways we can give out the cones *using all k flavors* is

$$k! \left\{ {n \atop k} \right\}.$$

6.4.20 The previous problem should have reminded you of problem 6.3.15, which stated that number of ways to give n kids ice-cream cones, one cone per kid, using all of k different flavors available, is

$$k^n - \binom{k}{1}(k-1)^n + \binom{k}{2}(k-2)^n - \binom{k}{3}(k-3)^n + \cdots + (-1)^k \binom{k}{k} 0^n.$$

Please do this now (using PIE), if you haven't already.

---

[7] See [9], Chapter 6, for information about Stirling numbers of the *first* kind.

6.4.21 Combine the two previous problems, and you get a nice formula for the Stirling numbers of the second kind. It is not closed-form, but who's complaining?

$$\left\{ {n \atop k} \right\} = \frac{1}{k!} \sum_{r=0}^{k} (-1)^r \binom{k}{r} (k-r)^n.$$

6.4.22 Returning to the ice-cream cone problem, now consider the case where we don't care whether we use all k flavors.

(a) Show that now, there are k^n different ways to feed the kids (this is an easy encoding problem that you should have done before).

(b) Partition these k^n possibilities by the exact number of flavors used. For example, let A_1 denote the set of "feedings" in which just one flavor is used; clearly $|A_1| = k$. Show that in general,

$$|A_r| = \left\{ {n \atop r} \right\} P(k, r).$$

6.4.23 *The Stirling Number "Dual" of the Binomial Theorem.* For positive integers r, define

$$x^{\underline{r}} := x(x-1)\cdots(x-r+1).$$

(Thus $x^{\underline{r}}$ is the product of r terms). Use the previous problems to show that

$$x^n = \left\{ {n \atop 1} \right\} x^{\underline{1}} + \left\{ {n \atop 2} \right\} x^{\underline{2}} + \cdots + \left\{ {n \atop n} \right\} x^{\underline{n}}.$$

Chapter 7

Number Theory

Number theory, the study of the integers **Z**, is one of the oldest branches of mathematics, and is a particularly fertile source of interesting problems that are accessible at many levels. This chapter will very briefly explore a few of the most important topics in basic number theory, focusing especially on ideas crucial for problem solvers. The presentation is a mix of exposition and problems for you to solve as you read. It is self-contained and does not assume that you have studied the subject before, but if you haven't, you may want to consult an elementary text such as [31] to learn things in more depth or at a more leisurely pace.

Several number theory topics were discussed in earlier chapters. If you read them, great. If not, the text below will remind you to read (or reread!) them at the appropriate times.

We will denote the integers (positive and negative whole numbers including zero) by **Z**, and the natural numbers (positive whole numbers) by **N**. In this chapter we will follow the convention that any Roman letter (a through z) denotes an integer. Furthermore, we reserve the letters p and q for primes.

7.1   Primes and Divisibility

If a/b is an integer, we say that b divides a or b is a factor or divisor of a. We also say this with the notation

$$b|a,$$

which you read "b divides a." Equivalently, there exists an integer m such that $a = bm$.

The Fundamental Theorem of Arithmetic

Undoubtedly, you are familiar with the set of prime numbers. A number n is **prime** if it has no divisors other than 1 and n. Otherwise, n is called **composite**. By convention, 1 is considered neither prime nor composite, so the set of primes begins with

$$2, 3, 5, 7, 11, 13, 17, 23, 29, 31, \ldots.$$

It is not at all obvious whether this sequence is finite or not. In turns out that

There are infinitely many primes.

This was Problem 2.3.21 on page 56, but in case you did not solve it, here is the complete proof, a classic argument by contradiction known to the ancient Greeks and written down by Euclid. This proof is one of the gems of mathematics. Master it.

We start by assuming that there are only finitely many primes $p_1, p_2, p_3, \ldots, p_N$. Now (the ingenious crux move!) consider the number $Q := (p_1 p_2 p_3 \cdots p_N) + 1$. Either it is prime, or it is composite. The first case would contradict the hypothesis that p_N is the largest prime, for certainly Q is much greater than p_N. But the second case also generates a contradiction, for if Q is composite, it must have at least one prime divisor, which we will call p. Observe that p cannot equal $p_1, p_2, \ldots, p_N$, for if you divide Q by any of the p_i, the remainder is 1. Consequently, p is a *new* prime that was not in the list $p_1, p_2, \ldots, p_N$, contradicting the hypothesis that this was the complete list of all primes.   ∎

7.1.1 There is a tiny gap in our proof above. We made the blithe statement that Q had to have at least one prime factor. Why is this true? More generally, prove that every natural number greater than 1 can be factored completely into primes. For example, $360 = 2^3 \cdot 3^2 \cdot 5$. Suggestion: use strong induction, if you want to be formal.

In fact, this factorization is unique, up to the order that we write the primes. This property of unique factorization is called the **Fundamental Theorem of Arithmetic (FTA)**.[1] We call the grouping of factors into primes the **prime-power factorization (PPF)**. An ugly, but necessary, notation is sometimes to write a number n in "generic" PPF:

$$n = p_1^{e_1} p_2^{e_2} \cdots p_t^{e_t}.$$

You probably are thinking, "What is there to prove about the FTA? It's obvious." In that case, you probably also consider the following "obvious."

Let x, y be integers satisfying $5x = 3y$. Then $3 \mid x$ and $5 \mid y$.

But how do you prove this rigorously? You reason that $5x$ is a multiple of 3 and since 3 and 5 are different primes, x has to contain the 3 in its PPF. This reasoning depended on the FTA, for if we did not have unique factorization, then it may be quite possible for a number to be factored in one way and have a 5 as one of its primes with no 3, and factored another way and have a 3 as one of its primes, without a 5.

Example 7.1.2 Not convinced? Consider the set $E := \{0, \pm 2, \pm 4, \ldots\}$ which contains only the even integers.

- Notice that 2, 6, 10, 30 are all "primes" in E since they cannot be factored *in E*.
- Observe also that

$$60 = 2 \cdot 30 = 6 \cdot 10;$$

---

[1] "FTA" is also used to abbreviate the Fundamental Theorem of Algebra.

in other words, 60 has two completely different E-prime factorizations, and consequently, if $x, y \in E$ satisfy $2x = 10y$, even though 2 and 10 are both E-primes, we *cannot* conclude that $10|x, 2|y$. After all, x could equal 30, and y could equal 6. In this case, x is *not* a multiple of 10 in E, since 30 is an E-prime. Likewise y is not a multiple of 2.

This example is rather contrived, but it should convince you that the FTA has real content. Certain number systems have unique factorization, and others do not. The set $\mathbf{Z}$ possesses important properties which make the FTA true. We will discover these properties and construct a proof of the FTA in the course of this section, developing useful problem-solving tools along the way.

GCD, LCM, and the Division Algorithm

Given two natural numbers a, b, their **greatest common factor**, written (a, b) or sometimes as $\text{GCD}(a, b)$, is defined to be the largest integer which divides both a and b. For example, $(66, 150) = 6$ and $(100, 250) = 25$ and $(4096, 999) = 1$. If the GCD of two numbers is 1, we say that the two numbers are **relatively prime**. For example, $(p, q) = 1$ if p and q are different primes. We will frequently use the notation $a \perp b$ in place of $(a, b) = 1$. Likewise, we define the **least common multiple**, or LCM, of a and b to be the least positive integer which is a multiple of both a and b. We use the notation $[a, b]$ or $\text{LCM}(a, b)$.[2]

7.1.3 Here are some important facts about GCD and LCM that you should think about and verify. [You may assume the truth of the FTA, if you wish, for (a), (b), and (c).]

(a) $a \perp b$ is equivalent to saying that a and b share no common primes in their PPFs.

(b) If $a = p_1^{e_1} p_2^{e_2} \cdots p_t^{e_t}$ and $b = p_1^{f_1} p_2^{f_2} \cdots p_t^{f_t}$ (where some of the exponents may be zero), then

$$(a, b) = p_1^{\min(e_1, f_1)} p_2^{\min(e_2, f_2)} \cdots p_t^{\min(e_t, f_t)},$$
$$[a, b] = p_1^{\max(e_1, f_1)} p_2^{\max(e_2, f_2)} \cdots p_t^{\max(e_t, f_t)}.$$

For example, the GCD of $360 = 2^3 \cdot 3^2 \cdot 5$ and $1597050 = 2 \cdot 3^3 \cdot 5^2 \cdot 7 \cdot 13^2$ is $2^1 \cdot 3^2 \cdot 5^1 \cdot 7^0 \cdot 13^0 = 90$.

(c) $(a, b)[a, b] = ab$ for any positive integers a, b.

(d) If $g|a$ and $g|b$, then $g|ax + by$, where x and y can be any integers. We call a quantity like $ax + by$ a **linear combination** of a and b.

(e) Consecutive numbers are always relatively prime.

(f) If there exists x, y such that $ax + by = 1$, then $a \perp b$.

7.1.4 Recall the **division algorithm**, which you encountered in Problem 3.2.15 on page 92.

---

[2]We can also define GCD and LCM for more than two integers. For example, $(70, 100, 225) = 5$ and $[1, 2, 3, 4, 5, 6, 7, 8, 9] = 2520$.

Let a and b be positive integers, $b \geq a$. Then there exist integers q, r satisfying $q \geq 1$ and $0 \leq r < a$ such that

$$b = qa + r.$$

(a) If you haven't proven this yet, do so now (use the extreme principle).

(b) Show by example that the division algorithm does not hold in the number system E defined in Example 7.1.2 on page 244. Why does it fail? What is E missing that $\mathbf{Z}$ has?

7.1.5 An important consequence of the division algorithm, plus 7.1.3(d), is that

*The greatest common divisor of a and b is the **smallest** positive linear combination of a and b.*

For example, $(8, 10) = 2$, and sure enough, if x, y range through all integers (positive and negative), the smallest positive value of $8x + 10y$ is 2 (when, for example, $x = -1, y = 1$). Another example: $7 \perp 11$, which means there exist integers x, y such that $7x + 11y = 1$. It is easy to find possible values of x and y by trial-and-error; $x = -3, y = 2$ is one such pair.

Let's prove 7.1.5; it is rather dry, but a great showcase for the use of the extreme principle plus argument by contradiction. Define u to be the smallest positive linear combination of a and b and let $g := (a, b)$. We know from 7.1.3(d) that g divides any linear combination of a and b, and so certainly g divides u. This means that $g \leq u$. We would like to show that in fact, $g = u$. We will do this by showing that u is a common divisor of a and b. Since u is greater than or equal to g, and g is the *greatest* common divisor of a and b, this would force g and u to be equal.

Suppose, on the contrary, that u is not a common divisor of a and b. Without loss of generality, suppose that u does not divide a. Then by the division algorithm, there exists a quotient $k \geq 1$ and *positive* remainder $r < u$ such that

$$a = ku + r.$$

But then $r = a - ku$ is positive and *less than u*. This is a contradiction, because r is also a linear combination of a and b, yet u was defined to be the smallest positive linear combination! Consequently, u divides a, and likewise u divides b. So u is a common divisor; thus $u = g$. ∎

This linear combination characterization of the GCD is really quite remarkable, for at first one would think that PPF's are needed to compute the GCD of two numbers. But in fact, computing the GCD does not rely on PPF's; we don't need 7.1.3(b) but can use 7.1.5 instead. In other words, *we do not need to assume the truth of the FTA in order to compute the GCD.*

7.1.6 Here is another consequence of 7.1.5:

If p is a prime and p|ab, then p|a or p|b.

We shall prove this, *without using the FTA*. Let us argue by contradiction. Assume that p divides neither a nor b. If p does not divide a, then $p \perp a$, since p is a prime. Then there exist integers x, y such that $px + ay = 1$. Now we need to use the hypothesis that $p|ab$. Let's get ab into the picture by multiplying the last equality by b:

$$pxb + aby = b.$$

But now we have written b as the sum of two quantities, each of which are multiples of p. Consequently, b is a multiple of p, a contradiction. ∎

Finally, we can prove the FTA. Statement 7.1.6 is the key idea that we need. Let's avoid notational complexity by considering a concrete example. Let $n = 2^3 \cdot 7^6$. How do we show that this factorization is unique? First of all, we can show that n couldn't contain any other prime factors. For suppose that, say, $17|n$. Then repeated applications of 7.1.6 would force the conclusion that either $17|2$ or $17|7$, which is impossible, as no two different primes can divide one another. The other possibility is that n has 2 and 7 as factors, but there is a factorization with different exponents; for example, maybe it is also true that $n = 2^8 \cdot 7^3$. In this case we would have

$$2^3 \cdot 7^6 = 2^8 \cdot 7^3,$$

and after dividing out the largest exponents of each prime that we can from each side, we get

$$7^3 - 2^5.$$

This reduces to the first case: we cannot have two factorizations with different primes. You should check with a few more examples that this argument can be completely generalized. We can rest; FTA is true. ∎

This important property of integers was a consequence of the division algorithm, which in turn was a consequence of the well-ordering principle and the fact that 1 is an integer. That's all that we needed!

Here is a simple application of FTA-style reasoning—a solution to Problem 5.4.11 on page 188.

Example 7.1.7 Recall that polynomial with integer coefficients is called monic if the term with highest degree has coefficient equal to 1. Prove that if a monic polynomial has a rational zero, then this zero must in fact be an integer.

Solution: Let the polynomial be $f(x) := x^n + a_{n-1}x^{n-1} + \cdots + a_1x + a_0$. Let u/v be a zero of this polynomial. The crux move: *without loss of generality, assume that $u \perp v$.* Then we have

$$\frac{u^n}{v^n} + \frac{a_{n-1}u^{n-1}}{v^{n-1}} + \cdots + \frac{a_1u}{v} + a_0 = 0.$$

The natural step now is to get rid of fractions, by multiplying both sides by v^n. This gives us

$$u^n + a_{n-1}u^{n-1}v + \cdots + a_1uv^{n-1} + a_0v^n = 0,$$

or

$$u^n = -\left(a_{n-1}u^{n-1}v + \cdots + a_1 uv^{n-1} + a_0 v^n\right).$$

Since the right-hand side is a multiple of v, and v shares no factor with u^n, we must conclude that $v = \pm 1$; i.e., u/v is an integer.   ∎

Now that we understand GCD well, let us finish up an old problem which we started in Example 2.2.2 on page 31.

Example 7.1.8 (AIME 1985) The numbers in the sequence

$$101, 104, 109, 116, \ldots$$

are of the form $a_n = 100 + n^2$, where $n = 1, 2, 3, \ldots$. For each n, let d_n be the greatest common divisor of a_n and a_{n+1}. Find the maximum value of d_n as n ranges through the positive integers.

Solution: Recall that we considered the sequence $a_n := u + n^2$, where u was fixed, and after some experimentation discovered that a_{2u}, a_{2u+1} seemed fruitful, since $a_{2u} = u(4u + 1)$ and $a_{2u+1} = (u + 1)(4u + 1)$. This of course means that $4u + 1$ is a common factor of a_{2u} and a_{2u+1}. In fact, since consecutive numbers are relatively prime (7.1.3), we have

$$(a_{2u}, a_{2u+1}) = 4u + 1.$$

It remains to show that this is the largest possible GCD. Make the abbreviations

$$a := a_n = u + n^2, \quad b := a_{n+1} = u + (n + 1)^2, \quad g := (a, b).$$

Now we shall explore profitable linear combinations of a and b, with the hope that we will get $4u + 1$. We have

$$b - a = 2n + 1.$$

This is nice and simple, but how do we get rid of the n? Since $a = u + n^2$, let's build the n^2 term. We have

$$n(b - a) = 2n^2 + n, \quad 2a = 2u + 2n^2,$$

and thus

$$2a - n(b - a) = 2u - n.$$

Doubling this and adding it back to $b - a$ yields

$$2\left(2a - n(b - a)\right) + (b - a) = 2(2u - n) + (2n + 1) = 4u + 1.$$

In other words, we have produced a linear combination of a and b [specifically, $(3 + 2n)a + (1 - 2n)b$] that is equal to $4u + 1$. Thus $g|(4u + 1)$, no matter what n equals. Since this value has been achieved, we conclude that indeed $4u + 1$ is the largest possible GCD of consecutive terms.   ∎

Notice that we did not need clairvoyance to construct the linear combination in one fell swoop. All it took was some patient "massage" moving toward the goal of eliminating the n.

Problems and Exercises

7.1.9 Write out a formal proof of the FTA.

7.1.10 Review Example 3.2.4 on page 85, an interesting problem involving GCD, LCM and the extreme principle.

7.1.11 Prove that the fraction $(n^3 + 2n)/(n^4 + 3n^2 + 1)$ is in lowest terms for every positive integer n.

7.1.12 *The Euclidean Algorithm.* Repeated use of the division algorithm allows one to easily compute the GCD of two numbers. For example, we shall compute $(333, 51)$:

$$333 = 6 \cdot 51 + 27;$$
$$51 = 1 \cdot 27 + 24;$$
$$27 = 1 \cdot 24 + 3;$$
$$24 = 8 \cdot 3 + 0.$$

We start by dividing 333 by 51. Then we divide 51 by the remainder from the previous step. At each successive step, we divide the last remainder by the previous remainder. We do this until the remainder is zero, and our answer—the GCD—is the final non-zero remainder (in this case, 3).

Here is another example. To compute $(89, 24)$, we have

$$89 = 3 \cdot 24 + 17;$$
$$24 = 1 \cdot 17 + 7;$$
$$17 = 2 \cdot 7 + 3;$$
$$7 = 2 \cdot 3 + 1;$$

so the GCD is 1.

This method is called the **Euclidean Algorithm**. Explain why it works!

7.1.13 *Linear Diophantine Equations.* Since $17 \perp 11$, there exist integers x, y such that $17x + 11y = 1$. For example, $x = 2, y = -3$ work. Here is a neat trick for generating more integer solutions to $17x + 11y = 1$: Just let

$$x = 2 + 11t, \quad y = -3 - 17t,$$

where t is *any* integer.

(a) Verify that $x = 2 + 11t, y = -3 - 17t$ will be a solution to $17x + 11y = 1$, no matter what t is. This is a simple algebra exercise, and is really just a nice example of the add zero creatively tool.

(b) Show that *all* integer solutions to $17x + 11y = 1$ have this form; i.e., if x and y are integers satisfying $17x + 11y = 1$, then $x = 2 + 11t, y = -3 - 17t$ for some integer t.

(c) It was easy to find the solution $x = 2, y = -3$ by trial-and-error, but for larger numbers we can use the Euclidean algorithm *in reverse*. For example, use the example in (7.1.12) to find x, y such that $89x + 24y = 1$. Start by writing 1 as a linear combination of 3 and 7; then write 3 as a linear combination of 7 and 17; etc.

(d) Certainly if $x = 2$, $y = -3$ is a solution to $17x + 11y = 1$, then $x = 2u$, $y = -3u$ is a solution to $17x + 11y = u$. And as above, verify that *all* solutions are of the form $x = 2u + 11t$, $y = -3u - 17t$.

(e) This method can certainly be generalized to any linear equation of the form $ax + by = c$, where a, b, c are constants. First we divide both sides by the GCD of a and b; if this GCD is *not* a divisor of c there cannot be solutions. Then we find a single solution either by trial-and-error, or by using the Euclidean algorithm.

(f) To see another example of generating infinitely many solutions to a diophantine equation, look at the problems about Pell's Equation (7.4.17–7.4.22).

7.1.14 Consider the set $F := \{a + b\sqrt{-6}, a, b \in \mathbf{Z}\}$. Define the **norm** function N by $N(a + b\sqrt{-6}) = a^2 + 6b^2$. This is a "natural" definition; it is just the square of the magnitude of the complex number $a + bi\sqrt{6}$, and it is a useful thing to play around with, because its values are integers.

(a) Show that the "norm of the product is equal to the product of the norm"; i.e., if $\alpha, \beta \in F$, then $N(\alpha\beta) = N(\alpha)N(\beta)$.

(b) Observe that if $\alpha \in F$ is not an integer (i.e., if it has a non-zero imaginary part), then $N(\alpha) \geq 6$.

(c) An F-prime is an element of F which has no factors in F other than 1 and itself. Show that 2 and 5 are F-primes.

(d) Show that 7 and 31 are *not* F-primes.

(e) Likewise, show that $2 - \sqrt{-6}$ and $2 - \sqrt{-6}$ are F-primes.

(f) Conclude that F does not possess unique factorization.

7.1.15 Prove that consecutive Fibonacci numbers are always relatively prime.

7.1.16 It is possible to prove that $(a, b)[a, b] = ab$ without using PPF, but instead just using the definitions of GCD and LCM. Try it!

7.1.17 (USAMO 1972) Let $a, b, c \in \mathbf{N}$. Show that

$$\frac{[a, b, c]^2}{[a, b][b, c][c, a]} = \frac{(a, b, c)^2}{(a, b)(b, c)(c, a)}.$$

7.1.18 Show that the sum of two consecutive primes is never twice a prime.

7.1.19 Is it possible for 4 consecutive integers to be composite? How about 5? More than 5? Arbitrarily many?

7.1.20 Show that

$$1 + \frac{1}{2} + \frac{1}{3} + \cdots + \frac{1}{n}$$

can never be an integer.

7.1.21 Show that $\binom{p}{k}$ is a multiple of p for all $0 < k < p$.

7.1.22 Show that 1000! ends with 249 zeros. Generalize!

7.1.23 Let $r \in \mathbf{N}$. Show that $\binom{p^r}{k}$ is a multiple of p for $0 < k < p^r$.

7.1.24 Show that if n divides a single Fibonacci number, then it will divide infinitely many Fibonacci numbers.

7.1.25 Prove that there are infinitely many primes of the form $4k + 3$; i.e. the sequence $\{3, 7, 11, 19, \ldots\}$ is infinite.

7.1.26 Prove that there are infinitely many primes of the form $6n - 1$.

7.1.27 A polynomial with integer coefficients is called **primitive** if its coefficients are relatively prime. For example, $3x^2 + 9x + 7$ is primitive while $10x^2 - 5x + 15$ is not.

(a) Prove that the product of two primitive polynomials is primitive. Hint: extreme principle.

(b) Use this to prove **Gauss's Lemma**: If a polynomial with integer coefficients can be factored into polynomials with rational coefficients, it can also be factored into primitive polynomials with integer coefficients. (See also page 186.)

7.1.28 True or false and why:

(a) The product of two consecutive positive integers cannot be equal to a perfect square.

(b) The product of three consecutive positive integers cannot be equal to a perfect square.

7.1.29 (Russia, 1995) The sequence $a_1, a_2, \ldots$ of natural numbers satisfies

$$\mathrm{GCD}(a_i, a_j) = \mathrm{GCD}(i, j)$$

for all $i \neq j$. Prove that $a_i = i$ for all i.

7.1.30 (USAMO 1973) Show that the cube roots of three distinct prime numbers cannot be three terms (not necessarily consecutive) of an arithmetic progression.

7.2 Congruence

Congruence notation was introduced on page 49. Recall that if $a - b$ is a multiple of m, we write $a \equiv b \pmod{m}$ (read "a is congruent to b modulo m").

7.2.1 Here are several facts that you should verify immediately.

(a) If you divide a by b and get a remainder of r, that is equivalent to saying that $a \equiv r \pmod{b}$.

(b) There are only m "different" integers modulo m, since there are only m different remainders $0, 1, 2, \ldots, m - 1$. We call these m numbers the **integers modulo** m or Z_m. For example, in Z_6 we have $5 + 5 = 4$, $2^5 = 2$, etc. Another term that is used is **residue** modulo m. For example, one might say that 7 and 3 are different residues modulo 5, but equal residues modulo 4.

(c) The statement $a \equiv b \pmod{m}$ is equivalent to saying that there exists an integer k such that $a = b + mk$.

(d) If $a \equiv b \pmod{m}$ and $c \equiv d \pmod{m}$, then $a + c \equiv b + d \pmod{m}$ and $ac \equiv bd \pmod{m}$.

The last statement is especially useful. For example, suppose we wanted to find the remainder when we divide 2^{1000} by 17. Note that $2^4 = 16 \equiv -1 \pmod{17}$. Thus $2^{1000} = (2^4)^{250} \equiv (-1)^{250} \equiv 1$, so the remainder is 1.

7.2.2 Two more examples of this method yield the following well-known divisibility rules. Prove them and learn them!

(a) If a number is written in base 10, then it is congruent to the sum of its digits modulo 9 and modulo 3.

(b) If a number is written in base 10, then it is congruent modulo 11 to the units digit − tens digit + hundreds digit − thousands digit, etc.

Viewing a problem modulo m for a suitably chosen m is a wonderful simplification tactic because it reduces the infinite universe of integers to the finite world of Z_m. You have encountered this idea before with parity (page 104), which is just the case $m = 2$, as well as other values of m (page 110). Often (but not always) we turn to prime values of m, since primes are simpler, more "fundamental" objects which are generally easier to understand. In general,

> When beginning a number theory investigation, assume that the variables are prime or at least relatively prime. Often the general case follows from the prime case, with just a few "technical" details.

Example 7.2.3 *Fermat's Last Theorem.* Let $n \geq 3$. Prove that the equation

$$x^n + y^n = z^n$$

has no non-zero integer solutions.

We are not going to prove this; Fermat's Last Theorem was perhaps the most famous outstanding problem in all of mathematics. The French mathematician Fermat conjectured its truth over 300 years ago, and the problem remained unsolved until 1995. But we shall point out two simplifications.

• Without loss of generality, n is prime. For example, if $x^3 + y^3 = z^3$ has no non-zero integer solutions, the same will be true of $x^{12} + y^{12} = z^{12}$, since this latter equation can be rewritten as

$$(x^4)^3 + (y^4)^3 = (z^4)^3.$$

- Likewise, we may assume that x, y, z is a **primitive** solution; i.e., x, y, z have no common factor (other than 1). To see why, suppose that g is the greatest common divisor of x, y, z. Then $x = ga, y = gb, z = gc$ for some integers a, b, c. Notice that a, b, c have no common factor, and

$$x^n + y^n = z^n \iff (ga)^n + (gb)^n = (gc)^n \iff a^n + b^n = c^n,$$

where the third equality followed from the second after division by g^n.

What's So Good About Primes?

One reason that primes are so convenient is that unique "multiplicative inverses" exist. For example, in Z_6, the number 5 has a multiplicative inverse, namely itself, since $5 \cdot 5 \equiv 1 \pmod 6$. However, 2 has no multiplicative inverse, nor does 4. In contrast, all the nonzero elements of Z_7 have inverses, and they are unique. We have

$$1 \cdot 1 \equiv 2 \cdot 4 \equiv 3 \cdot 5 \equiv 6 \cdot 6 \equiv 1 \pmod 7,$$

so the inverses of $1, 2, 3, 4, 5, 6$ in Z_7 are respectively $1, 4, 5, 2, 3, 6$. In general,

If p is prime, and x is not a multiple of p, then there is a unique $y \in \{1, 2, 3, \ldots, p - 1\}$ such that $xy \equiv 1 \pmod p$.

This was proven in Example 2.3.4 on page 49. The penultimate step was the very useful fact that

If p is prime, and $x \not\equiv 0 \pmod p$, then the $(p - 1)$ numbers

$$x, 2x, 3x, \ldots, (p - 1)x$$

are distinct in Z_p.

Equivalently, if p is prime, and $x \not\equiv 0 \pmod p$, then in Z_p,

$$\{x, 2x, 3x, \ldots, (p - 1)x\} = \{1, 2, 3, \ldots, p - 1\}. \tag{1}$$

For example, if $p = 7$ and $x = 4$, we have

$$4 \cdot 1 \equiv 4, \quad 4 \cdot 2 \equiv 1, \quad 4 \cdot 3 \equiv 5, \quad 4 \cdot 4 \equiv 2, \quad 4 \cdot 5 \equiv 6, \quad 4 \cdot 6 \equiv 3,$$

verifying that the set of the non-zero multiples of 4 in Z_7 is just a permutation of the non-zero values of Z_7.

Fermat's Little Theorem

Let's derive a nice consequence of (1). Let p be a prime and let $a \perp p$. Since the sets

$$\{a, 2a, 3a, \ldots, (p - 1)a\} \quad \text{and} \quad \{1, 2, 3, \ldots, p - 1\}$$

are equal in Z_p, the products of their elements are equal. In other words,

$$a \cdot 2a \cdot 3a \cdot \cdots \cdot (p - 1)a \equiv 1 \cdot 2 \cdot 3 \cdot \cdots \cdot (p - 1) \pmod p,$$

which is equivalent to

$$a^{p-1}(p - 1)! \equiv (p - 1)! \pmod p.$$

Since p is a prime, $(p-1)! \perp p$ and consequently, we can "cancel out" the $(p-1)!$ from both sides,[3] obtaining **Fermat's little theorem**:

$$a^{p-1} \equiv 1 \pmod{p}.$$

(The word "little" is used to distinguish it from Fermat's Last Theorem.) We can also eliminate the hypothesis that a be non-zero[4] modulo p by multiplying by a, producing the equivalent statement that

$$a^p \equiv a \pmod{p}$$

for all a, if p is prime.

The next example shows how one can use Fermat's little theorem to create composite numbers.

Example 7.2.4 (Germany, 1995) Let a and b be positive integers and let the sequence $(x_n)_{n \geq 0}$ be defined by $x_0 = 1$ and $x_{n+1} = ax_n + b$ for all nonnegative integers n. Prove that for any choice of a and b, the sequence $(x_n)_{n \geq 0}$ contains infinitely many composite numbers.

Solution: First, let's experiment. Try $a = 5, b = 7$ and the sequence is

$$1, 12, 67, 342, 1717, 8592, \ldots.$$

Clearly, every other term will be even, and that will always hold if a and b are both odd. If a and b have opposite parity, for example, $a = 2, b = 3$, our sequence is

$$1, 5, 13, 29, 61, 125, 253, 509, 1021, 2045, \ldots.$$

Notice that, starting with $a_2 = 5$, every 4th term appears to be a multiple of 5. Can we prove this? Let $u = a_k$ be a multiple of 5. Then the next terms are

$$a_{k+1} = 2u + 3,$$
$$a_{k+2} = 2(2u + 3) + 3 = 4u + 9,$$
$$a_{k+3} = 2(4u + 9) + 3 = 8u + 21,$$
$$a_{k+4} = 2(8u + 21) + 3 = 16u + 45.$$

So indeed, a_{k+4} will be a multiple of 5. It seems that we can generate composite numbers in this sequence if we are careful. Let's try a formal argument by contradiction, using any given values of a and b.

Assume that the conclusion is false, i.e., that the sequence only contains finitely many composite numbers. This means that "eventually" the sequence is only primes; i.e., there exists some M such that a_n is prime for all $n > M$. This statement has "footholds," since prime numbers are often easier to deal with than composite numbers, and in particular, we have nice tools we can use with them, like Fermat's little theorem. Now we need to use our experimentation to produce a contradiction that works for any values of a and b!

[3] See Problem 7.2.5 on page 255.

[4] The phrases "non-zero modulo p," "relatively prime to p," "not a multiple of p," and "non-zero in Z_p" are all equivalent.

Let $n > M$ and let $x_n = p$, a prime. What happens later in the sequence? We have $x_{n+1} = ap + b$, $x_{n+2} = a^2 p + ab + b$, $x_{n+3} = a^3 p + a^2 b + ab + b$, etc. In general, for any k we have

$$x_{n+k} = a^n p + b(a^{n-1} + a^{n-2} + \cdots + 1) = a^n p + b \left(\frac{a^n - 1}{a - 1} \right).$$

It would be nice if we could show that x_{n+k} is *not* a prime for some k. Since p is already in the picture, let's use it: if we could show that $b \left(\dfrac{a^n - 1}{a - 1} \right)$ was a multiple of p, we'd be done! Now Fermat's little theorem comes to the rescue: As long as a is not a multiple of p, we have $a^{p-1} \equiv 1 \pmod{p}$, so if we choose $n = p - 1$, then $a^n - 1$ is a multiple of p. However, we are dividing this by $a - 1$. What if p divides $a - 1$? Then we'd be in trouble. So we won't worry about that for now! Just assume that p doesn't divide $a - 1$.

Recapping, if we assume that neither p divides neither a nor $a - 1$, then x_{n+p-1} will be a multiple of p, which is a contradiction. How do we ensure that p satisfies these two conditions? We are given a; it is fixed. Hence there are only finitely many primes p which either divide a or divide $a - 1$. But by assumption, for $n > M$, the values of a_n are all primes, and there will be infinitely many primes, since the sequence is increasing. Since we have infinitely many primes to pick from, just pick one that works! ∎

Problems and Exercises

7.2.5 Let $ar \equiv br \pmod{m}$, with $r \perp m$. Show that we can "cancel out" the r from both sides of this congruence and conclude that $a \equiv b \pmod{m}$. What happens if $(m, r) > 1$?

7.2.6 Show that if $a^2 + b^2 = c^2$, then $3 | ab$.

7.2.7 If $x^3 + y^3 = z^3$, show that one of the three must be a multiple of 7.

7.2.8 Find all x such that $x^2 + 2$ is a prime.

7.2.9 Let N be a number with 9 distinct non-zero digits, such that, for each k from 1 to 9 inclusive, the first k digits of N form a number which is divisible by k. Find N (there is only one answer).

7.2.10 Let $f(n)$ denote the sum of the digits of n.

(a) For any integer n, prove that eventually the sequence

$$f(n), f(f(n)), f(f(f(n))), \ldots$$

will become constant. This constant value is called the **digital sum** of n.

(b) Prove that the digital sum of the product of any two twin primes, other than 3 and 5, is 8. (Twin primes are primes that are consecutive odd numbers, such as 17 and 19.)

(c) (IMO 1975) Let $N = 4444^{4444}$. Find $f(f(f(n)))$, without a calculator.

7.2.11 The **order** of a modulo p is defined to be the smallest positive integer k such that $a^k \equiv 1 \pmod{p}$. Show that the order of a must divide $p - 1$, if p is a prime.

7.2.12 Let p be a prime. Show that if $x^k \equiv 1 \pmod{p}$ for all nonzero x, then $p - 1$ divides k.

7.2.13 Let $\{a_n\}_{n \geq 0}$ be a sequence of integers satisfying $a_{n+1} = 2a_n + 1$. Is there an a_0 so that the sequence consists entirely of prime numbers?

7.2.14 Prove Fermat's little theorem by induction on a (you'll need the binomial theorem).

7.2.15 Our discussion of Fermat's little theorem involved the quantity $(p-1)!$. Please reread the proof of Wilson's theorem (Example 3.1.9 on page 76), which states that if p is prime, then $(p-1)! \equiv -1 \pmod{p}$.

7.2.16 *The Chinese Remainder Theorem.* Consider the following simultaneous congruence.

$$x \equiv 3 \pmod{11},$$
$$x \equiv 5 \pmod{6}.$$

It is easy to find a solution, $x = 47$, by inspection. Here's another method. Since $6 \perp 11$, we can find a linear combination of 6 and 11 that equals one, for example, $(-1) \cdot 11 + 2 \cdot 6 = 1$. Now compute

$$5 \cdot (-1) \cdot 11 + 3 \cdot 2 \cdot 6 = -19.$$

This number is a solution, modulo $66 = 6 \cdot 11$. Indeed, $47 \equiv -19 \pmod{66}$.

(a) Why does this work?

(b) Note that the two moduli (which were 11 and 6 in the example) must be relatively prime. Show by example that there may not always be a solution to a simultaneous congruence if the two moduli share a factor.

(c) Let $m \perp n$, let a and b be arbitrary, and let x simultaneously satisfy the congruences $x \equiv a \pmod{m}$ and $x \equiv b \pmod{n}$. The algorithm described above will produce a solution for x. Show that this solution is *unique* modulo mn.

(d) Show that this algorithm can be extended to any finite number of simultaneous congruences, as long as the moduli are pairwise relatively prime.

(e) Show that there exist 3 consecutive numbers, each of which is divisible by the 1999th power of an integer.

(f) Show that there exist 1999 consecutive numbers, each of which is divisible by the cube of an integer.

7.2.17 (USAMO 1995) Let p be an odd prime. The sequence $(a_n)_{n \geq 0}$ is defined as follows: $a_0 = 0$, $a_1 = 1$, ..., $a_{p-2} = p - 2$ and, for all $n \geq p - 1$, a_n is the least integer greater than a_{n-1} that does not form an arithmetic sequence of length p with any of the preceding terms. Prove that, for all n, a_n is the number obtained by writing n in base $p - 1$ and reading the result in base p.

7.2.18 (Putnam 1995) The number $d_1 d_2 \ldots d_9$ has nine (not necessarily distinct) decimal digits. The number $e_1 e_2 \ldots e_9$ is such that each of the nine 9-digit numbers formed by replacing just one of the digits d_i is $d_1 d_2 \ldots d_9$ by the corresponding digit e_i $(1 \leq i \leq 9)$ is divisible by 7. The number $f_1 f_2 \ldots f_9$ is related to $e_1 e_2 \ldots e_9$ in the same way: that is, each of the nine numbers formed by replacing one of the e_i by the corresponding f_i is divisible by 7. Show that, for each i, $d_i - f_i$ is divisible by 7. (For example, if $d_1 d_2 \ldots d_9 = 199501996$, then e_6 may be 2 or 9, since 199502996 and 199509996 are multiples of 7.)

7.2.19 (Putnam 1996) Suppose a, b, c, d are integers with $0 \leq a \leq b \leq 99$, $0 \leq c \leq d \leq 99$. For any integer i, let $n_i = 101i + 1002^i$. Show that if $n_a + n_b$ is congruent to $n_c + n_d$ mod 10100, then $a = c$ and $b = d$.

7.3 Number Theoretic Functions

Of the infinitely many functions with domain $\mathbf{N}$, we will single out a few that are especially interesting. Most of these functions are **multiplicative**. A function f with this property satisfies

$$f(ab) = f(a)f(b)$$

whenever $a \perp b$.

7.3.1 Show that if $f : \mathbf{N} \to \mathbf{N}$ is multiplicative, then $f(1) = 1$.

7.3.2 If a function f is multiplicative, then in order to know all values of f, it is sufficient to know the values of $f(p^r)$ for each prime p and each $r \in \mathbf{N}$.

Divisor Sums

Define

$$\sigma_r(n) = \sum_{d | n} d^r,$$

where r can be any integer. For example,

$$\sigma_2(10) = 1^2 + 2^2 + 5^2 + 10^2 = 130.$$

In other words, $\sigma_r(n)$ is the sum of the rth power of the divisors of n. Although it is useful to define this function for any value of r, in practice we rarely consider values other than $r = 0$ and $r = 1$.

7.3.3 Notice that $\sigma_0(n)$ is equal to the number of divisors of n. This function is usually denoted by $d(n)$. You encountered it in Example 3.1 on page 75 and Problem 6.1.21 on page 212. Recall that if

$$n = p_1^{e_1} p_2^{e_2} \cdots p_t^{e_t}$$

is the prime factorization of n, then

$$d(n) = (e_1 + 1)(e_2 + 1) \cdots (e_t + 1).$$

From this formula we can conclude that $d(n)$ is a multiplicative function.

7.3.4 Show by examples that $d(ab)$ does not always equal $d(a)d(b)$ if a and b are *not* relatively prime. In fact, prove that $d(ab) < d(a)d(b)$ when a and b are not relatively prime.

7.3.5 The function $\sigma_1(n)$ is equal to the sum of the divisors of n and is usually denoted simply by $\sigma(n)$.

(a) Show that $\sigma(p^r) = \dfrac{p^{r+1} - 1}{p - 1}$ for prime p and positive r.

(b) Show that $\sigma(pq) = (p+1)(q+1)$ for distinct primes p, q.

7.3.6 *An Important Counting Principle.* Let $n = ab$, where $a \perp b$. Show that if $d|n$ then $d = uv$, where $u|a$ and $v|b$. Moreover, this is a 1-1 correspondence; each different pair of u, v satisfying $u|a$, $v|b$ produces a different $d := uv$ which divides n.

7.3.7 Use (7.3.6) to conclude that $\sigma(n)$ is a multiplicative function. For example, $12 = 3 \cdot 4$, with $3 \perp 4$, and we have

$$
\begin{aligned}
\sigma(12) &= 1 + 2 + 4 + 3 + 6 + 12 \\
&= (1 + 2 + 4) + (1 \cdot 3 + 2 \cdot 3 + 4 \cdot 3) \\
&= (1 + 2 + 4)(1 + 3) \\
&= \sigma(4)\sigma(3).
\end{aligned}
$$

7.3.8 Notice in fact, that (7.3.6) can be used to show that $\sigma_r(n)$ is multiplicative, no matter what r is.

7.3.9 In fact, we can do more. The counting principle that we used can be reformulated in the following way: Let $n = ab$ with $a \perp b$, and let f be any multiplicative function. Carefully verify (try several concrete examples) that

$$
\begin{aligned}
\sum_{d|n} f(d) &= \sum_{u|a} \left(\sum_{v|b} f(uv) \right) \\
&= \sum_{u|a} \left(\sum_{v|b} f(u)f(v) \right) \\
&= \left(\sum_{u|a} f(u) \right) \left(\sum_{v|b} f(v) \right).
\end{aligned}
$$

We have proven the following general fact.

Let

$$F(n) := \sum_{d|n} f(d).$$

If f is multiplicative then F will be multiplicative as well.

Phi and Mu

Define $\phi(n)$ to be the number of positive integers less than or equal to n which are relatively prime to n. For example, $\phi(12) = 4$, since $1, 5, 7, 11$ are relatively prime to 12.

We can use PIE (the principle of inclusion-exclusion; see section 6.3) to evaluate $\phi(n)$. For example, to compute $\phi(12)$, the only relevant properties to consider are divisibility by 2 or divisibility by 3 because 2 and 3 are the only primes that divide 12. As we have done many times, we shall count the complement; i.e., we will count how many integers between 1 and 12 (inclusive) share a factor with 12. If we let M_k denote the multiples of k up to 12, then we need to compute

$$|M_2 \cup M_3|,$$

since any number that shares a factor with 12 will either be a multiple of 2 or 3 (or both). Now PIE implies that

$$|M_2 \cup M_3| = |M_2| + |M_3| - |M_2 \cap M_3|.$$

Because $2 \perp 3$, we can rewrite $M_2 \cap M_3$ as M_6. Thus we have

$$\phi(12) = 12 - |M_2 \cup M_3| = 12 - (|M_2| + |M_3|) + |M_6| = 12 - (6+4) + 2 = 4.$$

7.3.10 Let p, q be distinct primes. Show that

(a) $\phi(p) = p - 1$,

(b) $\phi(p^r) = p^r - p^{r-1} = p^{r-1}(p-1) = p^r \left(1 - \dfrac{1}{p}\right)$,

(c) $\phi(pq) = (p-1)(q-1) = pq \left(1 - \dfrac{1}{p}\right) \left(1 - \dfrac{1}{q}\right)$,

(d) $\phi(p^r q^s) = p^{r-1}(p-1)q^{s-1}(q-1) = p^r q^s \left(1 - \dfrac{1}{p}\right) \left(1 - \dfrac{1}{q}\right)$.

These special cases above certainly suggest that ϕ is multiplicative. This is easy to verify with PIE. For example, suppose that n contains only the distinct primes p, q, w. If we let M_k denote the number of positive multiples of k less than or equal to n, we have

$$\phi(n) = n - \left(|M_p| + |M_q| + |M_w|\right) + \left(|M_{pq}| + |M_{pw}| + |M_{qw}|\right) - |M_{pqw}|.$$

In general,

$$M_k = \left\lfloor \frac{n}{k} \right\rfloor,$$

but since p, q, w all divide n, we can drop the brackets;

$$\phi(n) = n - \left(\frac{n}{p} + \frac{n}{q} + \frac{n}{w} \right) + \left(\frac{n}{pq} + \frac{n}{pw} + \frac{n}{qw} \right) - \frac{n}{pqw}, \tag{2}$$

and this factors beautifully as

$$\phi(n) = n \left(1 - \frac{1}{p} \right) \left(1 - \frac{1}{q} \right) \left(1 - \frac{1}{w} \right).$$

If we write $n = p^r q^s w^t$, we have

$$\phi(n) = p^r \left(1 - \frac{1}{p} \right) q^s \left(1 - \frac{1}{q} \right) w^t \left(1 - \frac{1}{w} \right) = \phi(p^r)\phi(q^s)\phi(w^t),$$

using the formulas in 7.3.10. This argument certainly generalizes to any number of distinct primes, so we have established that ϕ is multiplicative. And in the process, we developed an intuitively reasonable formula. For example, consider $360 = 2^3 \cdot 3^2 \cdot 5$. Our formula says that

$$\phi(360) = 360 \left(1 - \frac{1}{2} \right) \left(1 - \frac{1}{3} \right) \left(1 - \frac{1}{5} \right),$$

and this *makes sense*; for we could argue that half of the positive integers up to 360 are odd, and two-thirds of these are not multiples of three, and four- fifths of what are left are not multiples of five. The final fraction ($\frac{1}{2} \cdot \frac{2}{3} \cdot \frac{4}{5}$) of 360 will be the numbers that share no divisors with 360. This argument is not quite rigorous. It tacitly assumes that divisibility by different primes is in some sense "independent" in a probabilistic sense. This is true, and it can be made rigorous, but this is not the place for it.[5]

Let us pretend to change the subject for a moment by introducing the **Möbius** function $\mu(n)$. We define

$$\mu(n) = \begin{cases} 1 & \text{if } n = 1; \\ 0 & \text{if } p^2 \mid n \text{ for some prime } p; \\ (-1)^r & \text{if } n = p_1 p_2 \cdots p_r, \text{ each } p \text{ a distinct prime.} \end{cases}$$

This is a rather bizarre definition, but it turns out that the Möbius function very conveniently "encodes" PIE. Here is a table of the first few values of $\mu(n)$.

n	1	2	3	4	5	6	7	8	9	10	11	12	13	14	15
$\mu(n)$	1	-1	-1	0	-1	1	-1	0	0	1	-1	0	-1	1	1

7.3.11 Verify that μ is multiplicative.

[5]See [15] for a wonderful discussion of this and related issues.

7.3.12 Use 7.3.11 and 7.3.9 to show that

$$\sum_{d|n} \mu(d) = \begin{cases} 1 & \text{if } n = 1; \\ 0 & \text{if } n > 1. \end{cases}$$

The values of $\mu(n)$ alternate sign depending on the parity of the number of prime factors of n. This is what makes the Möbius function related to PIE. For example, we could rewrite equation (2) as

$$\phi(n) = \sum_{d|n} \mu(d) \frac{n}{d}. \tag{3}$$

This works because of μ's "filtering" properties. If a divisor d in the sum above contains powers of primes larger than 1, $\mu(d) = 0$, so the term is not present. If d is equal to a single prime, say p, the term will be $-\frac{n}{p}$. Likewise, if $d = pq$, the term becomes $+\frac{n}{pq}$, etc. And of course (3) is a general formula, true for any n.

Problems and Exercises

7.3.13 Make a table, either by hand or with the aid of a computer, of the values of $d(n), \phi(n), \sigma(n), \mu(n)$ for, say, $1 \leq n \leq 100$ or so.

7.3.14 Prove that $\phi(n) = 14$ has no solutions.

7.3.15 In 7.3.4, you showed that $d(ab) < d(a)d(b)$ whenever a and b are relatively prime. What can you say about the σ function in this case?

7.3.16 Find the smallest integer n for which $\phi(n) = 6$.

7.3.17 Find the smallest integer n for which $d(n) = 10$.

7.3.18 Find $n \in \mathbf{N}$ such that $\mu(n) + \mu(n+1) + \mu(n+2) = 3$.

7.3.19 Show that for all $n, r \in \mathbf{N}$,

$$\frac{\sigma_r(n)}{\sigma_{-r}(n)} = n^r.$$

7.3.20 For $n > 1$, define

$$\omega(n) = \sum_{p|n} 1,$$

where the p in the sum must be prime. For $n = 1$, let $\omega(n) = 1$. In other words, $\omega(n)$ is the number of *distinct* prime divisors of n. For example, $\omega(12) = 2$ and $\omega(7^{344}) = 1$.

(a) Compute $\omega(n)$ for $n = 1, \ldots, 25$.

(b) What is $\omega(17!)$?

(c) Is ω multiplicative? Explain.

7.3.21 Likewise, for $n > 1$, define

$$\Omega(n) = \sum_{p^e \| n} e,$$

where again, p must be prime. For $n = 1$, we define $\Omega(n) = 1$. Thus $\Omega(n)$ is the sum of all the exponents that appear in the prime-power factorization of n. For example, $\Omega(12) = 2 + 1 = 3$, because $12 = 2^2 3^1$.

(a) Compute $\Omega(n)$ for $n = 1, \ldots, 25$.

(b) Show, with a counterexample, that Ω is *not* multiplicative.

(c) However, there is a simple formula for $\Omega(ab)$, when $(a, b) = 1$. What is it? Explain.

7.3.22 Define

$$F(n) = \sum_{d|n} g(d),$$

where $g(1) = 1$ and $g(k) = (-1)^{\Omega(k)}$ if $k > 1$. Find a simple rule for the F.

7.3.23 There are two very different ways to prove 7.3.12. One method, which you probably used already, was to observe that $F(n) := \sum_{d|n} \mu(d)$ is a multiplicative function, and then calculate that $F(p^r) = 0$ for all primes. But here is another method: ponder the equation

$$\sum_{k=0}^{\omega(n)} \binom{\omega(n)}{k} (-1)^k = (1 - 1)^{\omega(n)} = 0,$$

where $\omega(n)$ was defined in 7.3.20. Explain why this equation is true, and also why it proves 7.3.12.

7.3.24 Prove that $\phi(n) + \sigma(n) = 2n$ if and only if n is prime.

7.3.25 *Euler's Extension of Fermat's Little Theorem.* Emulate the proof of Fermat's little theorem (page 253) to prove the following:

Let $m \in \mathbf{N}$, *not necessarily prime, and let* $a \perp m$. *Then*

$$a^{\phi(m)} \equiv 1 \pmod{m}.$$

7.3.26 Let $f(n)$ be a strictly increasing multiplicative function with positive integer range satisfying $f(1) = 1$ and $f(2) = 2$. Prove that $f(n) = n$ for all n.

7.3.27 Find the last two digits of 9^{99} without using a machine.

The Möbius Inversion Formula

Problems 7.3.28–7.3.31 explore the Möbius Inversion Formula, a remarkable way to "solve" the equation $F(n) = \sum_{d|n} f(d)$ for f.

7.3.28 In 7.3.9, we showed that if $F(n) := \sum_{d|n} f(d)$, and f is multiplicative, then F will be multiplicative as well. Prove the converse of this statement: Show that if $F(n) := \sum_{d|n} f(d)$, and F is multiplicative, then so f must be multiplicative as well. Suggestion: strong induction.

7.3.29 *Another Counting Principle.* Let $F(n) := \sum_{d|n} f(d)$, and let g be an arbitrary function. Consider the sum

$$\sum_{d|n} g(d) F\left(\frac{n}{d}\right) = \sum_{d|n} g(d) \sum_{k|(n/d)} f(k).$$

For each $k \le n$, how many of the terms in this sum will contain the factor $f(k)$? First observe that $k|n$. Then show that the terms containing $f(k)$ will be

$$f(k) \sum_{u|(n/k)} g(u).$$

Conclude that

$$\sum_{d|n} g(d) F\left(\frac{n}{d}\right) = \sum_{k|n} f(k) \sum_{u|(n/k)} g(u).$$

The above equations are pretty hairy; but they are not hard if *you get your hands dirty and work out several examples!*

7.3.30 The above equation used an arbitrary function g. If we replace g with μ, we can use some special properties such as 7.3.12. This leads to the **Möbius Inversion Formula**, which states that if $F(n) := \sum_{d|n} f(d)$, then

$$f(n) = \sum_{d|n} \mu(d) F\left(\frac{n}{d}\right).$$

7.3.31 *An Application.* Consider all possible n-letter "words" that use the 26-letter alphabet. Call a word "prime" if it cannot be expressed as a concatenation of identical smaller words. For example, booboo is not prime, while booby is prime. Let $p(n)$ denote the number of prime words of length n. Show that

$$p(n) = \sum_{d|n} \mu(d) 26^{n/d}.$$

For example, this formula shows that $p(1) = \mu(1) \cdot 26^1 = 26$, which makes sense, since every single-letter word is prime. Likewise, $p(2) = \mu(1) \cdot 26^2 + \mu(2) \cdot 26^1 = 26^2 - 26$, which also makes sense since there are 26^2 two-letter words, and all are prime except for the 26 words aa, bb, ..., zz.

7.4 Diophantine Equations

A **diophantine equation** is any equation whose variables only assume integral values. You encountered linear diophantine equations in Problem 7.1.13 on page 249, a class of equations for which there is a "complete theory." By this we mean that given any linear diophantine equation, one can determine if there are solutions or not, and if there are solutions, there is an algorithm for finding *all* solutions. For example, the linear diophantine equation $3x + 21y = 19$ has no solution, since $GCD(3, 21)$ does not divide 19. On the other hand, the equation $3x + 19y = 4$ has infinitely many solutions, namely $x = -24 + 19t, 4 - 3t$, as t ranges through all integers.

Most higher-degree diophantine equations do not possess complete theories. Instead, there is a menagerie of different types of problems with diverse methods for understanding them, and sometimes only partial understanding is possible. We will just scratch the surface of this rich and messy topic, concentrating on a few types of equations that can sometimes be understood, and a few useful tactics that you will use again and again on many sorts of problems.

General Strategy and Tactics

Given any diophantine equation, there are four questions that you must ask:

- *Is the problem in "simple" form?* Always make sure that you have divided out all common factors, or assume the variables share no common factors, etc. See Example 7.2.3 on page 252 for a brief discussion of this.
- *Do there exist solutions?* Sometimes you cannot actually solve the equation, but you can show that at least one solution exists.
- *Are there no solutions?* Quite frequently, this is the first question to ask. As with argument by contradiction, it is sometimes rather easy to prove that an equation has no solutions. It is always worth spending some time on this question when you begin your investigation.
- *Can we find all solutions?* Once one solution is found, we try to understand how we can generate more solutions. It is sometimes quite tricky to prove that the solutions found are the complete set.

Here is a simple example of a problem with a "complete" solution, illustrating one of the most importance tactics: factoring.

Example 7.4.1 Find all right triangles with integer sides such that the area and perimeter are equal.

Solution: Let x, y be the legs and let z be the hypotenuse. Then $z = \sqrt{x^2 + y^2}$ by the Pythagorean theorem. Equating area and perimeter yields

$$\frac{xy}{2} = x + y + \sqrt{x^2 + y^2}.$$

Basic algebraic strategy dictates that we eliminate the most obvious difficulties, which in this case are the fraction and the radical. Multiply by 2, isolate the radical, and square. This yields

$$(xy - 2(x + y))^2 = 4(x^2 + y^2),$$

or

$$x^2 y^2 - 4xy(x + y) + 4(x^2 + y^2 + 2xy) = 4(x^2 + y^2).$$

After we collect like terms, we have

$$x^2 y^2 - 4xy(x+y) + 8xy = 0.$$

Clearly, we should divide out xy, as it is never equal to zero. We get

$$xy - 4x - 4y + 8 = 0.$$

So far, everything was straightforward algebra. Now we do something clever: add 8 to both sides to make the left-hand side factor. We now have

$$(x-4)(y-4) = 8,$$

and since the variables are integers, there are only finitely many possibilities. The only solutions (x, y) are $(6, 8)$, $(8, 6)$, $(5, 12)$, $(12, 5)$, which yield just two right triangles, namely the 6-8-10 and 5-12-13 triangles. ∎

The only tricky step was finding the factorization. But this wasn't really hard, as it was clear that the original left-hand side "almost" factored. As long as you try to factor, it usually won't be hard to find the proper algebraic steps.

The factor tactic is essential for finding solutions. Another essential tactic is to "filter" the problem modulo n for a suitably chosen n. This tactic often helps to show that no solutions are possible, or that all solutions must satisfy a certain form.[6] You saw a bit of this already on page 252. Here is another example.

Example 7.4.2 Find all solutions to the diophantine equation $x^2 + y^2 = 1000003$.

Solution: Consider the problem modulo 4. The only **quadratic residues** in Z_4 are 0 and 1, because

$$0^2 \equiv 0, 1^2 \equiv 1, 2^2 \equiv 0, 3^2 \equiv 1 \pmod 4.$$

Hence the sum $x^2 + y^2$ can only equal 0, 1 or 2 in Z_4. Since $1000003 \equiv 3 \pmod 4$, we conclude that there are no solutions. In general, $x^2 + y^2 = n$ will have no solutions if $n \equiv 3 \pmod 4$. ∎

Now let's move on to a meatier problem: the complete theory of Pythagorean triples.

Example 7.4.3 Find all solutions to

$$x^2 + y^2 = z^2. \tag{4}$$

Solution: First, we make the basic simplifications. Without loss of generality, we assume of course that all variables are positive. In addition, we will assume that our solution is **primitive**, i.e., that the three variables share no common factor. Any

---

[6]Use of the division algorithm is closely related to the factor tactic. See Example 5.4.1 on page 180 for a nice illustration of this.

primitive solution will produce infinitely many non-primitive solutions by multiplying; for example $(3, 4, 5)$ gives rise to $(6, 8, 10)$, $(9, 12, 15)$, ...

In this particular case, the assumption of primitivity leads to something a bit stronger. If d is a common divisor of x and y, then $d^2|x^2 + y^2$; in other words, $d^2|z^2$ so $d|z$. Similar arguments show that if d is a common divisor of any two of the variables, then d also divides the third. Therefore we can assume that our solution is not just primitive, but *relatively prime in pairs*.

Next, a little parity analysis; i.e., let's look at things modulo 2. Always begin with parity. You never know what you will discover. Let's consider some cases for the parity of x and y.

- *Both are even.* This is impossible, since the variables are relatively prime in pairs.
- *Both are odd.* This is also impossible. By the same reasoning used in Example 7.4.2 above, if x and y are both odd, it will force $z^2 \equiv 2 \pmod{4}$, which cannot happen.

We conclude that if the solutions to (4) are primitive, then exactly one of x and y is even. Without loss of generality, assume that x is even.

Now we proceed like a seasoned problem solver. Wishful thinking impels us to try some of the tactics that worked earlier. Let's make the equation factor! The obvious step is to rewrite it as

$$x^2 = z^2 - y^2 = (z - y)(z + y). \tag{5}$$

In other words, the product of $z - y$ and $z + y$ is a perfect square. It would be nice to conclude that each of $z - y$ and $z + y$ are also perfect squares, but this is not true in general. For example, $6^2 = 3 \cdot 12$.

On the other hand, it is true that if $v \perp w$ and $u^2 = vw$, then v and w must be perfect squares (this is easy to check by looking at PPF's). So in our problem, as in many problems, we should now focus our attention on the GCD's of the critical quantities, which at this moment are $z + y$ and $z - y$.

Let $g := \text{GCD}(z + y, z - y)$. Since $z - y$ and $z + y$ are even, we have $2|g$. On the other hand, g must divide the sum and difference of $z - y$ and $z + y$. This means that $g|2z$ and $g|2y$. But $y \perp z$, so $g = 2$.

Returning to (5), what can we say about two numbers if their GCD is 2 and their product is a perfect square? Again, a simple analysis of the PPF's yields the conclusion that we can write

$$z + y = 2r^2, \quad z - y = 2s^2,$$

where $r \perp s$. Solving for y and z yields $y = r^2 - s^2, z = r^2 + s^2$. We're almost done, but there is one minor detail: if r and s are both odd, this would make y and z both even, which violates primitivity. So one of r, s must be even, one must be odd.

Finally, we can conclude that all primitive solutions to (4) are given by

$$x = 2rs, \, y = r^2 - s^2, z = r^2 + s^2,$$

where r and s are relatively prime integers, one odd, one even. ∎

Factoring, modulo n filtering (especially parity), and GCD analysis are at the heart of most diophantine equation investigations, but there are many other tools available. The next example involves inequalities, and a very disciplined use of the tool of comparing exponents of primes in PPF's. We will use a new notation. Let $p^t \| n$ mean that t is the greatest exponent of p which divides n. For example, $3^2 \| 360$.

Example 7.4.4 (Putnam 1992) For a given positive integer m, find all triples (n, x, y) of positive integers, with n relatively prime to m, which satisfy

$$(x^2 + y^2)^m = (xy)^n. \tag{6}$$

Solution: What follows is a complete solution to this problem, but we warn you that our narrative is rather long. It is a record of a "natural" course of investigation: we make several simplifications, get a few ideas which partially work, and then gradually eliminate certain possibilities. In the end, our originally promising methods (parity analysis, mostly) do not completely work, but the partial success points us in a completely new direction, one that yields a surprising conclusion.

OK, let's get going. One intimidating thing about this problem are the two exponents m and n. There are so many possibilities! The AM-GM inequality (see page 192) helps to eliminate some of them. We have

$$x^2 + y^2 \geq 2xy,$$

which means that

$$(xy)^n = (x^2 + y^2)^m \geq (2xy)^m = 2^m (xy)^m,$$

so we can conclude that $n > m$. That certainly helps. Let's consider one example, with, say, $m = 1, n = 2$. Our diophantine equation is now

$$x^2 + y^2 = x^2 y^2.$$

Factoring quickly establishes that there is no solution, for adding 1 to both sides yields

$$(x^2 - 1)(y^2 - 1) = 1,$$

which has no positive integer solutions. But factoring won't work (at least not in an obvious way) for other cases. For example, let's try $m = 3, n = 4$. We now have

$$(x^2 + y^2)^3 = (xy)^4. \tag{7}$$

The first thing to try is parity analysis. A quick perusal of the four cases shows that the only possibility is that both x and y must be even. So let's write them as $x = 2a$, $y = 2b$. Our equation now becomes, after some simplifying,

$$(a^2 + b^2)^3 = 4(ab)^4.$$

Ponder parity once more. Certainly a and b cannot be of opposite parity. But they cannot both be odd either, for in that case the left-hand side will be the cube of an even number, which makes it a multiple of 8. However, the right-hand side is equal to

4 times the fourth power of an odd number, a contradiction. Therefore a and b must both be even. Writing $a = 2u$, $b = 2v$ transforms our equation into

$$(u^2 + v^2)^3 = 16(uv)^4.$$

Let's try the kind of analysis as before. Once again u and v must have the same parity, and once again, they cannot both be odd. If they were odd, the right-hand side equals 16 times an odd number; in other words, 2^4 is the highest power of 2 which divides it. But the left-hand side is the cube of an even number, which means that the highest power of 2 which divides it will be 2^3 or 2^6 or 2^9, etc. Once again we have a contradiction, which forces u, v to both be even, etc.

It appears that we can produce an *infinite* chain of arguments showing that the variables can be successively divided by 2, yet still be even! This is an impossibility, for no finite integer has this property. But let's avoid the murkiness of infinity by using the extreme principle. Return to equation (7). Let r, s be the greatest exponents of 2 which divide x, y respectively. Then we can write $x = 2^r a$, $y = 2^s b$, where a and b are odd, and we know that r and s are both positive. There are two cases:

- Without loss of generality, assume that $r < s$. Then (7) becomes

$$\left(2^{2r}a^2 + 2^{2s}b^2\right)^3 = 2^{4r+4s}a^4 b^4,$$

and after dividing by 2^{6r}, we get

$$(a^2 + 2^{2r-2s}b^2)^3 = 2^{4s-2r}a^4 b^4.$$

Notice that the exponent $4s - 2r$ is positive, making the right-hand side even. But the left-hand side is the cube of an odd number, which is odd. This is an impossibility; there can be no solutions.

- Now assume that $r = s$. Then (7) becomes

$$(a^2 + b^2)^3 = 2^{2r}a^4 b^4. \tag{8}$$

It is true that both sides are even, but a more subtle analysis will yield a contradiction. Since a and b are both odd, $a^2 \equiv b^2 \equiv 1 \pmod 4$, so $a^2 + b^2 \equiv 2 \pmod 4$, which means that $2^1 \| a^2 + b^2$. Consequently, $2^3 \| (a^2 + b^2)^3$. On the other hand, $2^{2r} \| 2^{2r}a^4 b^4$, where r is a positive integer. It is impossible for $2r = 3$, so the left-hand and right-hand sides of equation (8) have different exponents of 2 in the PPF's, an impossibility.

We are finally ready to tackle the general case. It seems as though there may be no solutions, but let's keep an open mind.

Consider the equation $(x^2 + y^2)^m = (xy)^n$. We know that $n > m$ and that both x and y are even (using the same parity argument as before). Let $2^r \| x$, $2^s \| y$. We Consider the two cases:

1. Without loss of generality, assume that $r < s$. Then $2^{2rm} \| (x^2 + y^2)^m$ and $2^{nr+ns} \| (xy)^n$. This means that $2rm = nr + ns$, which is impossible, since m is strictly less than n.

2. Assume that $r = s$. Then we can write $x = 2^r a$, $y = 2^r b$, where a and b are both odd. Thus

$$(x^2 + y^2)^m = 2^{2rm}(a^2 + b^2)^m,$$

where $a^2 + b^2 \equiv 2 \pmod 4$ and consequently $2^{2rm+m} \| (x^2 + y^2)^m$. Since $2^{2nr} \| (xy)^n$, we equate

$$2rm + m = 2rn,$$

and surprisingly, now, this equation has solutions. For example, if $m = 6$, then $r = 1$ and $n = 9$ work. That doesn't mean that the original equation has solutions, but we certainly cannot rule out this possibility.

Now what? It looks like we need to investigate more cases. But first, let's think about other primes. In our parity analysis, could we have replaced 2 with an arbitrary prime p? In case 1 above, yes: Pick any prime p and let $p^u \| x$, $p^v \| y$. Now, if we assume that $u < v$, we can conclude that $p^{2um} \| (x^2 + y^2)^m$ and $p^{nu+nv} \| (xy)^n$, and this is impossible because $n > m$. What can we conclude? Well, if it is impossible that u and v be different, no matter what the prime is, then the only possibility is that u and v are always equal, for every prime. That means that x and y are equal!

In other words, we have shown that there are no solutions, except for the possible case where $x = y$. In this case, we have

$$\left(2x^2\right)^m = \left(x^2\right)^n,$$

so $2^m x^{2m} = x^{2n}$, or $x^{2n-2m} = 2^m$. Thus $x = 2^t$, and we have $2nt - 2mt = m$, or

$$(2t + 1)m = 2nt.$$

Finally, we use the hypothesis that $n \perp m$. Since $2t \perp 2t + 1$ as well, the only way that the above equation can be true is if $n = 2t + 1$ and $m = 2t$. And this finally produces infinitely many solutions. If $m = 2t$, and $n = m + 1$, then it is easy to check that $x = y = 2^t$ indeed satisfies $(x^2 + y^2)^m = (xy)^n$.

And these will be the only solutions. In other words, if m is odd, there are no solutions, and if m is even, then there is the single solution

$$n = m + 1, x = y = 2^{m/2}.$$

■

Problems and Exercises

7.4.5 Prove rigorously these two statements, which were used in Example 7.4.3:

(a) If $u \perp v$ and $uv = x^2$, then u and v must be perfect squares.

(b) If p is a prime and $GCD(u, v) = p$ and $uv = x^2$, then $u = pr^2$, $v = ps^2$, with $r \perp s$.

7.4.6 (Greece, 1995) Find all positive integers n for which $-5^4 + 5^5 + 5^n$ is a perfect square. Do the same for $2^4 + 2^7 + 2^n$.

7.4.7 (United Kingdom, 1995) Find all triples of positive integers (a, b, c) such that

$$\left(1 + \frac{1}{a}\right)\left(1 + \frac{1}{b}\right)\left(1 + \frac{1}{c}\right) = 2.$$

7.4.8 Show that there is exactly one integer n such that $2^8 + 2^{11} + 2^n$ is a perfect square.

7.4.9 Find the number of ordered pairs of positive integers (x, y) that satisfy

$$\frac{xy}{x + y} = n.$$

7.4.10 (USAMO 1979) Find all non-negative integral solutions $(n_1, n_2, \ldots, n_{14})$ to

$$n_1^4 + n_2^4 + \cdots + n_{14}^4 = 1,599.$$

7.4.11 Find all positive integer solutions to $abc - 2 = a + b + c$.

7.4.12 (Germany, 1995) Find all pairs of nonnegative integers (x, y) such that $x^3 + 8x^2 - 6x + 8 = y^3$.

7.4.13 (India, 1995) Find all positive integers x, y such that $7^x - 3^y = 4$.

7.4.14 Develop a complete theory for the equation $x^2 + 2y^2 = z^2$. Can you generalize this even further?

7.4.15 Twenty-three people, each with integral weight, decide to play football, separating into two teams of 11 people, plus a referee. To keep things fair, the teams chosen must have equal *total* weight. It turns out that no matter who is chosen to be the referee, this can always be done. Prove that the 23 people must all have the same weight.

7.4.16 (India, 1995) Find all positive integer solutions x, y, z, p, with p a prime, of the equation $x^p + y^p = p^z$.

Pell's Equation

The quadratic diophantine equation $x^2 - dy^2 = n$, where d and n are fixed, is called **Pell's equation**. Problems 7.4.17–7.4.22 will introduce you to a few properties and applications of this interesting equation. We will mostly restrict our attention to the cases where $n = \pm 1$. For a fuller treatment of this subject, including the relationship between Pell's equation and continued fractions, consult just about any number theory textbook.

7.4.17 Notice that if d is negative, then $x^2 - dy^2 = n$ has only finitely many solutions.

7.4.18 Likewise, if d is perfect square, then $x^2 - dy^2 = n$ has only finitely many solutions.

7.4.19 Consequently, the only "interesting" case is when d is positive and not a perfect square. Let us consider a concrete example: $x^2 - 2y^2 = 1$.

(a) It is easy to see by inspection that $(1, 0)$ and $(3, 2)$ are solutions. A bit more work yields the next solution: $(17, 12)$.

(b) Cover the next line so you can't read it! Now, see if you can find a simple linear recurrence that produces $(3, 2)$ from $(1, 0)$ and produces $(17, 12)$ from $(3, 2)$. Use this to produce a new solution, and check to see if it works.

(c) You discovered that if (u, v) is a solution to $x^2 - 2y^2 = 1$, then so is $(3u + 4v, 2u + 3v)$. Prove why this works. It is much easy to see why it works than to discover it in the first place, so don't feel bad if you "cheated" in (b).

(d) But now that you understand the lovely tool of **generating new solutions** from clever linear combinations of old solutions, you should try your hand at $x^2 - 8y^2 = 1$. In general, this method will furnish infinitely many solutions to Pell's equation for any positive non-square d.

7.4.20 Notice that $(3 + 2\sqrt{2})^2 = 17 + 12\sqrt{2}$. Is this a coincidence? Ponder, conjecture, generalize.

7.4.21 Try to find solutions to $x^2 - dy^2 = -1$, for a few positive non-square values of d.

7.4.22 An integer is called square-full if each of its prime factors occurs to at least the second power. Prove that there exist infinitely many pairs of consecutive square-full integers.

7.5   Miscellaneous Instructive Examples

The previous sections barely sampled the richness of number theory. We conclude the chapter with a few interesting examples. Each example either illustrates a new problem-solving idea or illuminates an old one. In particular, we present several "crossover" problems that show the deep interconnections between number theory and combinatorics.

Can a Polynomial Always Output Primes?

Example 7.5.1 Consider the polynomial $f(x) := x^2 + x + 41$, which you may remember from Problem 2.2.34 on page 44. Euler investigated this polynomial, and discovered that $f(n)$ is prime for all integers n from 1 to 40. The casual observer may suspect after plugging in a few values of n that this polynomial *always* outputs primes, but it takes no calculator to see that this cannot be: just let $x = 41$, and we have $f(41) = 41^2 + 41 + 41$, obviously a multiple of 41. So now we are confronted with the "interesting" case:

> *Does there exist a polynomial $f(x)$ with integral coefficients and constant term equal to ± 1, such that $f(n)$ is a prime for all $n \in \mathbf{N}$?*

Investigation: Let us write

$$f(x) = a_n x^n + a_{n-1} x^{n-1} + \cdots + a_0, \tag{9}$$

where the a_i are integers and $a_0 = \pm 1$. Notice that we cannot use the trick of "plugging in a_0" that worked with $x^2 + x + 41$. Since we are temporarily stumped, we do what all problem solvers do: experiment! Consider the example $f(x) := x^3 + x + 1$. Let's make a table:

n	1	2	3	4	5	6	7
$f(n)$	3	11	31	69	131	223	351

This polynomial doesn't output all primes; the first x-value at which it "fails" is $x = 4$. But we are just looking for patterns. Notice that $f(4) = 3 \cdot 23$, and the next composite value is $f(7) = 3^3 \cdot 13$. Notice that $4 = 1 + 3$ and $7 = 4 + 3$. At this point, we are ready to make a tentative guess that $f(7 + 3)$ will also be a multiple of 3. This seems almost too good to be true, yet

$$f(10) = 10^3 + 10 + 1 = 1011 = 3 \cdot 337.$$

Let's attempt to prove this conjecture, at least for this particular polynomial. Given that $3 \mid f(a)$, we'd like to show that $3 \mid f(a + 3)$. We have

$$f(a + 3) = (a + 3)^3 + (a + 3) + 1 = (a^3 + 9a^2 + 27a + 27) + (a + 3) + 1.$$

Don't even think of adding up the like terms—that would be mindless "simplification." Instead, incorporate the hypothesis that $f(a)$ is a multiple of 3. We then write

$$f(a + 3) = (a^3 + a + 1) + (9a^2 + 27a + 27 + 3),$$

and we are done, since the first expression in parenthesis is $f(a)$, which is a multiple of 3, and all of the coefficients of the second expression are multiples of 3.

Solution: Now we are ready to attempt a more general argument. Assume that $f(x)$ is the generic polynomial defined in (9). Let $f(u) = p$ for some integer a, where p is prime. We would like to show that $f(u + p)$ will be a multiple of p. We have

$$f(u + p) = a_n(u + p)^n + a_{n-1}(u + p)^{n-1} + \cdots + a_1(u + p) + a_0.$$

Before we faint at the complexity of this equation, let's think about it. If we expand each $(u + p)^k$ expression by the binomial theorem, the leading term will be u^k. So we can certainly extract $f(u)$. We need to look at what's left over; i.e.,

$$f(u + p) - f(u) = a_n \left((u + p)^n - u^n \right) + a_{n-1} \left((u + p)^{n-1} - u^{n-1} \right) + \cdots + a_1 p.$$

Now it suffices to show that

$$(u + p)^k - u^k$$

is divisible by p for all values of k. This is very easy to do, for example, by induction, or by expanding $(u + p)^k$ with the binomial theorem:

$$(u + p)^k = u^k + \binom{k}{1} u^{k-1} p + \binom{k}{2} u^{k-2} p^2 + \cdots + p^k.$$

When we subtract u^k from this, the remaining terms are all multiples of p, so we are almost done.

We aren't completely done, because of the possibility that $f(u+p) - f(u) = 0$. In this case, $f(u+p) = p$, which is not composite. If this happens, we keep adding multiples of p. By the same argument used, $f(u+2p)$, $f(u+2p)$, ... will all be multiples of p. Because $f(x)$ is a *polynomial*, it can only equal p (or $-p$) for finitely many values of x [for otherwise, the new polynomial $g(x) := f(x) - p$ would have infinitely many zeros, violating the Fundamental Theorem of Algebra]. ∎

Incidentally, we never used the fact that p was prime. So we obtained a "bonus" result:

> If $f(x)$ is a polynomial with integer coefficients, then for all integers a, $f(a + f(a))$ is a multiple of $f(a)$.

If You Can Count It, It's an Integer

Example 7.5.2 Let $k \in \mathbf{N}$. Show that the product of k consecutive integers is divisible by $k!$.

Solution: It is possible to solve this problem with "pure" number theoretic reasoning, but it is far simpler and much more enjoyable to simply observe that

$$\frac{m(m+1)(m+2)\cdots(m+k-1)}{k!} = \binom{m+k-1}{k},$$

and binomial coefficients are integers! ∎

The moral of the story: Keep your point of view flexible. Anything involving integers is fair game for combinatorial reasoning. The next example continues this idea.

A Combinatorial Proof of Fermat's Little Theorem

Recall that Fermat's little theorem (page 253) states that if p is prime, then

$$a^p \equiv a \pmod{p}$$

holds for all a. Equivalently, FLT says that $a^p - a$ is a multiple of p. The expression a^p has many simple combinatorial interpretations. For example, there are a^p different p-letter words possible using an alphabet with a letters.

Let's take the example of $a = 26$, $p = 7$, and consider the "dictionary" $\mathcal{D}$ of these 26^7 words. Define the **shift function** $s : \mathcal{D} \to \mathcal{D}$ to be the operation that moves the last (rightmost) letter of a word to the beginning position. For example, $s(\texttt{fermats}) = \texttt{sfermat}$. We will call two words in $\mathcal{D}$ "sisters" if it is possible to transform one into the other with finitely many applications of the shift function. For example, `integer` and `gerinte` are sisters, since $s^3(\texttt{integer}) = \texttt{gerinte}$. Let us call all of the sisters of a word its "sorority." Since any word is its own sister, the sorority containing `integer` consists of this word and

`rintege,   erinteg,   gerinte,   egerint,   tegerin,   ntegeri.`

7.5.3 Show that if $s(U) = U$, then U is a word all of whose letters are the same. There are of course exactly 26 such "boring" words,

`aaaaaaa, bbbbbbb, ..., zzzzzzz.`

7.5.4 Show additionally that if $s^r(U) = U$ where $0 < r < 7$, then U must be a boring word.

7.5.5 Show that all sororities have either 1 member or exactly 7 members.

7.5.6 Conclude that the $26^7 - 26$ non-boring words in $\mathcal{D}$ must be a multiple of 7.

7.5.7 Finally, generalize your argument so that it works for any prime p. What simple number theory principle is needed to carry out the proof?

Sums of Two Squares

We shall end the chapter with an exploration of the diophantine equation

$$x^2 + y^2 = n.$$

We won't produce a complete theory here (see [18] for a very readable exposition), but we will consider the case where n is a prime p. Our exploration will use several old strategic and tactical ideas, including the pigeonhole principle, Gaussian pairing, and drawing pictures. The narrative will meander a bit, but please read it slowly and carefully, because it is a model of how many different problem-solving techniques come together in the solution of a *hard* problem.

First recall that $x^2 + y^2 = p$ will have no solutions if $p \equiv 3 \pmod 4$ (Example 7.4.2 on page 265). The case $p = 2$ is pretty boring. So all that remains to investigate are primes which are congruent to 1 modulo 4.

7.5.8 Find solutions to $x^2 + y^2 = p$, for the cases $p = 5, 13, 17, 29, 37, 41$.

By now you are probably ready to guess that $x^2 + y^2 = p$ will always have a solution if $p \equiv 1 \pmod 4$. One approach is to ponder some of the experiments we just did, and see if we can deduce the solution "scientifically," rather than with trial-and-error. Let's try $p = 13$. We know that one solution (the only solution, if we don't count sign and permuting the variables) is $x = 3$, $y = 2$. But how do we "solve"

$$x^2 + y^2 = 13?$$

This is a diophantine equation, so we should try factoring. "But it doesn't factor," you say. Sure it does! We can write

$$(x + yi)(x - yi) = 13,$$

where i is of course equal to the square root of -1. The only problem, and it is a huge one, is that i is not an integer.

But let's stay loose, bend the rules a bit, and make the problem easier. It is true that the square root of -1 is not an integer, but what if we looked at the problem in Z_{13}? Notice that

$$5^2 = 25 \equiv -1 \pmod{13}.$$

In other words, i "makes sense" modulo 13. The square root of -1 is equal to 5 modulo 13.

If we look at our diophantine equation modulo 13, it becomes

$$x^2 + y^2 \equiv 0 \pmod{13},$$

but the left-hand side now factors beautifully. Observe that we can now write

$$x^2 + y^2 \equiv (x - 5y)(x + 5y) \pmod{13},$$

and therefore we can make $x^2 + y^2$ congruent to 0 modulo 13 as long as $x \equiv \pm 5y$ (mod 13). For example, $y = 1$, $x = 5$ is a solution (and sure enough, $x^2 + y^2 = 26 \equiv 0$ (mod 13). Another solution is $y = 2$, $x = 10$, in which case $x^2 + y^2 = 104$, which is also a multiple of 13. Yet another solution is $y = 3, x = 15$. If we reduce this modulo 13, it is equivalent to the solution $y = 3$, $x = 2$ and this satisfies $x^2 + y^2 = 13$.

This is promising. Here is an outline of a possible "algorithm" for solving $x^2 + y^2 = p$.

1. First find the square root of -1 in Z_p. Call it u.
2. Then we can factor $x^2 + y^2 \equiv (x - uy)(x + uy) \pmod{p}$, and consequently the pairs $y = k$, $x = uk$, for $k = 1, 2, \ldots$ will solve the congruence

$$x^2 + y^2 \equiv 0 \pmod{p}.$$

   In other words, $x^2 + y^2$ will be a multiple of p.
3. If we are lucky, when we reduce the values of these solutions modulo p, we may get a pair of x and y which are sufficiently small so that not only will $x^2 + y^2$ be a multiple of p, it will actually equal p.

There are two major hurdles. First, how do we know if we can always find a square root of -1 modulo p? And second, how do we ensure that the values of x and y will be small enough so that $x^2 + y^2$ will equal p rather than, say, $37p$?

Do some experiments with primes under 50, and a calculator or computer if you wish, to determine for which primes p will the square root of -1 exists modulo p. You should discover that you can always find $\sqrt{-1}$ in Z_p if $p \equiv 1 \pmod{4}$, but never if $p \equiv 3 \pmod{4}$.

Sure enough, it seems that it *may* be true that

$$x^2 \equiv -1 \pmod{p}$$

has a solution if and only if $p \equiv 1 \pmod{4}$. But of course, numerical experiments only suggest the truth. We still need to prove something.

For the time being, let's not worry about this, and just assume that we can find the square root of -1 modulo p whenever we need to. Let's return to the second hurdle. We can produce infinitely many pairs (x, y) such that $x^2 + y^2$ is a multiple of p. All we need to do is get the values small enough. Certainly, if x and y are both less than $\sqrt{p}$, then $x^2 + y^2 < 2p$ which forces $x^2 + y^2 = p$. That is helpful, for if $u < \sqrt{p}$, we are immediately done. The pair $y = 1$, $x = u$ will do the trick.

On the other hand, what if $u > \sqrt{p}$ (notice that it cannot equal $\sqrt{p}$, since p is prime and hence not a perfect square)? Let $t := \lfloor \sqrt{p} \rfloor$. Then the problem is reduced to showing that one of

$$\pm u, \pm 2u, \pm 3u, \ldots, \pm tu,$$

when reduced modulo p, will be less than $\sqrt{p}$ in absolute value.

Let's try an example for $p = 29$. We have $t := \lfloor \sqrt{29} \rfloor = 5$, and $u = 17$ (found by trial-and-error; verify that it works!). Since $u > t$, we need to look at the sequence

$$u, 2u, 3u, 4u, 5u,$$

reduced modulo 29. **Draw a picture**:

The small ticks are the integers from 0 to 29, while the long ticks are placed at the locations

$$\sqrt{p}, \quad 2\sqrt{p}, \quad 3\sqrt{p}, \quad 4\sqrt{p}, \quad 5\sqrt{p}.$$

Notice that both $2u = 34 \equiv 5 \pmod{29}$ and $5u = 85 \equiv -2 \pmod{29}$ both work, yielding respectively the solutions $y = 2$, $x = 5$ and $y = 5$, $x = -2$. But *why* did it work? In general, we will be dropping dots onto the number line, and would like to see a dot land either to the left of the first long tick, or to the right of the last long tick. The dots will never land exactly on these ticks, since they correspond to irrational values. We are assuming without loss of generality that $u > \sqrt{p}$. In general, there are t long ticks. The sequence $u, 2u, \ldots, tu$ consists of *distinct* values modulo p (why?), so we will be placing t different dots on the number line. There are three possibilities.

1. One of the dots lands to the left of the first long tick.
2. One of the dots lands to the right of the last long tick.
3. The dots don't land in either of the above locations.

In cases 1 or 2, we are immediately done. In case 3, we have t dots that lie in $t - 1$ intervals (separated by long ticks). By the pigeonhole principle, one of these intervals must contain 2 dots. Suppose mu and nu lie in one interval. This means that $(m - n)u$ will be an integer (perhaps negative), but smaller than $\sqrt{p}$ in absolute value. Since $|m - n| \leq t$, we can just choose $k := |m - n|$ and we are guaranteed that $y = k$ and $x = ku$ (when reduced modulo p) will both be less than $\sqrt{p}$. So we're done!

That was a fun application of the pigeonhole principle. All that remains is to prove that we can always obtain the square root of -1. We have one tool available, something that was proven earlier, which seemed a curiosity at that time: Wilson's theorem. Recall that Wilson's theorem (Example 3.1.9 on page 76) said that if p is a prime, then $(p-1)! \equiv -1 \pmod{p}$. A small dose of Gaussian pairing can reorganize $(p-1)!$ into a perfect square modulo p, provided that $p \equiv 1 \pmod 4$. In that case $p-1$ will be a multiple of 4, so the individual terms in $(p-1)!$ can be arranged in *an even number* of pairs. For example, let $p = 13$. We can then write

$$12! = (1 \cdot 12)(2 \cdot 11)(3 \cdot 10)(4 \cdot 9)(5 \cdot 8)(6 \cdot 7).$$

Each pair has the form $k \cdot (-k)$ modulo 13, so we have

$$12! \equiv (-1^2)(-2^2)(-3^2)(-4^2)(-5^2)(-6^2) \pmod{13},$$

and since there are an even number of minus signs in this product, the whole thing is congruent to $(6!)^2$ modulo 13. Combining this with Wilson's theorem yields

$$(6!)^2 \equiv -1 \pmod{13}.$$

The argument certainly generalizes to any prime p of the form $4k + 1$. Not only have we shown that the square root of -1 exists, we can compute it explicitly. Our general result is that

If p is prime and $p \equiv 1 \pmod 4$, then

$$\left(\frac{p-1}{2}\right)^2 \equiv -1 \pmod{p}.$$

This concludes our exploration of the equation $x^2 + y^2 = p$. We have succeeded in proving that solutions can always be found if $p \equiv 1 \pmod 4$. ∎

Problems and Exercises

Below are a wide variety of problems and exercises, arranged in roughly increasing order of difficulty.

7.5.9 Example 7.5.1 on page 271 stated that one could prove that $(u + p)^k - u^k$ is divisible by p for all values of k using induction.

(a) Do this induction proof. It is an easy exercise.

(b) Even easier: Think about the factorization of $x^n - y^n$. You should know this by heart, but if not, consult formula 5.2.7 on page 163.

7.5.10 Show that $(a + b)^p \equiv a^p + b^p \pmod{p}$ for any prime p.

7.5.11 Use the multinomial theorem (Problem 6.1.27 on page 213) to examine

$$\underbrace{(1 + 1 + \cdots + 1)}_{a\ 1's}{}^{p},$$

and thus derive yet another combinatorial proof of Fermat's little theorem. Explain why this proof is really equivalent to the one we discussed on page 273.

7.5.12 (Putnam 1983) How many positive integers n are there such that n is an exact divisor of at least one of the numbers $10^{40}, 20^{30}$?

7.5.13 (Russia, 1995) Let m and n be positive integers such that

$$\text{LCM}[m, n] + \text{GCD}[m, n] = m + n.$$

Prove that one of the two numbers is divisible by the other.

7.5.14 (Russia, 1995) Is it possible for the numbers $1, 2, 3, \ldots, 100$ to be the terms of 12 geometrical progressions?

7.5.15 (Kiran Kedlaya) Let p be an odd prime and $P(x)$ a polynomial of degree at most $p - 2$.

(a) Prove that if P has integer coefficients, then $P(n) + P(n+1) + \cdots + P(n+p-1)$ is an integer divisible by p for every integer n.

(b) If $P(n) + P(n+1) + \cdots + P(n+p-1)$ is an integer divisible by p for every integer n, must P have integer coefficients?

7.5.16 (IMO 1972) Let m and n be arbitrary non-negative integers. Show that

$$\frac{(2m)!(2n)!}{m!n!(m+n)!}$$

is an integer.

7.5.17 Find all integer solutions to

$$x^2 + y^2 + z^2 = 2xyz.$$

7.5.18 (USAMO 1981) The measure of a given angle is $180°/n$ where n is a positive integer not divisible by 3. Prove that the angle can be trisected by Euclidean means (straight edge and compasses).

7.5.19 Show that

$$\binom{n}{1}, \binom{n}{2}, \ldots, \binom{n}{n-1}$$

are all even if and only if n is a power of 2.

7.5.20 Use problem 7.5.19 as a starting point for the more interesting open-ended question: What can you say about the parity of the numbers in Pascal's triangle? Are there patterns? Can you find a formula or algorithm for the number of odd (or even) numbers in each row? And can you say anything meaningful about divisibility modulo m for other values of m?

7.5.21 Show that given p, there exists x such that $d(px) = x$ and also y such that $d(p^2 x) = x$.

7.5.22 Given n, the number of solutions to $d(nx) = n$ is 1 if and only if n equals 1, 4, or a prime, is finite if n is a product of distinct primes, and otherwise is infinite.

7.5.23 Show that the sum of the odd divisors of n is equal to $-\sum_{d|n}(-1)^{n/d}d$.

7.5.24 Let $\omega(n)$ be the number of distinct primes dividing n. Show that

$$\sum_{d|n} |\mu(d)| = 2^{\omega(n)}.$$

7.5.25 Does there exist an x such that

$$\mu(x) = \mu(x+1) = \mu(x+2) = \cdots = \mu(x+1996)?$$

7.5.26 (Putnam 1983) Let p be an odd prime and let

$$F(n) := 1 + 2n + 3n^2 + \cdots + (p-1)n^{p-2}.$$

Prove that if a, b are distinct integers in $\{0, 1, 2, \ldots, p-1\}$ then $F(a)$ and $F(b)$ are not congruent modulo p.

7.5.27 Recall the definition of the **inverse image** of a function (page 159). Show that for each $n \in \mathbf{N}$,

$$\sum_{k \in \phi^{-1}(n)} \mu(k) = 0.$$

For example, if $n = 4$, then $\phi^{-1}(n) = \{5, 8, 10, 12\}$ [of course, you need to verify why there are no other k such that $\phi(k) = 4$] and

$$\mu(5) + \mu(8) + \mu(10) + \mu(12) = -1 + 0 + 1 + 0 = 0.$$

7.5.28 Does there exist a row of Pascal's Triangle containing four distinct elements a, b, c and d such that $b = 2a$ and $d = 2c$?

7.5.29 (IMO 1974) Prove that the number

$$\sum_{k=0}^{n} \binom{2n+1}{2k+1} 2^{3k}$$

is not divisible by 5 for any integer $n \geq 0$.

7.5.30 (Romania, 1995) Let $f : \mathbf{N} - \{0, 1\} \to \mathbf{N}$ be the function defined by

$$f(n) = \text{LCM}[1, 2, \ldots, n].$$

(a) Prove that for all $n, n \geq 2$, there exist n consecutive numbers for which f is constant.

(b) Find the greatest number of elements of a set of consecutive integers on which f is strictly increasing, and determine all sets for which this maximum is realized.

7.5.31 For a deck containing an even number of cards, define a "perfect shuffle" as follows: divide the deck into two equal halves, the top half and the bottom half; then interleave the cards one by one between the two halves, starting with the top card of the bottom half, then the top card of the top half, etc. For example, if the deck has 6 cards, labeled "123456" from top to bottom, after a perfect shuffle the order of the cards will be "415263." Determine the minimum (positive) number of perfect shuffles needed to restore a 94-card deck to its original order. Can you generalize this to decks of arbitrary (even) size?

7.5.32 (Iran, 1995) Let $n > 3$ be an odd integer with prime factorization

$$n = p_1^{\alpha_1} \cdots p_k^{\alpha_k}.$$

If

$$m = n \left(1 - \frac{1}{p_1}\right) \left(1 - \frac{1}{p_2}\right) \cdots \left(1 - \frac{1}{p_k}\right),$$

prove that there is a prime p such that p divides $2^m - 1$, but does not divide m.

Perfect Numbers

Problems 7.5.33–7.5.35 explore some simple ideas about a topic that has fascinated and perplexed mathematicians for at least 2000 years.

7.5.33 Prove the following two facts about the σ function:

(a) A positive integer n is prime if and only if $\sigma(n) = n + 1$.

(b) If $\sigma(n) = n + a$ and $a | n$ and $a < n$, then a must equal 1.

7.5.34 An integer n is called **perfect** if $\sigma(n) = 2n$. For example, 6 is perfect, since $1 + 2 + 3 + 6 = 2 \cdot 6$.

(a) Show that if $2^k - 1$ is a prime, then $2^{k-1}(2^k - 1)$ is perfect. This fact was known to the ancient Greeks, who computed the perfect numbers $28, 496, 8128$.

(b) It was not until the 18th century that Euler proved a partial converse to this:

> *Every even perfect number must be of the form* $2^{k-1}(2^k - 1)$, *where* $2^k - 1$ *is a prime.*

Now you prove it.

7.5.35 What can you say about odd perfect numbers? (Incidentally, no one has ever found one, nor proven that they do not exist. But that doesn't mean that you can't say something meaningful about them.)

Primitive Roots of Unity and Cyclotomic Polynomials

Problems 7.5.36–7.5.41 explore some fascinating connections between polynomials, number theory, and complex numbers. You may want to read about roots of unity (page 137) before attempting these problems.

7.5.36 *Primitive nth Roots of Unity.* The complex number ζ is called a **primitive** nth root of unity if n is the smallest positive integer such that $\zeta^n = 1$. For example, the 4th roots of unity are $1, i, -1, -i$ but only i and $-i$ are primitive 4th roots of unity.

(a) If p is a prime, then there are $p - 1$ primitive pth roots of unity, namely all the pth roots of unity except for 1:

$$\zeta, \zeta^2, \zeta^3, \ldots, \zeta^{p-1},$$

where $\zeta = \text{Cis } \frac{2\pi}{p}$.

(b) If $\zeta = \text{Cis } \frac{2\pi}{n}$, then ζ^k is a primitive nth root of unity if and only if k and n are relatively prime. Consequently, there are $\phi(n)$ primitive nth roots of unity.

7.5.37 Define $\Phi_n(x)$ to be the polynomial with leading coefficient 1 and degree $\phi(n)$ whose roots are the $\phi(n)$ different *primitive* roots of unity. This polynomial is known as the nth **cyclotomic polynomial**. Compute $\Phi_1(x), \Phi_2(x), \ldots, \Phi_{12}(x)$, and $\Phi_p(x)$ and Φ_{p^2} (for p prime).

7.5.38 Prove that $x^n - 1 = \prod_{d|n} \Phi_d(x)$ for all positive integers n.

7.5.39 Prove that $\Phi_n(x) = \prod_{d|n} (x^d - 1)^{\mu(n/d)}$.

7.5.40 Prove that for each $n \in \mathbf{N}$, the sum of the primitive nth roots of unity is equal to $\mu(n)$. In other words, if $\zeta := \text{Cis } \frac{2\pi}{n}$, then

$$\sum_{\substack{a \perp n \\ 1 \leq a < n}} \zeta^a = \mu(n).$$

7.5.41 Prove that the coefficients of $\Phi_n(x)$ are integers for all n. Must the coefficients only be ± 1?

Chapter 8

Calculus

In this chapter, we take it for granted that you are familiar with basic calculus ideas like limits, continuity, differentiation, integration, and power series. On the other hand, we assume that you may have have heard of, but *not* mastered:

- Formal "$\delta - \epsilon$" proofs;
- Taylor series with "remainder;";
- The mean value theorem.

In contrast to, say, Chapter 7, this chapter is not a systematic, self-contained treatment. Instead, we concentrate on just a few important ideas that enhance your understanding of how calculus works. Our goal is twofold: to uncover the *practical* meaning of some of the things that you have already studied, by developing useful reformulations of old ideas; and to enhance your intuitive understanding of calculus, by showing you some useful, albeit non-rigorous "moving curtains." The meaning of this last phrase is best understood with an example.

8.1 The Fundamental Theorem of Calculus

To understand what a moving curtain is, we shall explore, in some detail, the most important idea of elementary calculus. This example also introduces a number of ideas that we will keep returning to throughout the chapter.

Example 8.1.1 What is the fundamental theorem of calculus (FTC), what does it mean, and why is it true?

Partial Solution: You have undoubtedly learned about the FTC. One formulation of it says that if f is a continuous function,[1] then

$$\int_a^b f(x)dx = F(b) - F(a), \tag{1}$$

where F is any **antiderivative** of f; i.e., $F'(x) = f(x)$. This is a remarkable statement. The left-hand side of (1) can be interpreted as the area under the graph of $y = f(x)$

---

[1] In this chapter we will assume that the domain and range of all functions are subsets of the real numbers.

bounded by the x-axis and the vertical lines $x = a$, $x = b$, while the right-hand side is related to $f(x)$ by differentiation, the computation of the slope of the tangent line to the graph of a function.

Stating it that way makes the FTC seem quite mysterious. Let us try to shed some light on it. On one level, the FTC is an amazing *algorithmic* statement, since in practice, antiderivatives are sometimes rather easy to compute. But that explains *what* it is, not *why* it is true. Understanding why it is true is a matter of choosing the proper interpretation of the entities in (1).

We start with the very useful **define a function** tool, which you have seen before (for example, 5.4.2). Let

$$g(t) := \int_a^t f(x)dx.$$

We chose the variable t on purpose, to make it easy to visualize $g(t)$ as a function of *time*. As t increases from a, the function $g(t)$ is computing the area of a "moving curtain" as seen below. Notice that $g(a) = 0$.

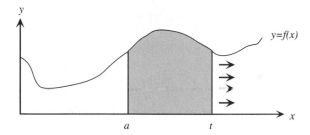

Differentiation is not just about tangent lines—it has a *dynamic* interpretation as instantaneous rate of change. Thus $g'(t)$ is equal to the rate of change of the area of the curtain at time t. With this in mind, look at the picture below: what does your intuition tell you the answer must be?

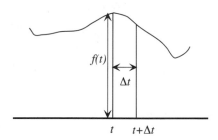

The area grows fast when the leading edge of the curtain is tall, and it grows slowly when the leading edge is short. It makes intuitive sense that

$$g'(t) = f(t), \tag{2}$$

since in a small interval of time Δt, the curtain's area will grow by approximately $f(t)\Delta t$. Equation (2) immediately yields the FTC, because if we define $F(t) := g(t) + C$, where C is any constant, we have $F'(t) = f(t)$ and

$$F(b) - F(a) = g(b) - g(a) = \int_a^b f(x)dx.$$

The crux move was to interpret the definite integral dynamically, and then observe the intuitive relationship between the speed that the area changes and the height of the curtain. This classic argument illustrates the critical importance of knowing as many possible alternate interpretations of both differentiation and integration.

You may argue that we have not proved FTC rigorously, and indeed (2) deserves a more careful treatment. After all, the curtain does not grow by exactly $f(t)\Delta t$. The exact amount is equal to

$$\int_t^{t+\Delta t} f(x)dx,$$

which is equal to $f(t)\Delta t + E(t)$, where $E(t)$ is the area of the "error," shown shaded below [note that $E(t)$ is negative in this picture].

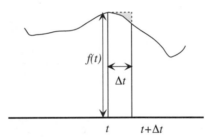

Everything hinges on showing that

$$\lim_{\Delta t \to 0} \left(\frac{E(t)}{\Delta t} \right) = 0. \tag{3}$$

This requires an understanding of continuity; we will prove (3) in 8.2.5.

8.2 Convergence and Continuity

You already have an intuitive understanding of concepts like limits and continuity, but in order to tackle interesting problems, you must develop a rigorous wisdom. Luckily, almost everything stems from one fundamental idea: convergence of sequences. If you understand this, you can handle limits, and continuity, and differentiation, and integration. Convergence of sequences is the theoretical foundation of calculus.

Convergence

We say that the real-valued sequence (a_n) **converges** to the limit L if

$$\lim_{n \to \infty} a_n = L.$$

What does this mean? That if we pick an arbitrary distance $\epsilon > 0$, eventually, *and forever after*, the a_i will get within ϵ of L. More specifically, for any $\epsilon > 0$ (think of ϵ as a really tiny number), there is an integer N (think of it as a really huge number, one that depends on ϵ) such that *all of the numbers*

$$a_N, a_{N+1}, a_{N+2}, \ldots$$

lie within ϵ of L. In other words, for all $n \geq N$,

$$|a_n - L| < \epsilon.$$

Sometimes, if the context is clear, we use the abbreviation $a_n \to L$ for $\lim_{n \to \infty} a_n = L$.

In practice, there are several possible methods of showing that a given sequence converges to a limit.

1. Somehow guess the limit L, and then show that the a_i get arbitrarily close to L.
2. Show that the a_i eventually get arbitrarily close to one another. More precisely, a sequence (a_n) possesses the **Cauchy** property if for any (very tiny) $\epsilon > 0$ there is a (huge) N such that

$$|a_m - a_n| < \epsilon$$

   for all $m, n \geq N$. If a sequence of real numbers has the Cauchy property, it converges.[2] The Cauchy property is often fairly easy to verify, but the disadvantage is that one doesn't get any information about the actual limiting value of the sequence.
3. Show that the sequence is **bounded** and **monotonic**. A sequence (a_n) is bounded if there is a finite number B such that $|a_n| \leq B$ for all n. The sequence is monotonic if it is either non-increasing or non-decreasing. For example, (a_n) is monotonically non-increasing if $a_{n+1} \leq a_n$ for all n.
   Bounded monotonic sequences are good, because they always converge. To see this, argue by contradiction: if the sequence did not converge, it would not have the Cauchy property, etc. ...
4. Show that the terms of the sequence are bounded above and below by the terms of two convergent sequences that converge to the same limit. For example, suppose that for all n, we have

$$0 < x_n < (0.9)^n.$$

   This forces $\lim_{n \to \infty} x_n = 0$. Conversely, if the terms of a sequence are greater in absolute value than the corresponding terms of a sequence that **diverges** (has infinite limit), then the sequence in question also diverges.

[2]See [23] for more information about this and other "foundational" issues regarding the real numbers.

5. Draw pictures whenever possible. Pictures rarely supply rigor, but often furnish the key ideas that make an argument both lucid and correct.

The next example illustrates some of these ideas.

Example 8.2.1 Fix $\alpha > 1$, and consider the sequence $(x_n)_{n \geq 0}$ defined by $x_0 = \alpha$, and

$$x_{n+1} = \frac{1}{2}\left(x_n + \frac{\alpha}{x_n}\right), \quad n = 0, 1, 2, \ldots.$$

Does this sequence converge, and if so, to what?

Solution: Let us try an example where $\alpha = 5$. Then we have

$$x_0 = 5,$$
$$x_1 = \frac{1}{2}\left(5 + \frac{5}{5}\right) = 3$$
$$x_2 = \frac{1}{2}\left(3 + \frac{5}{3}\right) = \frac{7}{3}.$$

Observe that the values (so far) are strictly decreasing. Will this always be the case? Let us visualize the evolution of the sequence. If we draw the graphs of $y = 5/x$ and $y = x$, we can construct a neat algorithm for producing the values of this sequence, for x_{n+1} is the *average* of the two numbers x_n and $5/x_n$. In the picture below, the y-coordinates of points B and A are respectively x_0 and $5/x_0$. Notice that the y-coordinate of the midpoint of the line segment AB is the average of these two numbers, which is equal to x_1.

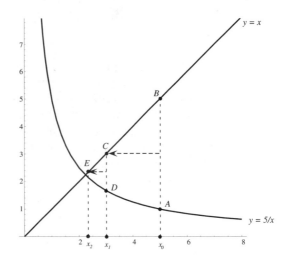

Next, draw a horizontal line left from this midpoint until it intersects the graph of $y = x$ (at C). The coordinates of C are (x_1, x_1), and we can drop a vertical line from C until it meets the graph of $y = 5/x$ (at D). By the same reasoning as before, x_2 is the y-coordinate of the midpoint of segment CD.

Continuing this process, we reach the point $E = (x_2, x_2)$, and it seems clear from this picture that if we keep going, we will converge to the intersection of the two graphs, which is the point $(\sqrt{5}, \sqrt{5})$.

Thus we conjecture that $\lim_{n\to\infty} x_n = \sqrt{5}$. However, the picture is not a rigorous proof, but an aid to reasoning. To show convergence with this picture, we would need to argue carefully why we will never "bounce" away from the convergence point. While it is possible to rigorize this, let's change gears and analyze the general problem algebraically.

The picture suggests two things: that the sequence decreases monotonically, and that it decreases to $\sqrt{\alpha}$. To prove monotonicity, we must show that $x_{n+1} \le x_n$. This is easy to do by computing the difference

$$x_n - x_{n+1} = x_n - \left(\frac{x_n^2 + \alpha}{2x_n}\right) = \frac{2x_n^2 - x_n^2 - \alpha}{2x_n} = \frac{x_n^2 - \alpha}{2x_n},$$

which is non-negative as long as $x_n^2 \ge \alpha$. And this last inequality is true; it is a simple consequence of the AM-GM inequality (see page 192):

$$\frac{1}{2}\left(x_n + \frac{\alpha}{x_n}\right) \ge \sqrt{x_n\left(\frac{\alpha}{x_n}\right)} = \sqrt{\alpha},$$

so $x_{n+1} \ge \sqrt{\alpha}$ no matter what x_n is equal to.[3] Since $x_0 = \alpha > \sqrt{\alpha}$, all terms of the sequence are greater than or equal to $\sqrt{\alpha}$.

Since the sequence is monotonic and bounded, it must converge. Now let us show that it converges to $\sqrt{\alpha}$. Since 0 is a much easier number to work with, let us define the sequence of "error" values E_n by

$$E_n := x_n - \sqrt{\alpha},$$

and show that $E_n \to 0$. Note that the E_n are all non-negative. Now we look at the ratio of E_{n+1} to E_n to see how the error changes, hoping that it decreases dramatically. We have (aren't you glad you studied factoring in Section 5.2?)

$$\begin{aligned} E_{n+1} &= x_{n+1} - \sqrt{\alpha} \\ &= \frac{1}{2}\left(x_n + \frac{\alpha}{x_n}\right) - \sqrt{\alpha} \\ &= \frac{x_n^2 + \alpha - 2x_n\sqrt{\alpha}}{2x_n} \\ &= \frac{(x_n - \sqrt{\alpha})^2}{2x_n} \\ &= \frac{E_n^2}{2x_n}. \end{aligned}$$

---

[3] Instead of studying the difference $x_n - x_{n+1}$, it is just as easy to look at the ratio x_{n+1}/x_n. This is always less than or equal to 1 (using a little algebra and the fact that $x_n \ge \sqrt{\alpha}$).

Thus

$$\frac{E_{n+1}}{E_n} = \frac{E_n}{2x_n} = \frac{x_n - \sqrt{\alpha}}{2x_n} < \frac{x_n}{2x_n} = \frac{1}{2}.$$

Since this ratio is also positive, we are guaranteed that $\lim_{n \to \infty} E_n = 0$, using method 4 on page 285.

We are done; we have shown that $x_n \to \sqrt{\alpha}$.   ∎

The trickiest part in the example above was guessing that the limit was $\sqrt{\alpha}$. What if we hadn't been lucky enough to have a nice picture? There is a simple but very productive tool that often works when a sequence is defined recursively. Let us apply it to the previous example. If $x_n \to L$, then for really large n, both x_n and x_{n+1} approach L. Thus, as n approaches infinity, the equation $x_{n+1} = (x_n + \alpha/x_n)/2$ becomes

$$L = \frac{1}{2}\left(L + \frac{\alpha}{L}\right),$$

and a tiny bit of algebra yields $L = \sqrt{\alpha}$. This **solve for the limit** tool does not prove that the limit exists, but it does show us what the limit must equal *if* it exists.

Continuity

Informally, a function is continuous if it is possible to draw its graph without lifting the pencil. Of the many equivalent formal definitions, the following is one of the easiest to use.

*Let $f : D \to \mathbf{R}$ and let $a \in D$. We say that f is **continuous** at a if*

$$\lim_{n \to \infty} f(x_n) = f(a)$$

for all sequences (x_n) in D with limit a.

We call f continuous on the set D if f is continuous at all points in D.

Continuity is a condition that you probably take for granted. This is because virtually every function that you have encountered (certainly most that can be written with a simple formula) are continuous.[4] For example, all elementary functions (finite combinations of polynomials, rational functions, trig and inverse trig functions, exponential and logarithmic functions, and radicals) are continuous at all points in their domains.

Consequently, we will concentrate on the many good properties that continuous functions possess. Here are two extremely useful ones.

Intermediate-Value Theorem (IVT)   If f is continuous on the closed interval $[a, b]$, then f assumes all values between $f(a)$ and $f(b)$. In other words, if y lies between $f(a)$ and $f(b)$, then there exists $x \in [a, b]$ such that $f(x) = y$.

Extreme-Value Theorem   If f is continuous on the closed interval $[a, b]$, then f attains minimum and maximum values on this interval. In other words, there exists $u, v \in [a, b]$ such that $f(u) \leq f(x)$ and $f(v) \geq f(x)$ for all $x \in [a, b]$.

---

[4]Notable exceptions are the floor and ceiling functions $\lfloor x \rfloor$ and $\lceil x \rceil$.

The extreme-value theorem seems almost without content, but examine the hypothesis carefully. If the domain is not a closed interval, it may not be true. For example, $f(x) := 1/x$ is continuous on $(0, 5)$, but achieves neither maximum nor minimum on this interval.

On the other hand, the IVT, while "obvious" (see problem 8.2.18 for hints about its proof), has many immediate applications. Here is one simple example. The crux move, defining a new function, is a typical tactic in problems of this kind.

Example 8.2.2 Let $f : [0, 1] \to [0, 1]$ be continuous. Prove that f has a **fixed point**; i.e., there exists $x \in [0, 1]$ such that $f(x) = x$.

Solution: Let $g(x) := f(x) - x$. Note that g is continuous, and that $g(0) = f(0) \geq 0$ and $g(1) = f(1) - 1 \leq 0$. By the IVT, there exists $u \in [0, 1]$ such that $g(u) = 0$. But this implies that $f(u) = u$. ∎

Uniform Continuity

Continuous functions on a closed interval (i.e., the domain is a closed interval) possess another important property, that of **uniform continuity**. Informally, this means that the amount of "wiggle" in the graph is constrained in the same way throughout the domain. More precisely,

> A function $f : A \to B$ is uniformly continuous on A, if, for each $\epsilon > 0$, there exists $\delta > 0$ such that if $x_1, x_2 \in A$ satisfy $|x_1 - x_2| < \delta$, then $|f(x_1) - f(x_2)| < \epsilon$.

The important thing in this definition is that the value of δ depends *only* on ϵ and not on the x-value. For each positive ϵ, there is a *single* δ which works everywhere on the domain. Because it is rather difficult to prove that all continuous functions on closed intervals are uniformly continuous, the concept of uniform continuity is not often introduced in elementary calculus classes. But it is such a useful idea that we will accept it, for now, on faith.[5]

Example 8.2.3 The function $f(x) = x^2$ is uniformly continuous on $[-3, 3]$. As long as $|x_1 - x_2| < \delta$, we are guaranteed that $|f(x_1) - f(x_2) < 6\delta$. It is easy to see why: For any $x_1, x_2 \in [-3, 3]$, the largest possible value for $|x_1 + x_2|$ is 6, and then

$$|f(x_1) - f(x_2)| = |x_1 + x_2| \cdot |x_1 - x_2| \leq 6|x_1 - x_2|.$$

Consequently, if we want to be sure that the function values are within ϵ, we need only require that the x-values be within $\epsilon/6$.

Example 8.2.4 The function $f(x) = 1/x$ defined on $(0, \infty)$ is not uniformly continuous. For x-values close to 0, the function changes too fast. Given an ϵ, no single δ will do, if the x-values are sufficiently close to 0. Note, however, that on any closed

---

[5]Consult any of the excellent texts by Boas [2], Spivak [27], or Apostol [1] for more information. The book by Boas stands out in particular, because it is less than 200 pages long!

interval, $f(x)$ is uniformly continuous. For example, verify that if we are restricting our attention to $x \in [2, 1000]$, then the "δ response" to the "ϵ challenge" is $\delta = 4\epsilon$. In other words, if we are challenged to constrain the f-values to be within ϵ of each other, we need only choose x-values within 4ϵ of one another.

Uniform continuity is just what we need to complete our proof of the FTC.

Example 8.2.5 Show that

$$\lim_{\Delta t \to 0} \left(\frac{E(t)}{\Delta t} \right) = 0,$$

where $E(t)$ was defined in the diagram on page 284.

Solution: Since f is defined on the closed interval $[a, b]$ and is continuous, it is uniformly continuous. Pick $\epsilon > 0$. By uniform continuity, there is a small enough Δt so that *no matter what t is*, the range of values between $f(t)$ and $f(t + \Delta t)$ is less than ϵ. In other words,

$$|f(t + \Delta t) - f(t)| < \epsilon.$$

Thus the area of the "error" $E(t)$ (in absolute value) is at most $\epsilon \cdot \Delta t$, and hence

$$\frac{|E(t)|}{\Delta t} < \frac{\epsilon \cdot \Delta t}{\Delta t} = \epsilon.$$

In other words, no matter how small we pick ϵ, we can pick a small enough Δt to guarantee that $E(t)/\Delta t$ is less than ϵ. Hence the limit is 0, and we have proven the FTC. ∎

Uniform continuity, as you see, is a powerful *technical* tool. But remember, the crux idea in our proof of the FTC was to picture a moving curtain. This simple picture is easy to remember and immediately leads to a one-sentence "proof" which is missing technical details. The details are important, but the picture—or at least, the idea behind the picture—is fundamental.

Problems and Exercises

8.2.6 Define the sequence (a_n) by $a_1 = 1$ and $a_n = 1 + 1/a_{n-1}$ for $n \geq 1$. Discuss the convergence of this sequence.

8.2.7 Suppose that (a_n) and (a_n/b_n) both converge. What can you say about the convergence of (b_n)?

8.2.8 Interpret the meaning of

$$\sqrt{2 + \sqrt{2 + \sqrt{2 + \sqrt{2} + \cdots}}}.$$

8.2.9 Fix $\alpha > 1$, and consider the sequence $(x_n)_{n \geq 0}$ defined by $x_0 > \sqrt{\alpha}$ and

$$x_{n+1} = \frac{x_n + \alpha}{x_n + 1}, \quad n = 0, 1, 2, \ldots.$$

Does this sequence converge, and if so, to what? Relate this to Example 8.2.1 on page 286.

8.2.10 Let (a_n) be a (possibly infinite) sequence of positive integers. A creature like

$$a_0 + \cfrac{1}{a_1 + \cfrac{1}{a_2 + \cfrac{1}{a_3 + \cfrac{1}{\ddots}}}}$$

is called a **continued fraction** and is sometimes denoted by $[a_0, a_1, a_2, \ldots]$.

(a) Give a rigorous interpretation of the number $[a_0, a_1, a_2, \ldots]$.

(b) Evaluate $[1, 1, 1, \ldots]$.

(c) Evaluate $[1, 2, 1, 2, \ldots]$.

(d) Find a sequence (a_n) of positive integers such that $\sqrt{2} = [a_0, a_1, a_2, \ldots]$. Can there be more than one sequence?

(e) Show that there does not exist a repeating sequence (u_n) such that

$$\sqrt[3]{2} = [a_0, a_1, a_2, \ldots].$$

8.2.11 Carefully prove the assertion stated on page 285, that all bounded monotonic sequences converge.

8.2.12 *Dense Sets.* A subset S of the real numbers is called **dense** if, given any real number x, there are elements of S which are arbitrarily close to x. For example, $\mathbf{Q}$ is dense, since any real number can be approximated arbitrarily well with fractions (look at decimal approximations). Here is a formal definition:

S is dense if, given any $x \in \mathbf{R}$ and $\epsilon > 0$, there exists $s \in S$ such that $|s - x| < \epsilon$.

More generally, we say that the set S is dense in the set T if any $t \in T$ can be approximated arbitrarily well by elements of S. For example, the set of positive fractions with denominator greater than numerator is dense in the unit interval $[0, 1]$.

(a) Observe that "S is dense in T" is equivalent to saying that for each $t \in T$, there is an infinite sequence (s_k) of elements in S such that $\lim_{k \to \infty} s_k = t$.

(b) Show that the set of real numbers that are zeros of quadratic equations with integer coefficients is dense.

(c) Let D be the set of **dyadic rationals**, the rational numbers whose denominators are powers of two ($1/2, 5/8, 1037/256$ are examples of dyadic rationals). Show that D is dense.

(d) Let S be the set of real numbers in $[0, 1]$ whose decimal representation contains no 3's. Show that S is not dense in $[0, 1]$.

8.2.13 Define $\langle x \rangle := x - \lfloor x \rfloor$. In other words, $\langle x \rangle$ is the "fractional part" of x; for example $\langle \pi \rangle = 0.14159 \ldots$. Let

$$a = 0.123456789101112131415\ldots;$$

in other words, a_0 is the number formed by writing every positive integer in order after the decimal point. Show that the set

$$\{\langle a \rangle, \langle 10a \rangle, \langle 10^2 a \rangle, \ldots\}$$

is dense in $[0, 1]$ (see 8.2.12 for the definition of "dense").

8.2.14 Let α be irrational. Show that the set

$$\{\langle \alpha \rangle, \langle 2\alpha \rangle, \langle 3\alpha \rangle, \ldots\}$$

is dense in $[0, 1]$. (See 8.2.12 for the definition of "dense" and 8.2.13 for the definition of $\langle x \rangle$.)

8.2.15 Consider a circle with radius 3 and center at the origin. Points A and B have coordinates $(2, 0)$ and $(-2, 0)$, respectively. If a dart is thrown at the circle, assuming a uniform distribution, it is clear that the probabilities of the two events.

- The dart is closer to A than to B.
- The dart is closer to B than to A.

are equal. We call the two points "fairly placed." Is is possible to have a third point C on this circle, so that A, B, C are all fairly placed?

8.2.16 Define the sequence (a_n) by $a_0 = \alpha$ and $a_{n+1} = a_n - a_n^2$ for $n \geq 1$. Discuss the converge of this sequence (it will depend on the initial value α).

8.2.17 Draw two nonintersecting circles on a piece of paper. Show that it is possible to draw a straight line which divides each circle into equal halves. That was easy. Next,

(a) What if the two shapes were arbitrary nonintersecting rectangles, instead of circles?

(b) What if the two shapes were arbitrary nonintersecting convex polygons, instead of circles?

(c) What if the two shapes were arbitrary nonintersecting "amoebas," possibly not convex?

8.2.18 Here is an allegory which should provide you with a strategy (the "repeated bisection method") for a rigorous proof of the intermediate value theorem. Consider the concrete problem of trying to find the square of 2 with a calculator that doesn't have a square-root key. Since $1^2 = 1$ and $2^2 = 4$, we figure that $\sqrt{2}$, if it exists, is between 1 and 2. So now we guess 1.5. But $1.5^2 > 2$, so our mystery number lies between 1 and 1.5. We try 1.25. This proves to be too small, so next we try $\frac{1}{2}(1.25 + 1.5)$, etc. We thus construct a sequence of successive approximations, alternating between too big and too small, but each approximation gets better and better.

8.2.19 (Leningrad Mathematical Olympiad, 1991) Let f be continuous and monotonically increasing, with $f(0) = 0$ and $f(1) = 1$. Prove that

$$f\left(\frac{1}{10}\right) + f\left(\frac{2}{10}\right) + \cdots + f\left(\frac{9}{10}\right) +$$
$$f^{-1}\left(\frac{1}{10}\right) + f^{-1}\left(\frac{2}{10}\right) + \cdots + f^{-1}\left(\frac{9}{10}\right) \leq \frac{99}{10}.$$

8.2.20 (Leningrad Mathematical Olympiad, 1988) Let $f : \mathbf{R} \to \mathbf{R}$ be continuous, with $f(x) \cdot f(f(x)) = 1$ for all $x \in \mathbf{R}$. If $f(1000) = 999$, find $f(500)$.

8.2.21 Let $f : [0, 1] \to \mathbf{R}$ be continuous, and suppose that $f(0) = f(1)$. Show that there is a value $x \in [0, 1998/1999]$ satisfying $f(x) = f(x + 1/1999)$.

8.2.22 (Putnam 1990) Is $\sqrt{2}$ the limit of a sequence of numbers of the form $\sqrt[3]{n} - \sqrt[3]{m}$ $(n, m = 0, 1, 2, \ldots)$?

8.2.23 (Putnam 1992) For any pair (x, y) of real numbers, a sequence $(a_n(x, y))_{n \geq 0}$ is defined as follows:

$$a_0(x, y) = x,$$
$$a_{n+1}(x, y) = \frac{(a_n(x, y))^2 + y^2}{2}, \qquad \text{for } n \geq 0.$$

Find the area of the region

$$\{(x, y) | (a_n(x, y))_{n \geq 0} \quad \text{converges}\}.$$

8.2.24 Let (x_n) be a sequence satisfying $\lim_{n \to \infty} (x_n - x_{n-1}) = 0$. Prove that

$$\lim_{n \to \infty} \frac{x_n}{n} = 0.$$

8.2.25 (Putnam 1970) Given a sequence (x_n) such that $\lim_{n \to \infty} (x_n - x_{n-2}) = 0$. Prove that

$$\lim_{n \to \infty} \frac{x_n - x_{n-1}}{n} = 0.$$

Infinite Series of Constant Terms

If (a_n) is a sequence of real numbers, we define the infinite series $\sum_{k=1}^{\infty} a_k$ to be the limit, if it exists, of the sequence of **partial sums** (s_n), where

$$s_n := \sum_{k=1}^{n} a_k.$$

Problems 8.2.26–8.2.30 below play around with infinite series of constant terms. (You may want to reread the section on infinite series in Chapter 5, and in particular Example 5.3.4 on page 176.)

8.2.26 Show *rigorously* that if $|r| < 1$, then

$$a + ar + ar^2 + ar^3 + \cdots = \frac{a}{1-r}.$$

8.2.27 Let $\sum_{k=1}^{\infty} a_k$ be a convergent infinite series, i.e., the partial sums converge. Prove that

(a) $\lim_{n \to \infty} a_n = 0$;

(b) $\lim_{n \to \infty} \sum_{k=n}^{\infty} a_k = 0.$

8.2.28 (Putnam 1994) Let (a_n) be a sequence of positive reals such that, for all n, $a_n \le a_{2n} + a_{2n+1}$. Prove that $\sum_{n=1}^{\infty} a_n$ diverges.

8.2.29 Let (a_n) be a sequence whose terms alternate in sign, and whose terms decrease monotonically to zero in absolute value. (For example, $1, -1/2, +1/3, -1/4, \ldots$.) Show that $\sum_{n=1}^{\infty} a_n$ converges.

8.2.30 Let (a_k) be the sequence used as an example in the previous problem. This problem showed that $\sum_{n=1}^{\infty} a_n$ converges. Notice also that $\sum_{n=1}^{\infty} |a_n|$ diverges (it is the harmonic series, after all). Use these two facts to show that given *any* real number x, it is possible to rearrange the terms of the sequence (a_n) so that the new sum converges to x. (By "rearrange" we mean reorder. For example, one rearrangement would be the series

$$\frac{1}{3} + \frac{1}{19} - \frac{1}{100} + \frac{1}{111} - 1 + \cdots.$$

8.3   Differentiation and Integration

Approximation and Curve Sketching

You certainly know that the derivative $f'(x)$ of the function $f(x)$ has two interpretations: a *dynamic* definition as rate of change [of $f(x)$ with respect to x], and a *geometric* definition as slope of the tangent line to the graph of $y = f(x)$ at the point $(x, f(x))$.

The rate-of-change definition is especially useful for understanding how functions grow. More elaborate information comes from the second derivative $f''(x)$, which of course measures how fast the derivative is changing. Sometimes just simple analysis of the signs of f' and f'' is enough to solve fairly complicated problems.

Example 8.3.1 Reread Example 2.2.7 on page 38, in which we studied the inequality $q(x) \ge q'(x)$ for a polynomial function q. Recall that we reduced the original problem to the following assertion:

> *Prove that if $p(x)$ is a polynomial with even degree with positive leading coefficient, and $p(x) - p''(x) \ge 0$ for all real x, then $p(x) \ge 0$ for all real x.*

Solution: The hypothesis that $p(x)$ has even degree with positive leading coefficient means that

$$\lim_{x \to -\infty} p(x) = \lim_{x \to +\infty} p(x) = +\infty;$$

therefore the minimum value of $p(x)$ is finite (since p is a polynomial, it only "blows up" as $x \to \pm\infty$). Now let us argue by contradiction, and assume that $p(x)$ is negative for some values of x. Let $p(a) < 0$ be the minimum value of the function. Recall that at relative minima, the second derivative is non-negative. Thus $p''(a) \geq 0$. But $p(a) \geq p''(a)$ by hypothesis, which contradicts $p(a) < 0$. ∎

Here is another polynomial example which adds analysis of the derivative to standard polynomial techniques.

Example 8.3.2 (United Kingdom 1995) Let a, b, c be real numbers satisfying $a < b < c$, $a + b + c = 6$ and $ab + bc + ca = 9$. Prove that $0 < a < 1 < b < 3 < c < 4$.

Solution: We are given information about $a + b + c$ and $ab + bc + ca$, which suggests that we look at the polynomial $P(x)$ whose zeros are a, b, c. We have

$$P(x) = x^3 - 6x^2 + 9x - k,$$

where $k = abc$, using the relationship between zeros and coefficients (see page 183). We must investigate the zeros of $P(x)$, and to this end we draw a rough sketch of the graph of this function. The graph of $P(x)$ must look something like the following picture.

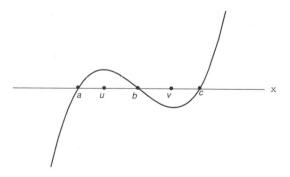

We have not included the y-axis, because we are not yet sure of the signs of a, b, c. But what we are sure of is that for sufficiently large negative values of x, $P(x)$ will be negative, since the leading term x^3 has this behavior, and it dominates the other terms of $P(x)$ if x is a large enough negative number. Likewise, for sufficiently large positive x, $P(x)$ will be positive. Since the zeros of $P(x)$ are $a < b < c$, $P(x)$ will have to be positive for x-values between a and b, with a relative maximum at $x = u$, and $P(x)$ will attain negative values when $b < x < c$, with a relative minimum at $x = v$.

We can find u and v by computing the derivative

$$P'(x) = 3x^2 - 12x + 9 = 3(x-1)(x-3).$$

Thus $u = 1$, $v = 3$ and we have $f(1) > 0$, $f(3) < 0$ so $a < 1 < b < 3 < c$.

It remains to show that $a > 0$ and $c < 4$. To do so, all we need to show is that $P(0) < 0$ and $P(4) > 0$. We will be able to determine the signs of these quantities if we can discover more about the unknown quantity k. But this is easy: $P(1) = 4 - k > 0$ and $P(3) = -k < 0$, so $0 < k < 4$. Therefore we have $P(0) = -k < 0$ and $P(4) = 4 - k > 0$, as desired. ∎

The tangent-line definition of the derivative stems from its formal definition as a limit. One of the first things you learned in your calculus class was the definition

$$f'(a) := \lim_{x \to a} \frac{f(x) - f(a)}{x - a} = \lim_{h \to 0} \frac{f(x+h) - f(x)}{h}.$$

The fractions in the definition compute the slope of "secant lines" which approach the tangent line in the limit. This suggests a useful, but less well-known, application of the derivative, the **tangent-line approximation** to the function. For example, suppose that $f(3) = 2$ and $f'(3) = 4$. Then

$$\lim_{h \to 0} \frac{f(3+h) - f(3)}{h} = 4.$$

Thus when h is small in absolute value, $(f(3+h) - f(3))/h$ will be close to 4; i.e.,

$$f(3+h) \approx f(3) + 4h = 2 + 4h.$$

In other words, the function $\ell(h) := 2 + 4h$ is the best *linear* approximation to $f(3 + h)$, because it is the only linear function $\ell(h)$ which satisfies

$$\lim_{h \to 0} (f(3+h) - \ell(h)) = 0.$$

In general, analyzing $f(a + h)$ with its tangent-line approximation $f(a) + hf'(a)$ is very useful, especially when combined with other geometric information, such as convexity.

Example 8.3.3 Prove **Bernoulli's Inequality**:

$$(1 + x)^\alpha \geq 1 + \alpha x,$$

for $x > -1$ and $\alpha \geq 1$, with equality when $x = 0$.

Solution: For integer α, this can be proven by induction, and indeed, this was Problem 2.3.33 on page 57. But induction won't work for arbitrary real α. Instead, define $f(u) := u^\alpha$, and note that $f'(u) = \alpha u^{\alpha-1}$ and $f''(u) = \alpha(\alpha-1)u^{\alpha-2}$. Thus $f(1) = 1$, $f'(1) = \alpha$ and $f''(u) > 0$ as long as $u > 0$ (provided, of course, that $\alpha \geq 1$). Thus the graph $y = f(u)$ is concave-up, for all $u \geq 0$, as shown below.

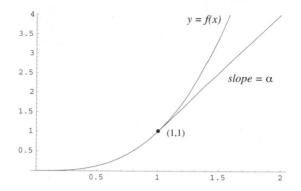

Therefore, the graph of $y = f(u)$ lies above the tangent line for all $u \geq 0$. Another way of saying this (make the substitution $x = u - 1$) is that $f(1 + x)$ is always strictly greater than its linear approximation $1 + \alpha x$, except when $x = 0$, in which case we have equality [corresponding to the point $(1, 1)$ on the graph]. We have established Bernoulli's inequality.[6]   ∎

The Mean Value Theorem

One difficulty that many beginners have with calculus problems is confusion over what should be rigorous and what can be assumed on faith as "intuitively obvious." This is not an easy issue to resolve, for some of the simplest, most "obvious" statements involve deep, hard-to-prove properties of the real numbers and differentiable functions.[7] We are not trying to be a real analysis textbook, and will not attempt to prove all of these statements. But we will present, with a "hand-waving" proof, one important theoretical tool which will allow you to begin to think more rigorously about many problems involving differentiable functions.

We begin with **Rolle's theorem**, which certainly falls into the "intuitively obvious" category.

> If $f(x)$ is continuous on $[a, b]$ and differentiable on (a, b), and $f(a) = f(b)$, then there is a point $u \in (a, b)$ at which $f'(u) = 0$.

The "proof" is a matter of drawing a picture. There will be a local minimum or maximum between a and b, at which the derivative will equal zero.

---

[6]The sophisticated reader may object that we need Bernoulli's inequality (or something like it) in the first place in order to compute $f'(u) = \alpha u^{\alpha-1}$ when α is not rational. This is not true; for example, see the brilliant treatment in [17], pp. 229–231, which uses the geometry of complex numbers in a surprising way.

[7]A function f is called differentiable on the open interval (a, b) if $f'(x)$ exists for all $x \in (a, b)$. We won't worry about differentiability at the endpoints a and b; there is a technical problem about how limits should be defined there.

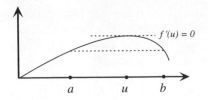

Rolle's theorem has an important generalization, the **mean value theorem**.

If $f(x)$ is continuous on $[a, b]$ and differentiable on (a, b), then there is a point $u \in (a, b)$ at which

$$f'(u) = \frac{f(b) - f(a)}{b - a}.$$

From the picture, it is clear that the mean value theorem asserts that there is an x-value $u \in (a, b)$ at which the slope of the tangent line at $(u, f(u))$ is parallel to the secant line joining $(a, f(a))$ and $(b, f(b))$. And the proof is just one sentence:

Tilt the picture for Rolle's theorem!

The mean value theorem connects a "global" property of a function (its values at the endpoints a and b) with a "local" property (the value of its derivative at a specific point) and is thus a deeper and more useful fact than is apparent at first glance. Here is an example.

Example 8.3.4 Suppose f is differentiable on $(-\infty, \infty)$ and there is a constant $k < 1$ such that $|f'(x)| \le k$ for all real x. Show that f has a fixed point.

Solution: Recall from Example 8.2.2 on page 289 that a fixed point is a point x such that $f(x) = x$. Thus we must show that the graphs of $y = f(x)$ and $y = x$ will intersect. Without loss of generality, suppose that $f(0) = v > 0$ as shown.

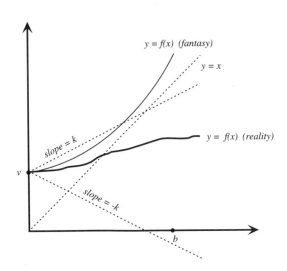

The picture gives us a vague idea. Since the derivative is at most k in absolute value, and since $k < 1$, the graph of $y = f(x)$ to the right of the y-axis will be trapped within the dotted-line "cone," and will eventually "catch up" with the graph of $y = x$. The mean value theorem lets us prove this in a satisfying way. Suppose that for all $x \geq 0$, we have $f(x) \neq x$. Then (IVT) we must have $f(x) > x$. Pick $b > 0$ (think large). By the mean value theorem, there is a $u \in (0, b)$ such that

$$f'(u) = \frac{f(b) - f(0)}{b - 0} = \frac{f(b) - v}{b}.$$

Since $f(b) > b$, we have

$$f'(u) > \frac{b - v}{b} = 1 - \frac{v}{b}.$$

Since b can be arbitrarily large, we can arrange things so that $f'(u)$ becomes arbitrarily close to 1. But this contradicts $|f'(u)| \leq k < 1$. Thus $f(x)$ must equal x for some $x > 0$.

If we supposed that $f(0) < 0$, the argument would be identical, except that we would draw our "cone" to the left of the y-axis, etc. ∎

The satisfying thing about this argument was the role that the mean value theorem played in guaranteeing exactly the right derivative values to get the desired contradiction.

The next example is a rather tricky problem that uses Rolle's theorem infinitely many times.

Example 8.3.5 (Putnam 1992) Let f be an infinitely differentiable real-valued function defined on the real numbers. If

$$f\left(\frac{1}{n}\right) = \frac{n^2}{n^2 + 1}, \qquad n = 1, 2, 3, \ldots,$$

compute the values of the derivatives $f^{(k)}(0), k = 1, 2, 3, \ldots$. (We are using the notation $f^{(k)}$ for the k-th derivative of f.)

Partial Solution: At first you might guess that we can let $n = 1/x$ and get

$$f(x) = \frac{\frac{1}{x^2}}{\frac{1}{x^2} + 1} = \frac{1}{x^2 + 1}$$

for all x. The trouble with this is that it is only valid for those values of x for which $1/x$ is an integer! So we know *nothing* at all about the behavior of $f(x)$ except at the points $x = 1, 1/2, 1/3, \ldots$.

But wait! The limit of the sequence $1, 1/2, 1/3, \ldots$ is 0, and the problem is only asking for the behavior of $f(x)$ at $x = 0$. So the strategy is clear: wishful thinking suggests that $f(x)$ and its derivatives agree with the behavior of the function $w(x) := 1/(x^2 + 1)$ at $x = 0$.

In other words, we want to show that the function

$$v(x) := f(x) - w(x)$$

satisfies

$$v^{(k)}(0) = 0, \quad k = 1, 2, 3, \ldots .$$

This isn't too hard to show, since $v(x)$ is "almost" equal to 0 and gets more like 0 as x approaches 0 from the right. More precisely, we have

$$0 = v(1) = v(1/2) = v(1/3) = \cdots . \qquad (4)$$

Since $v(x)$ is continuous, this means that $v(0) = 0$. Here's why: Let $x_1 = 1, x_2 = 1/2, x_3 = 1/3, \ldots$. Then $\lim_{n \to \infty} x_n = 0$ and

$$\lim_{n \to \infty} v(x_n) = \lim_{n \to \infty} 0 = 0,$$

and $v(x) = 0$ by the definition of continuity (see page 288).

Now you complete the argument! Use Rolle's theorem to get information about the derivative, as $x \to 0$, etc.[8]

A Useful Tool

We will conclude our discussion of differentiation with two examples that illustrate a useful idea inspired by logarithmic differentiation.

Example 8.3.6 *Logarithmic Differentiation.* Let $f(x) = \displaystyle\prod_{k=0}^{n} (x + k)$. Find $f'(1)$.

Solution: Differentiating a product is not that hard, but a more elegant method is to convert to a sum first by taking logarithms. We have

$$\log(f(x)) = \log x + \log(x + 1) + \cdots + \log(x + n),$$

and differentiation yields

$$\frac{f'(x)}{f(x)} = \frac{1}{x} + \frac{1}{x+1} + \cdots + \frac{1}{x+n}.$$

Thus

$$f'(1) = (n+1)! \left(1 + \frac{1}{2} + \cdots + \frac{1}{n+1} \right). \qquad \blacksquare$$

Logarithmic differentiation is not just a tool for computing derivatives. It is part of a larger idea: developing a bank of useful derivatives of "functions of a function" that you can recognize to analyze the original function. If a problem contains or can be made to contain the quantity $f'(x)/f(x)$, then antidifferentiation will yield the logarithm of $f(x)$, which in turn sheds light on $f(x)$. Here is another example of this style of reasoning.

----

[8]See Example 8.4.3 on page 311 for a neat way to compute the derivatives of $1/(x^2 + 1)$ at 0.

Example 8.3.7 (Putnam 1997) Let f be a twice-differentiable real-valued function satisfying

$$f(x) + f''(x) = -xg(x)f'(x), \tag{5}$$

where $g(x) \geq 0$ for all real x. Prove that $|f(x)|$ is bounded, i.e., show that there exists a constant C such that $|f(x)| \leq C$ for all x.

Partial Solution: The differential equation cannot easily be solved for $f(x)$, and integration likewise doesn't seem to help. However, the left-hand side of (5) is similar to the derivative of something familiar. Observe that

$$\frac{d}{dx}(f(x)^2) = 2f(x)f'(x).$$

This suggests that we multiply both sides of (5) by $f'(x)$, getting

$$f(x)f'(x) + f'(x)''(x) = -xg(x)(f'(x))^2.$$

Thus

$$(f(x)^2 + f'(x)^2)' = -xg(x)(f'(x))^2. \tag{6}$$

Now let $x \geq 0$. The right-hand side of (6) will be nonpositive, which means that $f(x)^2 + f'(x)^2$ is non-increasing for $x \geq 0$. Hence

$$f(x)^2 + f'(x)^2 \leq f(0)^2 + f'(0)^2$$

for all $x \geq 0$. This certainly implies that $f(x)^2 \leq f(0)^2 + f'(0)^2$. Thus there is a constant

$$C := \sqrt{f(0)^2 + f'(0)^2},$$

for which $|f(x)| \leq C$ for all $x \geq 0$. We will be done if we can do the argument for $x < 0$. We leave that as an exercise.

Integration

The fundamental theorem of calculus gives us a method for computing definite integrals. We are not concerned here with the process of antidifferentiation—we assume that you are well versed in the various techniques—but rather a better understanding of the different ways to view definite integrals. There is a lot of interplay between summation, integration, and inequalities; many problems exploit this.

Example 8.3.8 Compute $\displaystyle\lim_{n\to\infty} \sum_{k=1}^{n} \frac{n}{k^2 + n^2}$.

Solution: The problem is impenetrable until we realize that we are not faced with a sum, but the limit of a sum; and that is exactly what definite integrals are. So let us try to work backwards and construct a definite integral whose value is the given limit.

Recall that we can approximate the definite integral $\displaystyle\int_a^b f(x)dx$ by the sum

$$s_n := \frac{1}{n}(f(a) + f(a+\Delta) + f(a+2\Delta) + \cdots + f(b-\Delta)),$$

where $\Delta = \dfrac{b-a}{n}$. Indeed,

$$\lim_{n \to \infty} s_n = \int_a^b f(x)dx.$$

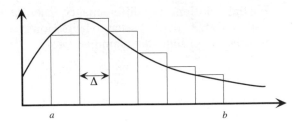

Now it is just a matter of getting $\displaystyle\sum_{k=1}^{n} \dfrac{n}{k^2+n^2}$ to look like s_n for appropriately chosen $f(x)$, a, and b. The crux move is to extract something that looks like $(b-a)/n$. Observe that

$$\frac{n}{k^2+n^2} = \frac{1}{n}\left(\frac{n^2}{k^2+n^2}\right) = \frac{1}{n}\left(\frac{1}{k^2/n^2+1}\right).$$

If k ranges from 1 to n, then k^2/n^2 ranges from $1/n^2$ to 1, which suggests that $a = 0$, $b = 1$, and $f(x) = 1/(x^2+1)$. It is easy to verify that this works; i.e.,

$$\frac{1}{n}\sum_{k=1}^{n} \frac{n}{k^2+n^2} = \frac{1}{n}\left(f\left(\frac{1}{n}\right) + f\left(\frac{2}{n}\right) + \cdots + f\left(\frac{n}{n}\right)\right).$$

Thus the given limit is equal to

$$\int_0^1 f(x)dx = \int_0^1 \frac{dx}{1+x^2} = \arctan x \Big]_0^1 = \frac{\pi}{4} - 0 = \frac{\pi}{4}. \qquad \blacksquare$$

Even finite sums can be analyzed with integrals. If the functions involved are monotonic, it is possible to relate integrals and sums with inequalities, as in the next example.

Example 8.3.9 (Putnam 1996) Show that for every positive integer n,

$$\left(\frac{2n-1}{e}\right)^{\frac{2n-1}{2}} < 1 \cdot 3 \cdot 5 \cdots (2n-1) < \left(\frac{2n+1}{e}\right)^{\frac{2n+1}{2}}.$$

Solution: The e's in the denominator along with the ugly exponents strongly suggest a simplifying strategy: take logarithms! This transforms the alleged inequality into

$$\frac{2n-1}{2}(\log(2n-1)-1) < S < \frac{2n+1}{2}(\log(2n+1)-1),$$

where

$$S = \log 1 + \log 3 + \log 5 + \cdots + \log(2n - 1).$$

Let us take a closer look at S. Because $\log x$ is monotonically increasing, it is apparent from the picture (here $\Delta = 2$) that

$$\int_1^{2n-1} \log x \, dx < 2S < \int_1^{2n+1} \log x \, dx.$$

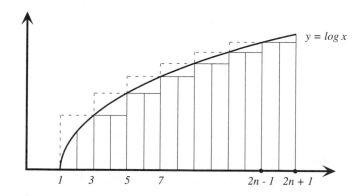

The inequality now follows, since an easy application of integration by parts yields $\int \log x \, dx = x \log x - x + C.$ ∎

Problems and Exercises

8.3.10 Prove that if the polynomial $P(x)$ and its derivative $P'(x)$ share the zero $x = r$, then $x = r$ is zero of multiplicity greater than 1. [A zero r of $P(x)$ has multiplicity m if $(x - r)$ appears m times in the factorization of $P(x)$. For example, $x = 1$ is a zero of multiplicity 2 of the polynomial $x^2 - 2x + 1$.]

8.3.11 Let $a, b, c, d, e \in \mathbf{R}$ such that

$$a + \frac{b}{2} + \frac{c}{3} + \frac{d}{4} + \frac{e}{5} = 0.$$

Show that the polynomial $a + bx + cx^2 + dx^3 + ex^5$ has at least one real zero.

8.3.12 *A Fable.* The following story was told by Doug Jungreis to his calculus class at UCLA. It is not completely true.

> A couple of years ago, I drove up to the Bay Area, which is 400 miles, and I drove fast, so it took me 5 hours. At the end of the trip, I slowed down, because I didn't want to get a ticket, and when I got off the freeway, I was traveling at the speed limit. Then a police officer pulled me over,

and he said, "You don't look like no Mario Andretti," and then he said, "You were going a little fast there." I said I was going the speed limit, but he responded, "Maybe you were a little while ago, but earlier, you were speeding." I asked how he knew that, and he said, "Son, by the mean value theorem of calculus, at some moment in the last 5 hours, you were going at *exactly* 80 m.p.h."

I took the ticket to court, and when push came to shove, the officer was unable to prove the mean value theorem beyond a reasonable doubt.

(a) Assuming that the officer could prove the mean value theorem, would his statement have been correct? Explain.

(b) Let us change the ending of the story so that the officer said, "I can't prove the mean value theorem, your Honor, but I can prove the intermediate value theorem, and using this, I can show that there was a time interval of exactly one minute during which the defendant drove at an average speed of 80 miles per hour." Explain his reasoning.

8.3.13 Finish up Example 8.3.7 by discussing the $x < 0$ case.

8.3.14 (Putnam 1994) Find all c such that the graph of the function $x^4 + 9x^3 + cx^2 + ax + b$ meets some line in four distinct points.

8.3.15 Let $f(x)$ be a differentiable function which satisfies

$$f(x+y) = f(x)f(y)$$

for all $x, y \in \mathbf{R}$. If $f'(0) = 3$, find $f(x)$.

8.3.16 (Putnam 1946) Let $f(x) := ax^2 + bx + c$, where $a, b, c \in \mathbf{R}$. If $|f(x)| \leq 1$ for $|x| \leq 1$, prove that $|f'(x)| \leq 4$ for $|x| \leq 1$.

8.3.17 *More About the Mean Value Theorem.*

(a) The "proof" of the mean value theorem on page 298 was simply to "tilt" the picture for Rolle's theorem. Prove the mean value theorem slightly more rigorously now, by assuming the truth of Rolle's theorem, and defining a new function in such a way that Rolle's theorem applied to this new function yields the mean value theorem. Use the tilting-picture idea as your guide.

(b) If you succeeded in (a), you may still grumble that you merely did an algebra exercise, but really did nothing new, and certainly achieved no insight better than tilting the picture. This is true, but the algebraic method is easy to generalize. Use it to prove the **generalized mean value theorem**, which involves two functions:

Let $f(x)$ and $g(x)$ be continuous on $[a, b]$ and differentiable on (a, b). Then there is a point $u \in (a, b)$ such that

$$f'(u)(g(b) - g(a)) = g'(u)(f(b) - f(a)).$$

(c) The regular mean value theorem is a special case of the generalized mean value theorem. Explain why.

(d) It is still worthwhile to understand the generalized mean value theorem pictorially. Draw a picture to illustrate what it says. Can you develop a pictorial proof, similar to the one we did for the mean value theorem?

8.3.18 *The Mean Value Theorem for Integrals.* This theorem states the following:

Let $f(x)$ be continuous on (a, b). Then there is a point $u \in (a, b)$ at which

$$f(u)(b - a) = \int_a^b f(x)dx.$$

(a) Draw a picture to see why this theorem is plausible, in fact "obvious."

(b) Use the regular mean value theorem to prove it (define a function cleverly, etc.).

8.3.19 Let $f(x)$ be a differentiable function which satisfies

$$f(xy) = f(x) + f(y)$$

for all $x, y > 0$. If $f'(e) = 3$, find $f(x)$.

8.3.20 Find and prove (using induction) a nice formula for the nth derivative of the product $f(x)g(x)$.

8.3.21 (Putnam 1993) The horizontal line $y = c$ intersects the curve $y = 2x - 3x^3$ in the first quadrant as in the figure. Find c so that the areas of the two shaded regions are equal.

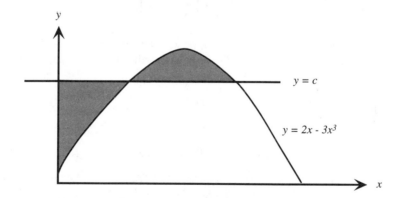

8.3.22 (Putnam 1964) Find all continuous functions $f(x)$, for $0 \leq x \leq 1$, such that

$$\int_0^1 f(x)dx = 1$$

$$\int_0^1 xf(x)dx = \alpha$$

$$\int_0^1 x^2 f(x)dx = \alpha^2,$$

where α is a given real number.

8.3.23 (Bratislava 1994) Define $f : [0, 1] \to [0, 1]$ by

$$f(x) = \begin{cases} 2x & 0 \le x \le 1/2, \\ -2x + 2 & 1/2 < x \le 1. \end{cases}$$

Next, define a sequence f_n of functions from $[0, 1]$ to $[0, 1]$ as follows: Let $f_1(x) = f(x)$ and let $f_n(x) = f(f_{n-1}(x))$ for $n > 1$. Prove that for each n,

$$\int_0^1 f_n(x)\,dx = 1/2.$$

8.3.24 Compute $\displaystyle\lim_{n\to\infty} \frac{1}{n} + \frac{1}{n+1} + \frac{1}{n+2} + \cdots + \frac{1}{2n-1}$.

8.3.25 In Example 8.3.4, there was a constant k which was strictly less than 1, and $|f'(x)| \le k$. Notice that this is *not* the same as saying that $|f'(x)| \le 1$. What would happen in this case?

8.3.26 (Putnam 1994) Let $f(x)$ be a positive-valued function over the reals such that $f'(x) > f(x)$ for all x. For what k must there exist N such that $f(x) > e^{kx}$ for $x > N$?

8.3.27 Let f be differentiable on $(-\infty, \infty)$ and suppose that $f'(x) \ne 1$ for all real x. Show that f can have either zero or one fixed point (but not more than one).

8.3.28 Compute $\displaystyle\lim_{n\to\infty} \left(\prod_{k=1}^{n} \left(1 + \frac{k}{n} \right) \right)^{1/n}$.

8.3.29 (Putnam 1991) Suppose f and g are non-constant, differentiable, real-valued functions defined on $(-\infty, \infty)$. Furthermore, suppose that for each pair of real numbers x and y,

$$f(x + y) = f(x)f(y) - g(x)g(y),$$
$$g(x + y) = f(x)g(y) + g(x)f(y).$$

If $f'(0) = 0$, prove that $(f(x))^2 + (g(x))^2 = 1$ for all x.

8.3.30 Let $f : [0, 1] \to \mathbf{R}$ satisfy $|f(x) - f(y)| \le (x - y)^2$ for all $x, y \in [0, 1]$. Furthermore, suppose $f(0) = 0$. Find all solutions to the equation $f(x) = 0$. Hint: assume that $f(x)$ is differentiable if you must, but this isn't really a differentiation problem.

8.3.31 Complete the proof started in Example 8.3.5 on page 299.

8.3.32 (Putnam 1976) Evaluate

$$\lim_{n\to\infty} \frac{1}{n} \sum_{k=1}^{n} \left(\left\lfloor \frac{2n}{k} \right\rfloor - 2 \left\lfloor \frac{n}{k} \right\rfloor \right).$$

Express your answer in the form $\log a - b$, with a, b both positive integers, and the logarithm to base e.

8.3.33 (Putnam 1970) Evaluate

$$\lim_{n\to\infty} \frac{1}{n^4} \prod_{j=1}^{2n} (n^2 + j^2)^{1/n}.$$

8.3.34 (Putnam 1939) Prove that

$$\int_1^a \lfloor x \rfloor f'(x)dx = \lfloor a \rfloor f(a) - (f(1) + \cdots + f(\lfloor a \rfloor)),$$

where $a > 1$.

8.3.35 (Turkey, 1996) Given real numbers

$$0 = x_1 < x_2 < \cdots < x_{2n} < x_{2n+1} = 1$$

with $x_{i+1} - x_i \leq h$ for $1 \leq i \leq 2n$, show that

$$\frac{1-h}{2} < \sum_{i=1}^{n} x_{2i}(x_{2i+1} - x_{2i-1}) < \frac{1+h}{2}.$$

8.3.36 *The Schwarz Inequality.* The Cauchy-Schwarz inequality has many generalizations. Here is one for integrals, known as the **Schwarz inequality**:

Let $f(x), g(x)$ be nonnegative continuous functions defined on the interval $[a, b]$. Then

$$\left(\int_a^b f(x)g(x)dx \right)^2 \leq \int_a^b (f(x))^2 dx \int_a^b (g(x))^2 dx.$$

(a) First, examine the Cauchy-Schwarz inequality (see page 198) to verify that the integral inequality above is a very plausible "version," since after all, integrals are "basically" sums.

(b) Proving the Schwarz inequality is another matter. Using the "integral-as- a-sum" idea is problematic, since limits are involved. However, we presented two other alternate proofs of Cauchy- Schwarz in Problems 5.5.35–5.5.36. Use one (or both) of these to come up with a nice proof of Schwarz's inequality.

(c) Generalize the inequality a tiny bit: remove the hypothesis that f, g are nonnegative, and replace the conclusion with

$$\left(\int_a^b |f(x)g(x)|dx \right)^2 \leq \int_a^b (f(x))^2 dx \int_a^b (g(x))^2 dx.$$

(d) Under what circumstances is Schwarz's inequality actually an equality?

8.4   Power Series and Eulerian Mathematics

Don't Worry!

In this final section of the book, we take a brief look at infinite series whose terms are not constants, but functions. This very quickly leads to technical questions of convergence.

Example 8.4.1 Interpret the meaning of the infinite series

$$\frac{x^2}{1+x^2} + \frac{x^2}{(1+x^2)^2} + \frac{x^2}{(1+x^2)^3} + \cdots,$$

where x can be any real number.

Solution: The only sensible interpretation is one that is consistent with our definition of series of constant terms. Thus we let

$$a_n(x) := \frac{x^2}{(1+x^2)^n}, \quad n = 1, 2, 3, \ldots,$$

and define the function $S(x)$ to be the limit of the partial sum functions. In other words, if

$$S_n(x) := \sum_{k=1}^{n} a_k(x),$$

then for each $x \in \mathbf{R}$, we define

$$S(x) := \lim_{n \to \infty} S_n(x),$$

provided that this limit exists.

Note that the $a_n(x)$ are all defined for all real x, so the same is true for each $S_n(x)$. In fact, we can explicitly compute $S_n(x)$, using the formula for a geometric series (see page 173); we have

$$S_n(x) = \frac{x^2}{1+x^2} + \frac{x^2}{(1+x^2)^2} + \cdots + \frac{x^2}{(1+x^2)^n}$$

$$= \frac{\dfrac{x^2}{(1+x^2)} - \dfrac{x^2}{(1+x^2)^{n+1}}}{1 - \dfrac{1}{1+x^2}}.$$

Multiply numerator and denominator by $1 + x^2$ and we get

$$S_n(x) = \frac{x^2 - \dfrac{x^2}{(1+x^2)^n}}{x^2}.$$

This formula is not defined if $x = 0$. As long as $x \neq 0$, we can simplify further to get

$$S_n(x) = 1 - \frac{1}{(1+x^2)^n}.$$

Fix a real number $x \neq 0$. As $n \to \infty$, the second term above will vanish, no matter what x is. Therefore $S(x) = 1$ for all nonzero x. But if $x = 0$, then each $a_n(0) = 0$ which forces $S_n(0) = 0$ and we conclude that $S(0) = 0$. ∎

That wasn't too bad, but something disturbing happened. Each $a_n(x)$ is continuous (in fact, differentiable), yet the infinite sum of these functions is discontinuous. This example warns us that infinite series of functions cannot be treated like finite series. There are plenty of other "pathologies," for example, a function $f(x)$ defined to be the infinite sum of $f_i(x)$, yet $f'(x)$ is *not* equal to the sum of the $f_i'(x)$.[9] The basic reason behind these troubles is the fact that properties like continuity, differentiation, etc. involve taking limits, as does finding the sum of a series. It is not always the case that a "limit of a limit" is unchanged when you interchange the order.

Luckily, there is one key property that prevents most of these pathologies: **uniform convergence**, which is defined in the same spirit as uniform continuity (see page 289). We say that the sequence of functions $(f_n(x))$ converge uniformly to $f(x)$ if the "N response" to the "ϵ challenge" is independent of x. We shall not discuss many details here (see [23] for a clear and concise treatment, and [17] for a fresh and intuitive discussion) because the punch line is essentially "don't worry." Here's why.

- If $f_n(x) \to f(x)$ uniformly, and the f_n are continuous, then $f(x)$ will also be continuous.

- If $f_n(x) \to f(x)$ uniformly, then $\int f_n(x)dx \to \int f(x)dx$.

- Uniformly convergent power series can be differentiated and integrated term by term.

The last item is important. A **power series** is a special case of a series of functions, namely one where each term has the form $a_n(x - c)^n$. From your elementary calculus courses, you learned about the radius of convergence.[10] For example, the series

$$1 + x + x^2 + x^3 + \cdots$$

converges to $\dfrac{1}{1 - x}$, provided that $|x| < 1$. In other words, the values of x for which the series converge lie within 1 unit of 0. The center of convergence is 0 and the radius of convergence is 1. What makes power series so useful is the fact that they converge uniformly as long as you contract the radius of convergence a bit. More formally,

Let

$$a_0 + a_1 x + a_2 x^2 + \cdots = f(x)$$

for all x such that $|x - c| < R$. Then for any positive ϵ, the convergence is uniform for all x such that $|x - c| \leq R - \epsilon$.

---

[9]See Chapter 7 of [23] for a nice discussion of these issues. (Our Example 8.4.1 was adapted from this chapter.)

[10]To *really* understand radius of convergence, you need to look at the complex plane. See [17] for an illuminating discussion.

Thus, once you are in possession of a uniformly convergent power series, you can abuse it quite a bit without fear of mathematical repercussions. You can differentiate or integrate term by term, multiply it by other well-behaving power series, etc., and be sure that what you get will behave as you think it should.

Taylor Series with Remainder

Most calculus textbooks present Taylor series, but the proof is rarely mentioned, or relegated to a technical appendix that is never read. This is a shame, because it is as easy as it is important. Let us derive the familiar Taylor series formula (including the remainder term) in a way that is both easy to understand and remember, with a simple example.

Example 8.4.2 Find the second-degree Taylor polynomial for $f(x)$, plus the remainder.

Solution: Assume that $f(x)$ is infinitely differentiable on its domain D, and all derivatives are bounded. In other words, for each $k \geq 1$ there is a positive number M_k such that $|f^{(k)}(x)| \leq M - k$ for all x in the domain. We shall construct the second-degree Taylor polynomial about $x = a$ (where $a \in D$). To do this, we start with the *third* derivative. All that we know for sure is that

$$-M_3 \leq f'''(t) \leq M_3$$

for all $t \in D$. Integrating this with respect to t from a to x yields[11]

$$-M_3(x - a) \leq \int_a^x f'''(t)dt \leq M_3(x - a)^3.$$

By the fundamental theorem of calculus, the integral of f''' is f'', so the above becomes

$$-M_3(x - a) \leq f''(x) - f''(a) \leq M_3(x - a)^3.$$

Now let us replace x with the variable t and integrate with respect to t once again from a to x. We have

$$-M_3 \frac{(x - a)^2}{2} \leq f'(x) - f'(a) - f''(a)(x - a) \leq M_3 \frac{(x - a)^2}{2}.$$

Repeat the process once more to get

$$-M_3 \frac{(x - a)^3}{2 \cdot 3} \leq f(x) - f(a) - f'(a)(x - a) - f''(a)\frac{(x - a)^2}{2} \leq M_3 \frac{(x - a)^3}{2 \cdot 3}.$$

Finally, we conclude that

$$f(x) = f(a) + f'(a)(x - a) + f''(a)\frac{(x - a)^2}{2}$$

---

[11] We are using the fact that $u(x) \leq v(x)$ implies $\int_a^b u(x)dx \leq \int_a^b v(x)dx$.

plus an error term which is at most equal to

$$M_3 \frac{(x-a)^3}{6}$$

in absolute magnitude. ∎

The general method is simple; just keep integrating the inequality $|f^{(n+1)}(x)| \leq M_n$ until you get the nth-degree Taylor polynomial. The general formula is

$$f(x) = f(a) + \sum_{i=1}^{n} f^{(i)}(a) \frac{(x-a)^i}{i!} + R_{n+1}, \tag{7}$$

where

$$|R_{n+1}| \leq M_{n+1} \frac{(x-a)^{n+1}}{(n+1)!}.$$

From this remainder formula it is clear that if the bounds on the derivatives are reasonable (for example, M_k does not grow exponentially in k), then the power series will converge. And that is an amazing thing. For example, consider the familiar series

$$\sin x = x - \frac{x^3}{3!} + \frac{x^5}{5!} - \cdots,$$

which converges (verify!) for all real x. Yet the coefficients for this "global" series come only from knowledge about the value of $\sin x$ and its derivatives at $x = 0$. In other words, complete "local" information yields complete global information! This is worth pondering.

In practice, it is not always necessary to use (7). As long as you know (or suspect) that the series exists, you can opportunistically extract terms of a series.

Example 8.4.3 Expand $\dfrac{1}{x^2 + 1}$ into a power series about $x = 0$.

Solution: We simply use the geometric series tool (see page 144):

$$\frac{1}{x^2 + 1} = \frac{1}{1 - (-x^2)} = 1 - x^2 + (x^2)^2 - (x^2)^3 + \cdots,$$

and thus

$$\frac{1}{x^2 + 1} = 1 - x^2 + x^4 - x^6 + \cdots. \qquad ∎$$

Example 8.4.4 Expand e^{x^2} into a power series about $x = 0$.

Solution: Just substitute $t = x^2$ into the familiar series

$$e^t = 1 + t + \frac{t^2}{2!} + \frac{t^3}{3!} + \cdots. \qquad ∎$$

You may wonder about these last two examples, asking, "Yes, we got a power series, but how do we know that we actually got the Taylor series that we would have gotten from (7)?" Once again, don't worry, for the power series expansion is unique. The essential reason is just a generalization of the "derivative is the best linear approximation" idea mentioned on page 296. For example, let $P_2(x)$ denote the second-degree Taylor polynomial for $f(x)$ about $x = a$. We claim that $P_2(x)$ is the best quadratic approximation to $f(x)$ for the simple reason that

$$\lim_{x \to a} (f(x) - P_2(x)) = 0,$$

while if $Q(x) \neq P_2(x)$ is any other quadratic polynomial, then

$$\lim_{x \to a} (f(x) - Q(x)) \neq 0. \tag{8}$$

This is one reason why power series are so important. Not only are they easy to manipulate, but they provide "ideal" information about the way the function grows.

Eulerian Mathematics

In the last few pages, we have been deliberately cavalier about rigor, partly because the technical issues involved are quite difficult, but mostly because we feel that too much attention to rigor and technical issues can inhibit creative thinking, especially at two times:

- The early stages of any investigation,
- The early stages of any person's mathematical education.

We certainly don't mean that rigor is evil, but we do wish to stress that lack of rigor is *not* the same as nonsense. A fuzzy, yet inspired idea may eventually produce a rigorous proof; and sometimes a rigorous proof completely obscures the essence of an argument.

There is, of course, a fine line between a brilliant, non-rigorous argument and poorly thought-out silliness. To make our point, we will give a few examples of "Eulerian mathematics," which we define as non-rigorous reasoning which may even be (in some sense) incorrect, yet which leads to an interesting mathematical truth. We name it in honor of the 18th-century Swiss mathematician Leonhard Euler, who was a pioneer of graph theory and generatingfunctionology, among other things. Euler's arguments were not always rigorous or correct by modern standards, but many of his ideas were incredibly fertile and illuminating.

Most of Euler's "Eulerian" proofs are notable for their clever algebraic manipulations, but that is not the case for all of the examples below. Sometimes a very simple yet "wrong" idea can help solve a problem.[12] We will begin with two examples that help to solve earlier problems. They are excellent illustrations of the "bend the rules" strategy discussed on page 23.

---

[12]The 20th-century king of algebraic Eulerian thinking was the self-educated Indian mathematician S. Ramanujan who did not use any rigor, yet made many incredible discoveries in number theory and analysis. See [26] for details. Recently, mathematics has begun to see a movement away from rigor and toward intuition and visualization; perhaps the most eloquent proponent of this approach is [17].

Example 8.4.5 Solve Problem 1.3.8 on page 11:

> *For any sequence of real numbers $A = (a_1, a_2, a_3, \ldots)$, define ΔA to be the sequence $(a_2 - a_1, a_3 - a_2, a_4 - a_3, \ldots)$, whose nth term is $a_{n+1} - a_n$. Suppose that all of the terms of the sequence $\Delta(\Delta A)$ are 1, and that $a_{19} = a_{94} = 0$. Find a_1.*

Partial Solution: Even though this is not a calculus problem—the variables are discrete, so notions of limit make no sense—we can apply calculus-style ideas. Think of A as a function of the subscript n. The Δ operation is reminiscent of differentiation; thus the equation

$$\Delta(\Delta A) = (1, 1, 1, \ldots)$$

suggests the differential equation

$$\frac{d^2 A}{dn^2} = 1.$$

Solving this (pretending that it makes sense) yields a quadratic function for n. None of this was "correct," yet it inspires us to try guessing that a_n is a quadratic function of n. And this guess turns out to be correct!

Example 8.4.6 Solve Problem 1.3.12 on page 11:

> *Determine, with proof, the largest number which is the product of positive integers whose sum is 1976.*

Partial Solution: Once again, we shall inappropriately apply calculus to a discrete problem. It makes intuitive sense for the numbers whose sum is 1976 to be equal (see the discussion of the AM-GM inequality in Section 5.5). But how large should these parts be? Consider the optimization question of finding the maximum value of

$$f(x) := \left(\frac{S}{x}\right)^x,$$

where S is a positive constant. An exercise in logarithmic differentiation (do it!) shows that $S/x = e$. Thus, if the sum is S each part should equal e and there should be S/e parts.

Now this really makes no sense if $S = 1976$ and the parts must be integers, and having a non-integral number of parts makes even less sense. But it at least focuses our attention on parts whose size is close to $e = 2.71828 \ldots$. Once we start looking at parts of size 2 and 3, the problem is close to solution (use an algorithmic approach similar to our proof of the AM-GM inequality).

The next example, due to Euler, is a generatingfunctionological proof of the infinitude of primes. The argument is interesting and ingenious and has many applications and generalizations (some of which you will see in the problems and exercises below). It can be rigorized pretty easily by considering partial sums and products, but that obscures the inspiration and removes the fun.

Example 8.4.7 Prove that the number of primes is infinite.

Solution: Consider the harmonic series

$$1 + \frac{1}{2} + \frac{1}{3} + \frac{1}{4} + \frac{1}{5} + \cdots.$$

Let us try to factor this sum, which doesn't really make sense, since it is infinite. Define

$$S_k := 1 + \frac{1}{k} + \frac{1}{k^2} + \frac{1}{k^3} + \cdots,$$

and consider the infinite product

$$S_2 S_3 S_5 S_7 S_{11} \cdots,$$

where the subscripts run through all primes. The first few factors are

$$\left(1 + \frac{1}{2} + \frac{1}{2^2} + \cdots\right)\left(1 + \frac{1}{3} + \frac{1}{3^2} + \cdots\right)\left(1 + \frac{1}{5} + \frac{1}{5^2} + \cdots\right)\cdots.$$

When we expand this infinite product, the first few terms will be

$$1 + \frac{1}{2} + \frac{1}{3} + \frac{1}{4} + \frac{1}{5} + \frac{1}{6} + \cdots,$$

and we realize that for each n, the term $1/n$ will appear. For example, if $n = 360$, then $1/n$ will appear when we multiply the terms $1/2^3$, $1/3^2$, and $1/5$. Moreover, $1/n$ will appear exactly once, because prime factorization is *unique* (see the discussion of the fundamental theorem of arithmetic in Section 7.1). Thus

$$1 + \frac{1}{2} + \frac{1}{3} + \frac{1}{4} + \frac{1}{5} + \cdots = S_2 S_3 S_5 S_7 S_{11} \cdots.$$

For each k, we have $S_k = 1/(1 - k)$, which is finite. But the harmonic series is infinite. So it cannot be a product of finitely many S_k. We conclude that there are infinitely many primes! ∎

Our final example is also due to Euler. Here the tables are turned: ideas from polynomial algebra are inappropriately applied to a calculus problem, resulting in a wonderful and correct evaluation of an infinite series (although in this case, complete rigorization is much more complicated). Recall that the zeta function (see page 177) is defined by the infinite series

$$\zeta(s) := \frac{1}{1^s} + \frac{1}{2^s} + \frac{1}{3^s} + \cdots.$$

Example 8.4.8 Is there a simple expression for $\zeta(2) = 1 + \frac{1}{2^2} + \frac{1}{3^2} + \cdots$?

Solution: Euler's wonderful, crazy idea was inspired by the relationship between zeros and coefficients (see Section 5.4) which says that the sum of the zeros of the monic polynomial

$$x^n + a_{n-1}x^{n-1}x^{n-1} + \cdots + a_1 x + a_0$$

is equal to $-a_{n-1}$; this follows from an easy argument that examines the factorization of the polynomial into terms of the form $(x - r_i)$, where each r_i is a zero.

Why not try this with functions that have *infinitely many* zeros? A natural candidate to start with is $\sin x$, because its zeros are $x = k\pi$ for all integers k. But we are focusing on squares, so let us modify our candidate to $\sin \sqrt{x}$. The zeros of this function are $0, \pi^2, 4\pi^2, 9\pi^2, \ldots$. Since we will ultimately take reciprocals, we need to remove the 0 from this list. This leads to our final candidate, the function

$$f(x) := \frac{\sin \sqrt{x}}{\sqrt{x}},$$

which has the zeros $\pi^2, 4\pi^2, 9\pi^2, \ldots$. Using the methods of Example 8.4.4, we easily discover that

$$f(x) = 1 - \frac{x}{3!} + \frac{x^2}{5!} - \frac{x^3}{7!} + \cdots. \tag{9}$$

Since we know all the zeros of $f(x)$, we can pretend that the factor theorem for polynomials applies, and write $f(x)$ as an infinite product of terms of the form $x - n^2\pi^2$. But we need to be careful. The product

$$(x - \pi^2)(x - 4\pi^2)(x - 9\pi^2) \cdots$$

won't work; the constant term is infinite, which is horribly wrong. The power series (9) tells us that the constant term must equal 1. The way out of this difficulty is to write the product as

$$f(x) = \left(1 - \frac{x}{\pi^2}\right)\left(1 - \frac{x}{4\pi^2}\right)\left(1 - \frac{x}{9\pi^2}\right)\cdots, \tag{10}$$

for now the constant term is 1, and when each factor $1 - x/(n^2\pi^2)$ is set equal to zero, we get $x = n^2\pi^2$, just what we want.

Now it is a simple matter of comparing coefficients. It is easy to see that the coefficient of the x-term in the infinite product (10) is

$$-\left(\frac{1}{\pi^2} + \frac{1}{4\pi^2} + \frac{1}{9\pi^2} + \cdots\right).$$

But the corresponding coefficient in the power series (9) is $-1/3!$. Equating the two, we have

$$\frac{1}{\pi^2} + \frac{1}{4\pi^2} + \frac{1}{9\pi^2} + \cdots = \frac{1}{6},$$

and thus

$$1 + \frac{1}{2^2} + \frac{1}{3^2} + \cdots = \frac{\pi^2}{6}. \qquad \blacksquare$$

Problems and Exercises

8.4.9 Verify that the sum in Example 8.4.1 does not converge uniformly (look at what happens near $x = 0$).

8.4.10 Consider the series $1 + x + x^2 + x^3 + \cdots = \dfrac{1}{1-x}$, which converges for $|x| < 1$.

(a) Show that this series does not converge uniformly.

(b) Show that this series does converge uniformly for $|x| \leq 0.9999$.

8.4.11 Prove an important generalization of the binomial theorem, which states that

$$(1+x)^\alpha = 1 + \binom{\alpha}{1}x + \binom{\alpha}{2}x^2 + \binom{\alpha}{3}x^3 + \cdots,$$

where

$$\binom{\alpha}{r} := \frac{\alpha(\alpha-1)\cdots\alpha-r+1}{r!}, \quad r = 1, 2, 3, \ldots.$$

Notice that this definition of binomial coefficient agrees with the combinatorial one which is defined only for positive integral α. Also note that the series above will terminate if α is a positive integer. Discuss convergence. Does it depend on α?

8.4.12 (Putnam 1992) Define $C(\alpha)$ to be the coefficient of x^{1992} in the power series about $x = 0$ of $(1+x)^\alpha$. Evaluate

$$\int_0^1 \left(C(-y-1) \sum_{k=1}^{1992} \frac{1}{y+k} \right) dy.$$

8.4.13 Prove the assertion on page 312 that concluded with (8), which stated that the second-degree Taylor polynomial is the "best" quadratic approximation to a function.

8.4.14 Let $x > 1$. Evaluate the sum

$$\frac{x}{x+1} + \frac{x^2}{(x+1)(x^2+1)} + \frac{x^4}{(x+1)(x^2+1)(x^4+1)} + \cdots.$$

8.4.15 Use power series to prove that $e^{x+y} = e^x e^y$.

8.4.16 (Putnam 1990) Prove that for $|x| < 1$, $|z| > 1$,

$$1 + \sum_{j=1}^\infty (1+x^j)P_j = 0,$$

where P_j is

$$\frac{(1-z)(1-zx)(1-zx^2)\cdots(1-zx^{j-1})}{(z-x)(z-x^2)(z-x^3)\cdots(z-x^j)}.$$

8.4.17 (Putnam 1997) Evaluate

$$\int_0^\infty \left(x - \frac{x^3}{2} + \frac{x^5}{2 \cdot 4} - \frac{x^7}{2 \cdot 4 \cdot 6} + \cdots \right)\left(1 + \frac{x^2}{2^2} + \frac{x^4}{2^2 \cdot 4^2} + \frac{x^6}{2^2 \cdot 4^2 \cdot 6^2} + \cdots \right).$$

8.4.18 (Putnam 1990) Is there an infinite sequence $a_0, a_1, a_2, \ldots$ of nonzero real numbers such that for $n = 1, 2, 3, \ldots$ the polynomial

$$p_n(x) = a_0 + a_1 x + a_2 x^2 + \cdots + a_n x^n$$

has exactly n distinct real roots?

8.4.19 Let S be the set of integers whose prime factorizations only include the primes 3, 5, and 7. Does the sum of the reciprocals of the elements of S converge, and if so, to what?

8.4.20 Consider the argument used in Example 8.4.7.

(a) Did this argument really require the fundamental theorem of arithmetic (unique factorization)?

(b) Make this argument rigorous, by considering only finite partial sums and products.

8.4.21 Show that $\zeta(s) = \displaystyle\prod_{p \text{ prime}} \frac{p^s}{1 - p^s}$.

8.4.22 Evaluate

$$\frac{1}{1^2} + \frac{1}{3^2} + \frac{1}{5^3} + \frac{1}{7^2} + \cdots.$$

8.4.23 Compute $\zeta(2) + \zeta(3) + \zeta(4) + \cdots$.

8.4.24 Prove that

$$\frac{\sin x}{x} = \prod_{n=1}^{\infty} \cos\left(\frac{x}{2^n}\right),$$

(a) using telescoping;

(b) using power series.

8.4.25 Let $P = \{4, 8, 9, 16, \ldots\}$ be the set of perfect powers, i.e., the set of positive integers of the form a^b, where a and b are integers greater than 1. Prove that

$$\sum_{j \in P} \frac{1}{j - 1} = 1.$$

Eulerian Mathematics and Number Theory

The following challenging problems are somewhat interrelated, all involving manipulations similar to Example 8.4.7. You may need to reread the combinatorics and number theory chapters, and some familiarity with probability is helpful for the last two problems.

8.4.26 Find a sequence $n_1, n_2, \ldots$ such that

$$\lim_{k \to \infty} \frac{\phi(n_k)}{n_k} = 0.$$

8.4.27 Prove that

$$\frac{1}{\zeta(s)} = \sum_{n=1}^{\infty} \frac{\mu(n)}{n^s}.$$

8.4.28 Fix a positive integer n. Let $p_1, \ldots, p_k$ be the primes less than or equal to $\sqrt{n}$. Let $Q_n := p_1 \cdot p_2 \cdots p_k$. Let $\pi(x)$ denote the number of primes less than or equal to x. Show that

$$\pi(n) = -1 + \pi(\sqrt{n}) + \sum_{d \mid Q_n} \mu(d) \left\lfloor \frac{n}{d} \right\rfloor.$$

8.4.29 For $n \in \mathbb{N}, x \in \mathbb{R}$, define $\phi(n, x)$ to equal the number of positive integers less than or equal to x which are relatively prime to n. For example, $\phi(n, n)$ is just plain old $\phi(n)$. Find a formula for computing $\phi(n, x)$.

8.4.30 Show that the number of pairs (x, y) where x and y are relatively prime integers between 1 and n inclusive, is

$$\sum_{1 \le r \le n} \mu(r) \left\lfloor \frac{n}{r} \right\rfloor^2.$$

8.4.31 Show that the probability that two randomly chosen positive integers are relatively prime is $\dfrac{6}{\pi^2}$.

8.4.32 Analogous to the concept of perfect numbers (see Problems 7.5.33–7.5.35) are the abundant numbers. The natural number n is considered **abundant** if $\sigma(n) > 2n$.

(a) How abundant can a number get? In other words, what is the largest possible value for the ratio $\sigma(n)/n$?

(b) What is the expected value of this "abundancy quotient" $\sigma(n)/n$? In other words, if you pick an integer n at random, and compute the value of $\sigma(n)/n$, what limiting average value do we get if we repeat this experiment indefinitely?

(c) What relative fraction of positive integers is abundant?

Appendix A

Hints to Selected Problems

We include here some very brief hints to some of the problems. Occasionally we include a full solution, but mostly we only include a suggestive word or two to spur on your investigations. For more complete solutions to most of the problems, consult the *Instructor's Resource Manual*. You can find more information about this at the web site www.wiley.com/college/zeitz.

Chapter 2

2.1.22 Conservation of mass.

2.1.27 (d) Try to actually build the shape.

2.2.11 Any time a problem involves doubling, like this one, you should think about writing numbers in binary (base 2).

2.2.14 It's too hard to draw pictures for $n \geq 4$. Try to represent the situation 2-dimensionally. If mapmakers can do it, so can you. The answer, by the way, is not a power of 2. It is $n^2 - n + 2$.

2.2.24 The exponents are numbers whose base-3 representation only contains the digits 0 and 1. This looks like base-2 numbers!

2.2.27 Once again, think about base-2 representation.

2.2.34 (a) What if $n = 41$?

2.3.14 You need the fundamental theorem of arithmetic (see Section 7.1).

2.3.22 If the conclusion were false, then there exist t_1, t_2, u_1, u_2 such that $t_1 t_2 \in U$ and $u_1 u_2 \in T$. Use these to get a contradiction by finding a product of several numbers which lies in both U and T.

2.3.23 Try replacing 1995 with 3. Then try 4. Think about parity!

2.3.25 It is helpful to think *dynamically*; imagine slowly drawing in the $(n+1)$st line and keep a running count of all the new areas that it creates.

2.3.31 Equivalently, you want to show that $7^n = 1 + 6N$, where N is an integer.

2.3.35 Some of these problems require strong induction. Another idea of "strengthening" is to try to prove *more* than what is asked (see Example 2.3.10 on page 54). For example, with problems (f) and/or (g), you may need to prove two different statements, not one, and use each statement to help the other.

2.4.7 Draw a distance-versus-time graph.

2.4.8 Draw a distance-versus-time graph.

2.4.12 Whenever you see sums of squares, think about distance in a coordinate system [since the distance between (x_1, y_1) and (x_2, y_2) is $\sqrt{(x_1 - x_2)^2 + (y_1 - y_2)^2}$].

2.4.13 Three dimensions are too hard!

2.4.14 The algebra is too hard. Try a picture. If the "floor" brackets confuse you at first, temporarily pretend that they aren't there.

8.2.19 Draw a careful graph. Make sure that you understand the graphical relationship between a function and its inverse function (the graph of $y = f^{-1}(x)$ is the reflection of the graph of $y = f(x)$ about the line $y = x$).

2.4.16 See the hint for 2.4.12.

2.4.18 This is a good "backburner" problem. Consult the sections on symmetry and invariants (Sections 3.1, 3.4, respectively) for ideas.

Chapter 3

3.1.13 You will need to use the reflection tool (Example 3.1.5 on page 71) *twice*.

3.1.17 After the bugs have turned, say, 1 degree, they are still at the vertices of a square, and they still haven't crashed into one another. So they will turn another 1 degree, and then, . . .

3.1.18 The center of the table is a "distinguished" point.

3.1.19 Virtually any problem involving actual reflection will benefit from the reflection tool (Example 3.1.5 on page 71). The strategy is to try to turn a jagged path into a nice straight line.

3.1.23 The temperature function $T(x, y, z)$, unfortunately, is not symmetric in the three variables. So *impose* symmetry: consider the new function

$$T(x, y, z) + T(y, z, x) + T(z, x, y).$$

Reread Example 3.1.7 on page 73 for inspiration.

3.1.24 See 3.1.23 above.

3.2.7 Look at a square which contains the smallest value.

3.2.12 Look at the longest string of consecutive zeros which appears in the number.

3.2.10 Consider the two people whose distance apart is the smallest.

3.2.13 At "worst," 1 and n^2 are diagonally opposite. In all other cases there is a shorter path connecting these two numbers.

3.2.15 (d) *Solution*: Let t be chosen so that $r := b - at$ is the smallest possible nonnegative value (t is a positive integer). We claim that $0 \leq r < a$. Assume not; since $r \geq 0$, then we would have $r \geq a$. But then $r - a = b - a(t + 1)$ is nonnegative, yet smaller, contradicting the minimality of r.   ∎

3.3.11 There seem to be not enough pigeons, but you can bypass this problem by looking at cases: What if someone knows no one? What if everyone knows at least someone?

3.3.16 Penultimate step: look at *rectangles* of size 0.02.

3.3.18 Penultimate step: there are many ways to force numbers to be relatively prime. Think of some simple ways.

3.3.19 How "far" is each person from his correct entrée?

3.3.20 Think about the parity of exponents.

3.3.26 Think about diagonals and things that "act like" diagonals.

3.4.17 First, try some examples! Given the consecutive numbers, look at the number which is the "most" even; i.e., has the largest power of two dividing it. Must this number be unique?

3.4.20 Try some examples. Look at divisibility.

3.4.23 The pair of numbers a, b can be nicely represented as a point in the plane. Try drawing some very careful pictures to keep track of your experiments.

3.4.25 Reading from left to right, keep track of changes.

3.4.28 First show that every person's weight must have the same parity.

3.4.30 Three dimensions are too hard. Make it easier!

Chapter 4

4.1.8 See Problem 3.3.11.

4.1.9 An equivalent problem: Prove that if you color the edges of a K_6 with two colors, then there will be a monochromatic triangle. To see this, start with an arbitrary vertex. By pigeonhole, at least 3 of the 5 edges emanating from this vertex are, say, blue. Then what?

4.1.11 Argue by contradiction and use pigeonhole-style thinking. Consider, for example, when $v = 12$. If the graph were not connected, then there would have to be a "clique" of with 6 or less vertices (a clique being a subset of vertices none of which are neighbors of any vertices outside the clique).

4.1.16 (a) Use extreme principle. Consider the the longest such "oriented path," and show that it must include all players.

(b) The equivalent statement: a directed complete graph possesses a Hamilton path.

4.1.18 Penultimate step: can we recast the problem in such a way that an Eulerian path would solve it? Or might a Hamilton path do the trick? Could either work? Be flexible about which entities play the role of vertices or edges. For example, one interpretation makes each of the 28 dominos a different vertex. What does an edge mean, then? Another possibility is to make each of the 7 numbers $0, 1, 2, \ldots, 6$ a vertex ...

4.1.22 Analyze degree numbers; use handshake lemma.

4.1.23 Devise an algorithm, where you travel first on a 1-cube, then a 2- cube, etc. Prove that your algorithm works using induction.

4.2.9 (a) Use a picture to show that $z\bar{z} = |z|^2$.

(c) The four points $z, 1 - z, -1, 0$ form a rhombus (a parallelogram with 4 equal side lengths). Recall (and prove!) that the diagonals of a rhombus are perpendicular and bisect one another.

(e) Verify that if $\text{Re}(w/z) = 0$, then w and z are perpendicular as vectors.

(f) Take absolute values of both sides of the equation $(z-1)^{10} = z^{10}$.

4.2.19 If z, w lie on the unit circle, then the four points $0, z, w, z + w$ form a rhombus. See Problem 4.2.9(c) as well.

4.2.22 Show that two of the zeros of this polynomial are ω and ω^2, where $\omega = e^{2\pi i/3}$ is a cube root of unity. This implies that a certain quadratic polynomial divides $z^5 + z + 1$.

4.2.24 Two approaches: One is to replace $\sin t$ with $(e^{it} - e^{-it})/2i$. The other is to use 4.2.9(c). You may need the factor theorem (see Section 5.4).

4.2.26 The polynomial $x^4 + x^3 + x^2 + x + 1$ is shouting at you, "5th roots of unity!"

4.2.33 Penultimate step: if the line segments are the vectors z, w, then $z = \pm iw$ is equivalent to saying that z, w are perpendicular and of equal length.

4.3.10 You will need the technique of partial fractions (see footnote on page 146).

4.3.13 Note that $1 + x + x^2 + x^3$ factors.

4.3.17 Use the geometric series tool.

4.3.18 Let a_n be the number of partitions of a positive integer n into parts that are not multiples of three and let b_n be the number of partitions of n in which there are at most two repeats, and let the generating functions for $(a_n), (b_n)$ be $A(x), B(x)$, respectively. Show that

$$A(x) = (1 + x + x^2 + \cdots)(1 + x^2 + x^4 + \cdots)(1 + x^4 + x^8 + \cdots)(1 + x^5 + x^{10} + \cdots) \cdots$$

and

$$B(x) = (1 + x + x^2)(1 + x^2 + x^4)(1 + x^3 + x$$

4.3.20 Use the geometric series tool. Think about dice.

Chapter 5

5.1.10 Penultimate step: to show that $\lfloor u \rfloor = \lfloor v \rfloor$, show that $N \leq u, v <$ some integer N. Note that $4n + 2$ is never a perfect square.

5.1.13 Many approaches work, including induction. There are a number of methods, but here is a brute-force approach that isn't too hard, but instructive: write in binary (base 2).

5.2.27 It is helpful to know that all answers to AIME problems are integers between 0 and 999. Let the number in question be N. Notice that N must end in a 2, so we can write $N = 10a + 2$, where a is a one- or two-digit number.

5.2.28 Use 5.2.10.

5.2.35 The fact that $(16x^2 - 9) + (9x^2 - 16) = (25x^2 - 25)$ is *not* a coincidence.

5.3.14 Use the catalyst tool (see Example 5.3.3).

5.3.19 The harmonic series diverges.

5.3.20 Get an upper bound on the sum by estimating the size of the $1/k$ terms when k has a one-digit, two-digit, etc. Geometric series are easy to use, so play around with them.

5.4.9 Plug in the zeros of $x^3 - x$.

5.4.10 Start by letting $x = \sqrt{2} + \sqrt{5}$, and then do smart algebraic things until you have a polynomial with integral coefficients.

5.4.11 See Example 7.1.7 on page 247 for a solution.

5.4.12 Think about parity.

5.4.15 Plug in $x = a, b, c$. If it factors, then one factor is linear, the other is quadratic.

5.4.21 Note that the converse to Problem 5.4.8 is also true.

5.5.27 Rationalize the numerator.

5.5.28 Use AMGM.

5.5.29 Verify that if any term in the product is greater than 3, the product can be increased by breaking up this term into two terms, one of which is 2 or 3.

5.5.34 Think about coordinates on the plane.

46 Notice that Cauchy-Schwarz implies that if $r_1, r_2, \ldots, r_n$ are real, then

$$n \sum r_i^2 \geq \left(\sum r_i \right)^2.$$

In addition, use the relationship between roots and coefficients (see page 183).

r 6

6.1.20 (a) Consider the problem of counting the number of ways you can choose two people from a pool of n men and n women. What are the cases?

6.1.23 See page 214 for a solution.

6.1.24 For (a), see page 214. For ideas about (b), look at Section 6.3.

6.1.26 Classify the color schemes according to symmetry. Different things will be overcounted by different amounts. Check your ideas with a much smaller case, such as a 3×3 board.

6.1.27 Here is the answer, but ponder it until you understand *why* it is true. The key tool used is the mississippi formula (6.1.10):

$$(x_1 + x_2 + \cdots + x_n)^r = \sum \frac{r!}{a_1! a_2! \cdots a_n!} x_1^{a_1} x_2^{a_2} \cdots x_n^{a_n},$$

where the indices a_i range through all non-negative values such that $a_1 + a_2 + \cdots + a_n = r$. For example, if we let $n = 3, r = 4$, we get (replacing x_1, x_2, x_3 with x, y, z)

$$(x + y + z)^4 = x^4 + \frac{4!}{3!1!} x^3 y + \frac{4!}{3!1!} x^3 z + \frac{4!}{3!1!} y^3 z + \cdots + \frac{4!}{2!2!} x^2 y^2 + \cdots.$$

6.2.9 Notice that there are four possibilities per person: no treat, just ice cream, just cookie, both treats. Then we have some correcting to do, since not all possibilities should be counted.

6.2.10 Where does the number 232 come from?

6.2.17 It is helpful to count the number of arrangements for which the sport- utility vehicle cannot park: next to each compact car, there will either be a single vacant space, or not.

6.2.26 It is easiest to assume that the trees are distinguishable; i.e., order matters. Thus there are 12! different ways of arranging the trees, and hence 12! will be the denominator of our probability. Then, when we count the number of arrangements for which no two birches are adjacent, we also assume that trees are distinguishable. This problem is not too different, then, from 6.2.17.

6.2.31 Look at a concrete example. Consider selecting 8 people from a pool of 20, of which 13 are men and 7 are women. What are the different cases?

6.2.34 The fact that $c(n, m)$ is a sum suggests a particular tactic. Which one? Show that if $A \subset B$, then $f(A) \le f(B)$. If you know just a few crucial values of f, that will determine f completely.

6.3.10 It is helpful to compute the number of of n-digit strings whose product is not a multiple of 10. Then the properties that matter are *not* being even, and *not* being equal to 5.

6.3.12 As with many PIE problems, it is helpful to count the complement. The permutations to focus on are not the ones that fix exactly k elements, but rather, the ones which fix element #k. These will overlap, but that's what PIE is designed for.

6.3.15 Once again, count the complement: in how many ways can we give the cones out, and not use all k flavors? Focus on arrangements for which flavor #i is not used.

6.4.4 Either a word begins with a single-letter word, or it does not.

6.4.9 You may want to consider an auxiliary problem: how many legal sequences of n pairs of parentheses are "prime," in the sense that when you read from left to right, no substring is legal? For example, "(())" is prime, but "()()" is not.

Chapter 7

7.1.3 (e) Notice that (d) implies that if $g|a$ and $g|b$, then $g|a - b$.

7.1.11 Study Example 7.1.8 on page 248. Another approach is to use the Euclidean algorithm (Problem 7.1.12).

7.1.13 (b) Imagine that (a, b) and (u, v) are two different solutions to $17x + 11y = 1$. Show that $a - u$ must equal a multiple of 11, while $b - v$ is a multiple of 17. At some point you will use FTA.

7.1.16 Consider $ab/(a, b)$ Show that this must be a multiple of both a and b and thus be at least $[a, b]$. Now look at $ab/[a, b]$.

7.1.20 Look at the parity of the numerator and denominator of this sum (in lowest terms).

7.1.23 First try Problem 7.1.22.

7.1.25 Notice that a number of the form $4k + 3$ must have at least some prime factors of this form. Why?

7.1.28 What else do you know about consecutive numbers?

7.1.29 First show that $k|a_k$ for each k.

7.2.6 Look at perfect squares modulo 3.

7.2.13 The answer is "no." Use Fermat's little theorem.

7.3.19 Use the Gaussian pairing tool.

7.3.23 The left-hand side is a PIE statement.

7.4.7 Without loss of generality, assume $a \geq b \geq c$. Note that we must have $2 \leq (1 + 1/c)^3$. What does this tell you about c?

7.4.10 The number 1,599 should force you to think hard about viewing the problem mod m, for a well-chosen m.

7.4.15 This already appeared as Problem 3.4.28 on page 117.

7.4.22 The pair 8, 9 is suggestive.

7.5.16 Use the methods of Problem 7.1.22 to count prime powers in numerator and denominator.

7.5.17 There are no non-zero solutions. Look at parity.

7.5.18 Two numbers are floating around: 3, and n, where $n \perp 3$. What do you do when you see relatively prime numbers?

7.5.25 Don't forget the Chinese remainder theorem (7.2.16).

7.5.31 Let the number of cards be $2n$. Look at things modulo $2n + 1$.

7.5.40 Define $f(n)$ to be the sum of the zeros of $\Phi_n(x)$. Then, by 7.5.38, we see that $\sum_{d|n} f(d)$ equals the sum of the zeros of $x^n - 1$, and this is equal to 0, unless $n = 1$, in which case it equals 1. In other words. But by the MIF, we already know of a function with this property: μ. Recall that $\sum_{d|n} \mu(d)$ equals 0 unless $n = 1$, in which case it equals 1. So $f(n) = \mu(n)$.

Chapter 8

8.2.6 It turns out that $a_n \to \phi$, where $\phi := (1 + \sqrt{5})/2$. To see this, try drawing a picture (see Example 8.2.1 on page 286), or define the "error" sequence (e_n) by $e_n := a_n - \phi$, and then try to express e_n in terms of e_{n-1} with the goal of showing $e_n \to 0$. You will use, somewhere along the way, the fact that $\phi^2 - \phi - 1 = 0$.

8.2.8 Let $a_1 = \sqrt{2}$ and $a_{n+1} = \sqrt{2 + a_n}$. Since the (only positive) solution to $x = \sqrt{2 + x}$ is $x = 2$, you should try to prove that $a_n \to 2$.

8.2.14 Without loss of generality, try to get within 0.1 of an arbitrary point $x \in [0, 1]$. Divide $[0, 1]$ into 10 equal parts and use the pigeonhole principle. It doesn't work immediately, but don't give up!

8.2.20 Think about $f^{-1}(500)$.

8.2.21 Define $g(x) := f(x + 1/1999) - f(x)$. It suffices to show that $g(x)$ changes sign for two values of x in the interval $[0, 1998/1999]$. If the endpoints don't work out, be creative.

8.2.22 Is there anything special about $\sqrt{2}$? Is there a more general statement?

8.2.24 Pick $\epsilon > 0$. Then there is an N such that $|x_n - x_{n-1}| < \epsilon$ for all $n \geq N$. Now get a handle on the size of x_n for $n \geq N$ by telescoping, and using the triangle-inequality (see hint for 2.3.29).

8.2.25 Do Problem 8.2.24 first.

8.2.28 Look at Example 5.3.4 on page 176.

8.3.11 Rolles theorem.

8.3.15 Use the definition $f'(x) = \lim\limits_{h \to 0} \dfrac{f(x+h) - f(x)}{h}$.

8.3.21 Let $f(x) := 2x - 3x^3$. It's too hard to solve for the values $x = a, b$ at which $f(x) = c$, so just call these values a and b. Then our problem involves two integrals, one from a to b, and the other from 0 to a. Can these two integrals be combined?

8.3.22 This problem is practically begging you to look at $(x + 1)^2$.

8.3.26 Take some time to figure out what the problem is asking. Your intuition tells you that it is plausible for $f(x)$ to be bigger than e^x. Might $f(x)$ exceed, say, e^{2x} for large enough x? Logarithmic differentiation will help.

8.3.29 Look at the hint for 8.3.15 above. Also, recall that one way to prove that something equals a constant is to show that its derivative is zero.

8.3.30 The triangle inequality (see hint for 2.3.29) may be useful.

8.4.12 Use the generalized binomial theorem (8.4.11). Logarithmic differentiation may also be helpful.

8.4.14 Look at partial sums. There are geometric series to be summed.

8.4.18 What kind of function does $p_n(x)$ approach as $n \to \infty$?

8.4.19 Use the ideas of Example 8.4.7 on page 313.

8.4.26 The basic idea: look at n_k that have many prime factors.

8.4.28 Use PIE.

References

1. Tom M. Apostol. *Calculus*. Blaisdell, second edition, 1967–69.

2. Ralph Boas. *A Primer of Real Functions*. The Mathematical Association of America, second edition, 1972.

3. H. S. M. Coxeter and S. L. Greitzer. *Geometry Revisited*. The Mathematical Association of America, 1967.

4. N. G. de Bruijn. Filling boxes with bricks. *American Mathematical Monthly*, 76:37–40, 1969.

5. Heinrich Dörrie. *100 Great Problems of Elementary Mathematics*. Dover, 1965.

6. Martin Gardner. *The Unexpected Hanging*. Simon and Schuster, 1969.

7. Martin Gardner. *Wheels, Life and Other Mathematical Amusements*. Freeman, 1983.

8. Edgar G. Goodaire and Michael M. Parmenter. *Discrete Mathematics with Graph Theory*. Prentice Hall, 1998.

9. Ronald L. Graham, Donald E. Knuth, and Oren Patashnik. *Concrete Mathematics*. Addison-Wesley, 1989.

10. Richard K. Guy. The strong law of small numbers. *American Mathematical Monthly*, 95:697–712, 1988.

11. Nora Hartsfield and Gerhard Ringel. *Pearls in Graph Theory*. Academic Press, revised edition, 1994.

12. I. N. Herstein. *Topics in Algebra*. Blaisdell, 1964.

13. D. Hilbert and S. Cohn-Vossen. *Geometry and the Imagination*. Chelsea, second edition, 1952.

14. Ross Honsberger. *Mathematical Gems II*. The Mathematical Association of America, 1976.

15. Mark Kac. *Statistical Independence in Probability, Analysis and Number Theory*. The Mathematical Association of America, 1959.

16. Nicholas D. Kazarinoff. *Analytic Inequalities*. Holt, Rinehart and Winston, 1961.

17. Tristan Needham. *Visual Complex Analysis*. Oxford University Press, 1997.

18. Ivan Niven, Herbert S. Zuckerman, and Hugh L. Montgomery. *An Introduction to the Theory of Numbers*. John Wiley & Sons, fifth edition, 1991.

19. Joseph O'Rourke. *Art Gallery Theorems and Algorithms*. Oxford University Press, 1976.

20. George Pólya. *How to Solve It*. Doubleday, second edition, 1957.

21. George Pólya. *Mathematical Discovery*, volume II. John Wiley & Sons, 1965.

22. George Pólya, Robert E. Tarjan, and Donald R. Woods. *Notes on Introductory Combinatorics*. Birkhäuser, 1983.

23. Walter Rudin. *Principles of Mathematical Analysis*. McGraw-Hill, third edition, 1976.

24. Will Shortz. *The Puzzlemaster Presents*. Random House, 1996.

25. Paul Sloane. *Lateral Thinking Puzzlers*. Sterling Publishing Co., 1992.

26. Alan Slomson. *An Introduction to Combinatorics*. Chapman and Hall, 1991.

27. Michael Spivak. *Calculus*. W. A. Benjamin, 1967.

28. John Stillwell. *Mathematics and Its History*. Springer-Verlag, 1989.

29. Clifford Stoll. *The Cuckoo's Egg: Tracking a Spy Through the Maze of Computer Espionage*. Pocket Books, 1990.

30. Alan Tucker. *Applied Combinatorics*. John Wiley & Sons, third edition, 1995.

31. Charles Vanden Eynden. *Elementary Number Theory*. McGraw-Hill, 1987.

32. Stan Wagon. Fourteen proofs of a result about tiling a rectangle. *American Mathematical Monthly*, 94:601–617, 1987.

33. Herbert S. Wilf. *generatingfunctionology*. Academic Press, 1994.

Index